New Paradigms in Technological Innovation Management

New Paradigms in Technological Innovation Management

Edited by Hans Roth

CLANRYE
INTERNATIONAL
www.clanryeinternational.com

Clanrye International,
750 Third Avenue, 9th Floor,
New York, NY 10017, USA

ISBN: 978-1-63240-835-8

Cataloging-in-Publication Data

New paradigms in technological innovation management / edited by Hans Roth.
p. cm.
Includes bibliographical references and index.
ISBN 978-1-63240-835-8
1. Technological innovations--Management. 2. Research, Industrial--Management.
3. Industrial management. I. Roth, Hans.
HD31.2 .N49 2019
658--dc23

For information on all Clanrye International publications
visit our website at www.clanryeinternational.com

CLANRYE
INTERNATIONAL

Printed in China.

Contents

Preface

In my initial years as a student, I used to run to the library at every possible instance to grab a book and learn something new. Books were my primary source of knowledge and I would not have come such a long way without all that I learnt from them. Thus, when I was approached to edit this book; I became understandably nostalgic. It was an absolute honor to be considered worthy of guiding the current generation as well as those to come. I put all my knowledge and hard work into making this book most beneficial for its readers.

Technological innovation system explains the nature and rate of technological change. It can be applied to three levels of analysis namely, to technology as a field of knowledge, to a product or artifact, or to a set of related products and artifacts developed with the aim to satisfy a specific purpose. The structures in a technological innovation system represent the static aspect of the system and are distinguished into three categories- actors, institutions and technological factors. The organizations that contribute to a technology are actors. These may also be developers, financiers, regulators, etc. The institutional structures form the core of the innovation system. These are the institutional rules and constraints that shape human interaction. Technological factors are essential for understanding the feedback mechanisms between technological and institutional change. The book studies, analyzes and upholds the pillars of technological innovation management with respect to business organizations and its utmost significance in modern times. It presents researches and studies performed by experts across the globe. It is appropriate for students seeking detailed information in this area as well as for experts.

I wish to thank my publisher for supporting me at every step. I would also like to thank all the authors who have contributed their researches in this book. I hope this book will be a valuable contribution to the progress of the field.

Editor

1

Governing Quintuple Helix Innovation: Urban Living Labs and Socio-Ecological Entrepreneurship

Bastiaan Baccarne, Sara Logghe, Dimitri Schuurman, and Lieven De Marez

> "*Cities are never random. No matter how chaotic they might seem, everything about them grows out of a need to solve a problem. In fact, a city is nothing more than a solution to a problem, that in turn creates more problems that need more solutions, until towers rise, roads widen, bridges are built, and millions of people are caught up in a mad race to feed the problem-solving, problem-creating frenzy.*"
>
> Neal Shusterman
> Author of young-adult fiction
> In *Downsiders* (2001)

Growing urbanization puts pressure on both social and ecological systems. This pressure raises complex and multi-facetted challenges that can only be tackled by collaborative and distributed innovation development processes. However, theoretical frameworks that assess such collaborations are often very conceptual, with little focus on the actual governance mechanisms that facilitate them. This article studies the urban living lab concept as an inter-organizational design and multi-stakeholder innovation development process to govern the quintuple helix model for innovation by means of an action research based multidimensional case study design, which focuses on the concepts of innovation democracy, mode 3 knowledge production, the innovation ecosystem as a system of societal subsystems, and socio-ecological transition. In this way, we provide a more profound understanding of such innovation processes to tackle socio-ecological challenges by means of public–private interactions driven by eco-entrepreneurship.

Introduction

Society is confronted with challenges of an increasingly complex and global nature. It is hard for a single societal actor to come up with the right solutions, given that knowledge and resources are distributed among a wide network of stakeholders (Bogers & West, 2012). Thus, innovation actors must reach out to external knowledge. Among the most pressing and interesting challenges are those involving public value and market failure. As urbanization continues at a rapid pace, socio-ecological systems are put under heavy pressure, inducing ecological issues such as global warming, decreasing air quality, increasing hazardous emissions, and geological instability. Although a sense of urgency for solutions is widespread, society is still struggling to find an adequate, sustainable, and agile way to react. It is clear, however, that these challenges need to be dealt with by a diverse ecosystem of private actors, universities, civil society, and politics.

Nevertheless, in innovation management theory, the question is not *why*, but rather *how* such challenges can be tackled. In the collaborative knowledge production and innovation management literature, one of the frameworks that attempt to take the natural environment into account is the "quintuple helix model" for innovation (Carayannis & Campbell, 2010). Although this rather recent analytical framework is very promising, only little empirical evidence exists that explores its possibilities and limitations. On top of that, this model is mainly applied to assess larger innovation ecosystems such as national or regional innovation systems, and to a lesser extent to the innovation development process.

Therefore, this article focuses on the specific governance mechanisms that can facilitate quintuple helix innovation at the level of the individual innovation development process. More specifically, the "urban living lab" concept is explored as an inter-organizational R&D design and multi-stakeholder innovation development process to govern the quintuple helix model for innovation. Hence, this article contributes to a more profound understanding of local collaborative innovation processes that are designed to tackle socio-ecological challenges by means of public–private interactions, driven by eco-entrepreneurship. In other words: How can urban living labs be a way to put quintuple helix innovation into practice?

First, we discuss the urbanization process to better understand the context in which these challenges occur. Next, we relate this evolution to collaborative innovation literature, and elaborate on the (urban) living lab concept as a way to put this into practice. Finally, we develop an analytical framework, which is structured along the concepts of innovation democracy, mode 3 knowledge production, the innovation ecosystem as a system of societal subsystems, and socio-ecological transition, and apply this framework to two urban living lab cases.

Background

Urbanization, socio-ecological challenges, and urban innovation

The speed of urbanization is overwhelming (Bocquier, 2005). This rapid evolution puts pressure on social, physical, and ecological systems as city populations continue to grow and more and more people are live in densely populated areas. This pressure, in combination with the associated emergence of grand societal challenges and rapid technological evolutions, forces cities to look for new ways to reinvent themselves (Atkinson, 1998; Foth, 2009; Viitanen & Kingston, 2014).

In practice, however, local governments often lack the capability and resources to tackle these challenges in a flexible way (O'Flynn, 2007). In the search for new ways to cope with this tension, transparency and close interaction with grassroots initiatives are increasingly put forward as solutions to overcome this gap (Buscher et al., 2010). This approach involves a wide variety of urban stakeholders (e.g., citizens, universities, enterprises, non-governmental organizations), thereby potentially leveraging the distributed knowledge in the urban environment.

Nevertheless, these interactions need to be governed and in some way be able to connect the traditional top-down approach with a grassroots or bottom-up approach. This strategy is in line with the open or distributed innovation approach (Bogers & West, 2012; Chesbrough, 2003), causing city governments to question the dominant paradigm of top-down innovation development, and implementation, and to experiment with innovation processes together with, and even by, citizens and other organizations in the urban environment (Paskaleva, 2011).

Understanding collaborative innovation processes

A useful framework for the analysis of such complex collaborative innovation networks is the "triple helix" model for innovation (Etzkowitz & Leydesdorff, 1995), which originally focused on collaboration and knowledge production in university–government–industry partnerships (Etzkowitz & Leydesdorff, 2000). It was later expanded with a fourth helix to incorporate civil society (Carayannis & Campbell, 2009).

However, from a socio-ecological systems point of view, the urban evolutions described above also need to be studied as changes in human–environment systems (Young et al., 2006).

Hence, a socio-ecological systems approach integrates social and ecological systems thinking in a holistic way to assess "system threats" (Berkes et al., 2000). Such socio-ecological systems can be considered complex units in which resources are exchanged and regulated by social and ecological systems (Berkes et al., 2002; Machlis et al., 1997), which makes them interesting conceptual frameworks to assess socio-ecological innovation in an urban environment. They also encourage the integration of this logic and analytical dimensions in the innovation ecosystem literature. Therefore, a fifth helix should be added to the quadruple helix model, to also take the natural environment into account (Carayannis & Campbell, 2010). This is what makes the quintuple helix model for innovation an interesting and valuable model to analyze innovation ecosystems.

To explain processes of knowledge exchange that take place in such collaborative innovation ecosystems, these models apply the concepts of "mode 3" knowledge production and "open innovation diplomacy". Mode 3 knowledge production is conceptualized as an extension of mode 1 knowledge production (traditional research by universities) (Godin & Gingras, 2000) and mode 2 knowledge production (knowledge that is generated when mode 1 knowledge is applied and put into practice) (Gibbons et al., 1994; Nowotny et al., 2003). Mode 3 adds a third component to this representation of knowledge production by highlighting the overarching system in which this knowledge is produced and exchanged (i.e., innovation networks and knowledge clusters). "Open innovation diplomacy", on the other hand, is used to describe the way in which different organizations and ecosystem are able to collaborate and bridge the divides that exist between traditionally separated domains (this can be social, organizational, cultural, or technological) (Carayannis & Campbell, 2011).

Living labs and the urban environment
Triple, quadruple, and quintuple helix models have a strong theoretical nature. One approach that tries to facilitate such models in a structured way is the "living lab" approach, which can be defined as an ecosystem approach in which end users and other stakeholders are involved in the development of an innovation over a long period of time, in a real-life environment, following an iterative process (Niitamo & Kulkki, 2006; Schuurman et al., 2012) applying multi-method, user-centric innovation research with a strong focus on user empowerment and real-world experimentation (Følstad, 2008; Schuurman et al., 2013; Ståhlbröst, 2008). Furthermore, it offers a structured process and environment to govern input from a wide variety of stakeholders and research methods (Eriksson et al., 2006; Ståhlbröst & Holst, 2012).

In the urban environment, living labs gain importance as a way to govern (complex) urban collaborative innovation processes (Buscher et al., 2010; Paskaleva, 2011). Although the process is similar, *urban* living labs have a distinct nature because the focus is on civic participation, and the output is aimed at increasing quality of life in the city rather than the development of a commercial product or service (Baccarne et al., 2014). As such, urban living labs are an instrument to include a wide variety of stakeholders (citizens, municipalities, entrepreneurs, etc.) in the search for innovations that meet local socio-ecological challenges (Franz, 2014). Juujärvi and Pesso (2013) define an urban living labs as “a physical region in which different stakeholders form public–private–people partnerships of public agencies, firms, universities, and users collaborate to create, prototype, validate, and test new technologies, services, products, and systems in real-life contexts”.

However, despite strong European support, this research and development concept is still struggling for an adequate and more profound theoretical anchoring and remains too much of a "practice-based" concept (Kviselius et al., 2008; Schuurman, 2015). Quintuple helix (-related) concepts provide potentially valuable tools and assumptions for the assessment and theoretical foundation of the more practical oriented living lab literature, embedded within a broader socio-ecological system. On the other hand, the living lab literature might provide a practical framework to put quintuple helix innovation into practice. Furthermore, exploring the quintuple helix model in the context of urban innovation contributes to a more profound understanding of urban innovation in relation with socio-ecological transition in urban areas.

Methodology

Research design
To investigate quintuple helix innovation in relation to urban living labs, we conducted an action research study (Reason & Bradbury, 2001) in which we purposefully designed and participated as researchers in two urban living lab projects. This approach allowed us direct access and control over the projects, as well as more profound insights on the observed phenomena. The analysis is structured following the principles of a multidimensional case study design (Eisenhardt, 1989; Yin, 1984). The unit of analysis in this design is the individual project-centric innovation ecosystem.

The two cases had to: i) focus on eco-entrepreneurship, ii) target the urban environment, iii) be open for the researcher team to shape and mold (cfr. action research), iv) be collaborative in nature, and v) encompass an innovation development process. The two selected research projects were instigated by start-up organizations with socio-ecological goals in the urban environment, and they were both incubated by a regional incubating organization (funded by the Flemish government in order to stimulate innovation in ICT). Both projects involved the set-up of an ad hoc collaborative network of stakeholders and a structured, semi-formal adaptive iterative product development process. Project A was set up around the development of an interactive platform to engage, collaborate, and communicate on the topic of air quality. Project B concerned the development of a peer-to-peer car sharing initiative for electric vehicles.

Both living lab processes were designed along principles whereby the living lab project had to: i) incorporate multiple iterations (Pierson & Lievens, 2005; Schaffers & Budweg, 2009), ii) involve multiple stakeholders (Feurstein et al., 2008; Frissen & Lieshout, 2004;

Juujärvi & Pesso, 2013), iii) be driven by multi-method research (De Moor et al., 2010; Eriksson et al., 2006; Niitamo & Kulkki, 2006), iv) involve real-world experimentation (Følstad, 2008; Niitamo & Kulkki, 2006), and v) be based on active co-creation by stakeholders (Følstad, 2008; Schumacher & Niitamo, 2008; Schuurman et al., 2012; Ståhlbröst, 2008). In line with the conceptualization of an urban living lab, the local government was a required stakeholder in the innovation development ecosystem. Between each iteration, steering committee meetings were held to evaluate the process and modify it if necessary. These steering committees consisted of the researchers, the project managers, and the project instigators (eco-entrepreneurs) (see also Rits et al., 2015).

Research framework
A beneficial aspect of case study design is the opportunity to include multiple sources of evidence, thus enhancing the validity of the analysis through data triangulation (Yin, 1984). The sources of evidence in this study include ethnographic field notes, in-depth interviews, email communications, meeting reports of steering committees, initial project proposals, project reports, and project deliverables. These sources of evidence were coded and interpreted by the author team following an analytical protocol (Yin, 1984) that encompassed the four dimensions that are key elements in the literature on innovation ecosystems, and in particular the quintuple helix model and socio-ecological systems:

1. *Innovation diplomacy:* the praxis of bridging barriers between traditionally separated actors and fields (Carayannis & Campbell, 2011).
   Assumption: properly targeted initiatives are able to connect know-how, tacit knowledge, creativity, and formal knowledge between different domains and nurture entrepreneurship.

2. *Mode 3 knowledge production:* based on a system-theoretic perspective in which knowledge is molded, remixed, shared, and applied within a knowledge-driven society.
   Assumption: quintuple helix ecosystems are knowledge production, distribution, and application systems in which knowledge is generated through the exchange of knowledge between actors in the ecosystem (Carayannis & Campbell, 2012).

3. *A system of subsystems:* the quintuple helix ecosystem encompasses the different domains that resonate and collaborate to solve mutual challenges.
   Assumption: each ecosystem actor provides unique resources to the ecosystem, but also mixes, translates, and processes resources from others. The quintuple helix model describes five societal subsystems (Carayannis et al., 2012): i) the educational system, which generates and disseminates new knowledge; ii) the economic system, which controls, possesses, and generates economic capital; iii) the political system, which has political and legal capital (e.g., laws, clearances, policy, public goods); iv) civil society, which has social capital, and is characterized by traditions, values, and behavioural patterns; and v) the natural environment, which has natural capital (e.g., natural resources, climate, air quality, geological stability).

4. Socio-ecological transition: the main contribution of the quintuple helix model is the integration of the natural environment, which is conceptualized as a contextualization of the four helices of the quadruple helix.
   Assumption: if this context is taken into account, it is possible to achieve sustainable socio-ecological transition, creating synergies between economy, society, and democracy (Carayannis & Campbell, 2011).

Analysis

Quintuple helix innovation in a structured process
An urban living lab follows a structured process in which a central problem, idea, concept, or prototype is at the heart of the collaboration. This innovation development process implements a combination of different methodologies (e.g., for Project A: interviews, focus groups, surveys, co-creation workshops, online crowdsourcing, and a field trial) to involve a wide variety of stakeholders (e.g., for Project A: different local government divisions, citizens, civil society organizations, large and small research organizations, and startups). As described earlier, this formal but flexible staged process was, in both cases, instigated by the eco-entrepreneurs, was financially supported by the Flemish government, and was managed by iMinds Living Labs, a semi-public living lab organization, embedded in an interdisciplinary and cross-university, technology-driven research centre. The formal living lab process structured the innovation development and governed stakeholder interaction, knowledge production, and knowledge transfer. For example, the process facilitated steering committees, safeguarded the overall planning and resources of the project, leveraged social capital within the different subsystems, provided coaching and implementation resources, and translated knowledge between ecosystem stakeholders.

Innovation diplomacy
The formal commitments within both cases were limited to the eco-entrepreneurs and the semi-public living lab management organization. However, both cases also had semi-formal commitments from their respective local governments. Both projects were unable to convince other key actors in the ecosystem to become a formal partner of the project because these other actors were not willing to commit themselves to an *uncertain* and *open* project. However, once the project gained momentum and tangibility, collaborations were still possible on an ad hoc basis. Not only were key stakeholders eventually attracted to the ecosystem, but also several unexpected smaller institutions (i.e., research institutes, companies, and civic organizations), who were very willing to contribute and share their knowledge. The ad hoc, open-ended, and semi-formal design of these temporal collaboration networks served as an *risk reducing* characteristic and helped to overcome collaboration barriers. As such, the collaborative design of the urban living lab stakeholder ecosystem acted as a *centripetal force* within the urban environment, as acupuncture to congregate urban actors with similar goals. Although every collaborator had their own agenda (e.g., leveraging their own business, connecting with the local government, pushing local change), this did not interfere with the overall goals of the project.

Finally, when it came to the involvement of the local government, the process facilitated access and interaction with different governmental levels. The deployment of a *temporal experimental window,* which was considered "neutral" and "external" to existing organizations, enabled individuals to detach themselves from restraining organizational structures and collaborate in a more agile environment, and to stretch the boundaries of what is generally possible (e.g., temporarily sharing data sources, talking freely outside the governmental organization, providing favourable exceptions on the use of public space).

Mode 3 knowledge production
In the air quality domain (Project A), a lot of knowledge is generated in mode 1. Traditionally, research institutes obtain grants to study atmospheric particulate matter (e.g., as PM2.5) or ozone concentrations. Most of these data remain hidden to the public. However, there are some initiatives that attempt to disseminate these data to civil society. Most of the time, these initiatives are built upon open data principles and are, for example, translated in dashboards showing air quality values. In theory, these initiatives distribute and apply mode 1 knowledge (potentially generating mode 2 knowledge). However, this information cannot be interpreted by actors outside the knowledge domain (e.g., regular citizens). Even if the raw numbers are translated in visual information (e.g., public visualizations of the air quality have been created in the city of Project A), the academic complexity was not interpretable for citizens and citizens had no idea how to act upon this information.

Through an iterative, multi-method research approach, the (tacit) knowledge of all stakeholders could be captured, exchanged, (re)combined, mixed, and molded. By distributing and translating fundamental chunks of knowledge to actors outside the original knowledge domain, unexpected but valuable interpretations and interactions occurred. However, such "sparks" and often volatile knowledge must be adequately captured and managed to contribute to the project goal. For Project A, this process of knowledge capture resulted in the development a conceptual model for socio-ecological change, which served as the basis for design requirements and the development of the project prototypes. Project B, on the other hand, focused on understanding end-user needs and frustrations in relation to technological evolutions and other stakeholder needs and knowledge, mainly to set-up a field trial experiment that matched the usage context. These insights were necessary for both practical and substantial reasons in order to be able to test the electric-car sharing system.

The ecosystem and circulation of knowledge
In both cases, various exchanges between different knowledge domains took place. Some examples include knowledge transfers from the political system to the economic system (i.e., knowledge regarding policy, internal procedures, the value network, and business model opportunities) and from civil society to both the educational system (e.g., the interpretation of complex data by citizens and the relation to their everyday behaviour) and the economic system (i.e., regarding needs and adoption potential). Furthermore, for the local government, the urban living lab projects also bridged different divisions and individuals *within* the organization (e.g., different divisions were working on open data and air quality for some time, but it took the project to connect them and expose the overlap between their efforts). This bridging can be conceptualized as an *intra-organizational centripetal force,* which is the mobilizing effect of an urban living lab *within* an organization to connect like-minded individuals beyond organizational structures. As such, the projects facilitated horizontal and agile collaboration and knowledge exchange on an ad hoc basis, largely bypassing traditional structures and processes.

Socio-ecological transition

As for the "natural context", both projects aimed to interact with the socio-ecological environment through their environmental goals. Against this backdrop, it makes sense to conceptually model the innovation ecosystem using the quintuple helix model. Using this framework, it is clear that (urban) innovation is related to issues regarding the socio-ecological context in which we live and (co-)develop our common socio-technological future. Neglecting this dimension is a failure to contextualize innovation development in a broader environment. Although this model fits with the theoretical understanding of urban living labs, they in turn offer an implementation approach for the quintuple helix model for innovation.

Concerning sustainability, both projects exposed the difficulty of designing viable business models in a "public" domain. Both civil society and the economic system consider it the duty of the political system to take responsibility. However, the political system is facing decreasing resources, which makes it hard to develop a sustainable business model. For Project B, this challenge resulted in the establishment of a cooperative organization with limited liability. Different actors who were previously involved in the urban living lab became members of this organization. As such, some of the social capital that was generated within the urban living lab was leveraged to help sustain the innovation.

Nevertheless, socio-ecological transition must also be considered in a broader sense. Through (series of) experiments, project-based quintuple helix innovation can foster long-term change on a more latent level, by inspiring and stimulating debate on contemporary urban challenges and solutions. Applied to the socio-ecological systems way of thinking, urban living labs contribute to the *resistance* of an urban socio-ecological system. More specifically, this goal is achieved through the facilitation of flexible and agile experimentation with possible solutions for issues related to "system stress" caused by urbanization itself, and by doing so, adding to a long-term and latent social transition that is closely interrelated with ecological concerns and associated knowledge.

Conclusions and Discussion

This article bridges the theoretical propositions of the quintuple helix model and the practice-based concept of urban living labs. More specifically, we discussed the concepts of innovation diplomacy, mode 3 knowledge production, the innovation ecosystem as a system of subsystems (related to the circulation of knowledge), and socio-ecological transition to analyze the urban living lab approach. Our findings largely support the theoretical assumptions of the quintuple helix model and elaborate on the urban living lab approach as a way to put this model into practice at the level of a single innovation development process. Urban living labs can be a way to work with ad hoc collectives, lowering the barriers for collaboration. The project-centric nature is a catalyst for knowledge exchange and collaborations within and outside the project and involved organizations.

A successful urban living lab can facilitate and balance top-down governance with bottom-up initiatives in the city. However, some challenges remain. Whereas experimental urban living lab activities activate and reinforce the quintuple helix ecosystem, it is still hard to harness the creation potential within the city in a sustainable way. Nevertheless, urban living labs facilitate urban transitions through an accumulation of experiments, which allow urban actors to experience change, leading to transitions in the long run. This logic suggests that urban living labs contribute to (long-term) sustainable socio-ecological transition, which is mainly facilitated by an interdisciplinary (and transdisciplinary) temporal experimental window that promotes collaborative learning and stakeholder engagement. However, although value is being created at the meso level (i.e., the project level), there is a need for a more formal value capture and retention processes at the macro level (i.e., the level of ecosystems and the overarching organization). In the urban context, it makes sense that local governments fulfil such a role. This is to some extent in line with the conceptualization of the government as a platform (O'Reilly, 2011).

The quintuple helix is a useful concept to understand and analyze how knowledge is created and exchanged in an urban environment, which can be studied as a collaborative innovation development ecosystem, while also taking the ecological context into account. Although such awareness is growing in most organizations, this dimension is not present in most distributed innovation theories and processes. An urban living lab thus can generate and evolve tacit and codified knowledge while focusing on the exchange of knowledge within a natural environment system. In this way, both the innovation outcomes and the urban socio-ecological transition can become more sustainable and recover ecological balance, thus ensuring the quality of life for future generations.

About the Authors

Bastiaan Baccarne is a Research and Teaching Assistant at iMinds-MICT, a research group within the communications department of Ghent University, where he works on the topic of user-centric innovation development in the context of (new) media and ICT. Bastiaan is also a PhD student working on citizen empowerment through the co-development of socio-technical innovations in an urban environment. Being part of iMinds Living Labs, Bastiaan works as a user researcher for several SME and startup projects, with an overall academic focus on the possibilities and limitations of user-centric innovation development ecosystems.

Sara Logghe obtained a master's degree in both History and Communication Sciences with specialization in New Media and Society at Ghent University. Her second master's thesis was *Cultural Communication on the Internet: A Study on the Potential of Social Media for Cultural Institutions.* A three-month internship at iMinds Living Labs encouraged her interest in research on living labs, and in May 2013, Sara joint iMinds-MICT as a Junior Researcher. Working on cultural projects within living lab research, Sara developed a specific interest in the changing library scene.

Dimitri Schuurman holds a PhD (2015) and Master's in Communication Sciences (2003) from Ghent University in Belgium. He joined the research group iMinds – MICT – Ghent University in 2005 and started working at iMinds Living Labs in 2009. Together with his iMinds colleagues, Dimitri developed a specific living lab offering targeted at startups and SMEs, in which he has managed over 50 innovation projects. As a senior researcher, Dimitri is currently responsible for the methodology and academic valorization of living lab projects. He also coordinates a dynamic team of living lab researchers from iMinds – MICT – Ghent University. His main interests and research topics are situated in the domains of open innovation, user innovation, and innovation management. In early 2015, he finished his PhD entitled *Bridging the Gap between Open and User Innovation? Exploring the Value of Living Labs as a Means to Structure User Contribution and Manage Distributed Innovation.*

Lieven De Marez is Research Director of the iMinds Media & ICT (MICT) research group and teaches on the topics of innovation research, media, market and ICT, and new communication technologies in the Department of Communication Sciences at Ghent University in Belgium. MICT is one of 16 research groups within iMinds, and Lieven is also part of the management team of iMinds Living Labs' facilitating infrastructure for living lab research.

References

Atkinson, R. 1998. Technological Change and Cities. *Cityscape,* 3(3): 129–170.
http://www.jstor.org/stable/20868462

Baccarne, B., Mechant, P., Schuurman, D., Colpaert, P., & De Marez, L. 2014. Urban Socio-Technical Innovations With and By Citizens. *Interdisciplinary Studies Journal,* 3(4): 143–156.

Berkes, F., Colding, J., & Folke, C. (Eds.). 2002. *Navigating Social–Ecological Systems.* Cambridge: Cambridge University Press.
http://dx.doi.org/10.1017/cbo9780511541957

Berkes, F., Folke, C., & Colding, J. (Eds.). 2000. *Linking Social and Ecological Systems: Management Practices and Social Mechanisms for Building Resilience.* Cambridge: Cambridge University Press.

Bocquier, P. 2005. World Urbanization Prospects. *Demographic Research,* 12: 197–236.
http://dx.doi.org/10.4054/demres.2005.12.9

Bogers, M., & West, J. 2012. Managing Distributed Innovation: Strategic Utilization of Open and User Innovation. *Creativity and Innovation Management,* 21(1): 61–75.
http://dx.doi.org/10.1111/j.1467-8691.2011.00622.x

Buscher, V., Tomordy, M., Ashley, G., & Tabet, M. 2010. *Smart Cities: Transforming the 21st Century City Via the Creative Use of Technology.* London: Arup IT & Communication Systems.

Carayannis, E. G., Barth, T. D., & Campbell, D. F. 2012. The Quintuple Helix Innovation Model: Global Warming as a Challenge and Driver for Innovation. *Journal of Innovation and Entrepreneurship,* 1(2): 2.
http://dx.doi.org/10.1186/2192-5372-1-2

Carayannis, E. G., & Campbell, D. F. 2009. Mode 3 and Quadruple Helix: Toward a 21st Century Fractal Innovation Ecosystem. *International Journal of Technology Management,* 46(3): 201–234.
http://dx.doi.org/10.1504/ijtm.2009.023374

Carayannis, E. G., & Campbell, D. F. 2010. Triple Helix, Quadruple Helix and Quintuple Helix and How Do Knowledge, Innovation and the Environment Relate To Each Other? A Proposed Framework for a Trans-disciplinary Analysis of Sustainable Development and Social Ecology. *International Journal of Social Ecology and Sustainable Development,* 1(1): 41–69.
http://dx.doi.org/10.4018/jsesd.2010010105

Carayannis, E. G., & Campbell, D. F. J. 2011. Open Innovation Diplomacy and a 21st Century Fractal Research, Education and Innovation (FREIE) Ecosystem: Building on the Quadruple and Quintuple Helix Innovation Concepts and the "Mode 3" Knowledge Production System. *Journal of the Knowledge Economy*, 2(3): 327–372.
http://dx.doi.org/10.1007/s13132-011-0058-3

Carayannis, E. G., & Campbell, D. F. J. 2012. *Mode 3 Knowledge Production in Quadruple Helix Innovation Systems*. New York, NY: Springer New York.
http://dx.doi.org/10.1007/978-1-4614-2062-0

Chesbrough, H. W. 2003. *Open Innovation: The New Imperative for Creating and Profiting from Technology*. Boston: Harvard Business School Press.

De Moor, K., Berte, K., De Marez, L., Joseph, W., Deryckere, T., & Martens, L. 2010. User-Driven Innovation? Challenges of User Involvement in Future Technology Analysis. *Science and Public Policy*, 37(1): 51–61.
http://dx.doi.org/10.3152/030234210x484775

Eisenhardt, K. 1989. Building Theories from Case Study Research. *Academy of Management Review*, 14(4): 532–550.
http://dx.doi.org/10.2307/258557

Eriksson, M., Niitamo, V., Kulkki, S., & Hribernik, K. 2006. *Living Labs as a Multi-Contextual R&D Methodology*. 12th International Conference on Concurrent Enterprising: Innovative Products and Services through Collaborative Networks, ICE 2006. Milan, Italy.

Etzkowitz, H., & Leydesdorff, L. 1995. The Triple Helix: University–Industry–Government Relations: A Laboratory for Knowledge-Based Economic Development. *EASST Review*, 14(1): 14–19.

Etzkowitz, H., & Leydesdorff, L. 2000. The Dynamics of Innovation: From National Systems and "Mode 2" to a Triple Helix of University–Industry–Government Relations. *Research Policy*, 29(2): 109–123.
http://dx.doi.org/10.1016/s0048-7333(99)00055-4

Feurstein, K., Hesmer, A., Hribernik, K., Thoben, T., & Schumacher, J. 2008. Living Labs: A New Development Strategy. In J. Schumacher & V. P. Niitamo (Eds.), *European Living Labs – A New Approach for Human Centric Regional Innovation*. Berlin, Germany: Wissenschaftlicher.

Følstad, A. 2008. Living Labs for Innovation and Development of Communication Technology: A Literature Review. *The Electronic Journal for Virtual Organisations and Networks*, 10: 99–131.

Foth, M. 2009. *Handbook of Research on Urban Informatics: The Practice and Promise of the Real-Time City*. Hershey, PA: IGI Global.
http://dx.doi.org/10.4018/978-1-60566-152-0

Franz, Y. 2014. Chances and Challenges for Social Urban Living Labs in Urban Research. *ENoLL OpenLivingLab Days 2014 Conference Proceedings:* 105–114. Amsterdam, Netherlands.

Frissen, V., & Lieshout, M. 2004. *To User-Centred Innovation Processes: The Role of Living Labs*. Delft, TNO-ICT.

Gibbons, M., Limoges, C., Nowotny, H., Schwartzman, S., Scott, P., & Trow, M. 1994. *The New Production of Knowledge: The Dynamics of Science and Research in Contemporary Societies*. London: Sage.
http://dx.doi.org/10.4135/9781446221853

Godin, B., & Gingras, Y. 2000. The Place of Universities in the System of Knowledge Production. *Research Policy*, 29(2): 273–278.
http://dx.doi.org/10.1016/s0048-7333(99)00065-7

Juujärvi, S., & Pesso, K. 2013. Actor Roles in an Urban Living Lab: What Can We Learn from Suurpelto, Finland? *Technology Innovation Management Review*, 11(3): 22–27.
http://timreview.ca/article/742

Kviselius, N., Ozan, H., Edenius, M., & Andersson, P. 2008. The Evolution of Living Labs–Propositions for Improved Design and Further Research. *5th International Conference on Innovation and Management (ICIM 2008):* 842–856. Maastricht, The Netherlands.

Machlis, G., Force, J., & Burch, W. J. 1997. The Human Ecosystem Part I: The Human Ecosystem as an Organizing Concept in Ecosystem Management. *Society & Natural Resources*, 10(4): 347–367.
http://dx.doi.org/10.1080/08941929709381034

Niitamo, V., & Kulkki, S. 2006. State-of-the-Art and Good Practice in the Field of Living Labs. *Proceedings of the 12th International Conference on Concurrent Enterprising: Innovative Products and Services through Collaborative Networks:* 349–357. Milan, Italy.

Nowotny, H., Scott, P., & Gibbons, M. 2003. Introduction: "Mode 2" Revisited: The New Production of Knowledge. *Minerva*, 41(3): 179–194.

O'Flynn, J. 2007. From New Public Management to Public Value: Paradigmatic Change and Managerial Implications. *Australian Journal of Public Administration*, 66(3): 353–366.
http://dx.doi.org/10.1111/j.1467-8500.2007.00545.x

O'Reilly, T. 2011. Government as a Platform. *Innovations: Technology, Governance, Globalization*, 6(1): 13–40.
http://dx.doi.org/10.1162/INOV_a_00056

Paskaleva, K. 2011. The Smart City: A Nexus for Open Innovation? *Intelligent Buildings International*, 3(3): 153–171.
http://dx.doi.org/10.1080/17508975.2011.586672

Pierson, J., & Lievens, B. 2005. Configuring Living Labs for a "Thick" Understanding of Innovation. *Ethnographic Praxis in Industry Conference Proceedings*, (1): 114–127.
http://dx.doi.org/10.1111/j.1559-8918.2005.tb00012.x

Reason, P., & Bradbury, H. 2001. *Handbook of Action Research: Participative Inquiry and Practice*. London: Sage Publications.

Rits, O., Schuurman, D., & Ballon, P. 2015. *Exploring the Benefits of Combining Business Modelling with Living Labs*. The XXVI ISPIM Conference – Shaping the Frontiers of Innovation Management, Budapest, Hungary on 14-17 June 2015, (June).

Schaffers, H., & Budweg, S. 2009. A Living Lab Approach for Enhancing Collaboration in Professional Communities. *Proceedings of the International Conference on Concurrent Enterprising*. Leiden, The Netherlands.

Schumacher, J., & Niitamo, V. P. (Eds.). 2008. *European Living Labs: A New Approach for Human Centric Regional Innovation*. Berlin, Germany: Fachhochschule Verlag Berlin.

Schuurman, D. 2015. *Bridging the Gap between Open and User Innovation? Exploring the Value of Living Labs as a Means to Structure User Contribution and Manage Distributed Innovation*. Doctoral dissertation, Ghent University, Belgium.

Schuurman, D., Baccarne, B., Kawsar, F., Seys, C., Veeckman, C., et al. 2013. *Living Labs as Quasi-experiments: Results from the Flemish LeYLab.* XXIV ISPIM Conference – Innovating in Global Markets: Challenges for Sustainable Growth, (June). Helsinki, Finland.

Schuurman, D., Lievens, B., De Marez, L., & Ballon, P. 2012. Towards Optimal User Involvement in Innovation Processes: A Panel-Centered Living Lab-approach. *Proceedings of PICMET'12,* 2046–2054.

Ståhlbröst, A. 2008. *Forming Future IT - The Living Lab Way of User Involvement.* Doctoral dissertation, Luleå University of Technology, Sweden.

Ståhlbröst, A., & Holst, M. 2012. *The Living Lab Methodology Handbook.* Luleå, Sweden: Luleå University of Technology and Centre for Distance-Spanning Technology (CDT).

Viitanen, J., & Kingston, R. 2014. Smart Cities and Green Growth: Outsourcing Democratic and Environmental Resilience to the Global Technology Sector. *Environment and Planning A,* 46(4): 803-819.
http://dx.doi.org/10.1068/a46242

Yin, R. 1984. *Case Study Research.* Beverly Hills, CA: Sage Publications.

Young, O. R., Berkhout, F., Gallopin, G. C., Janssen, M. A., Ostrom, E., & van der Leeuw, S. 2006. The Globalization of Socio-Ecological Systems: An Agenda for Scientific Research. *Global Environmental Change,* 16(3): 304–316.
http://doi.org/10.1016/j.gloenvcha.2006.03.004

Keywords: distributed innovation, quintuple helix innovation, social ecology, socio-ecological entrepreneurship, urban living labs

2

Enhancing Early Innovation in an Urban Living Lab: Lessons from Espoo, Finland

Soile Juujärvi and Virpi Lund

> *It is in the field of ignorance that the spark of something new is most often ignited.*
>
> Lotte Darsø
> Innovation researcher

Urban areas are often characterized by complex problems, such as social and economic deprivation, segregation, or bureaucratic administration. Urban living laboratories provide a promising approach to redefining and tackling such problems in novel ways by enabling bottom-up innovation with various actors. The present study examined an urban living lab initiative in a suburban area of Espoo, Finland, where guided workshops based on the Change Laboratory method were arranged. The findings show that, before development projects are launched, it is important to dedicate sufficient time to the early innovation process, which includes building relationships, sharing knowledge, exploring ignorance, and innovating new concepts. The study emphasizes the importance of distinguishing early innovation processes from later ones, which means separating the "preject" from the "project". We conclude that successful management of an urban living lab combines bottom-up and top-down approaches.

Introduction

Living laboratories have increasingly been used as platforms for innovation and experimentation in urban areas, involving key features of open innovation, a multi-stakeholder approach, real-life environments, and residents as users (Friedlich et al., 2013; Veeckman & Graaf, 2015). The goals of urban living labs can vary according to their environments, from small-scale experiments of new technology and services to large-scale social and economic improvement (Franz et al., 2015). In addition to complex problems in physical environments, there are social and economic problems that are difficult to understand and handle due to their multidimensional nature, such as stigmatization, unemployment, and segregation of ethnic minorities. There are also problems due to organized complexity: a multiplicity of organizations steering the region can result in competitive and overlapping systems of administration (Baynes, 2013; Wallin, 2013). Due to multi-layered problems, urban living labs call for practice-based innovation with diffuse and heterogeneous knowledge production, instead of homogenous accumulation of knowledge and clearly-defined problem solving (Melkas & Harmaakorpi, 2008). Thus, an urban living lab usually starts as a bottom-up process setting additional challenges for problem definition and the composition of actors.

The purpose of this article is to describe an urban living lab initiative in a suburban area by examining the early phase of its innovation process, which is also called the front-end phase in research literature. The front-end phase refers to the starting point of the project where opportunities are identified and concepts are created through adaptive interactions between participants. In our case, participants represented a range of living lab roles: enablers, providers, utilizers, and residents as users (Juujärvi & Pesso, 2013a; Leminen et al., 2012), whose further analysis is beyond the scope of this study. The present article is focused on advancing the urban living lab approach as an innovation method for urban development. We start by discussing the concept of the urban living lab and its implications for the early innovation process.

Urban Living Labs

The emerging interest in urban living labs calls for more precise definitions of the concept. An urban living lab has been defined as a forum for innovation that integrates residents and other stakeholders to develop and test new ideas, systems, and solutions in complex and real contexts (see Friedlich et al., 2013). Referring to Almirall and Wareham (2008), it can be seen as a specific type of open innovation network that acts as an intermediary between residents, public organizations, and private organizations to capture and codify user insights in their living environments. Franz, Tausz, and Thiel (2015) further distinguish technologically and socially oriented urban living labs, with the former ones focusing on co-developing new products and services and the latter ones dealing with the wider scope of urban and city development to improve living environments involving technological, social, and political questions. In socially oriented urban living labs, users have versatile roles as residents and citizens, and consequently, citizen participatory and co-creation processes are intertwined (Franz et al., 2015; Juujärvi & Pesso, 2013a). Consistent with this view, an urban living lab has been defined as a regional forum for innovation and dialogue focusing on solving challenges in the urban area (Friedlich et al., 2013).

Previous studies suggest that urban living labs may have various goals ranging in size and contents and calling for different forms of collaboration (see Leminen & Westerlund, 2015). In this article, we focus on socially oriented urban living labs, which are characterized by citizen participation, strong collaboration with local stakeholders, and the aim to create concepts and methodology that can be transferred into other contexts (see Franz et al., 2015). Proactive networking, experimentation as a bottom-up process, as well as commitment and longevity in development work has been previously suggested to be success factors for urban living labs (Juujärvi & Pesso, 2013a). Juujärvi and Pesso (2013a) have further elaborated actor roles for successful urban living labs. City representatives as enablers and public authorities bear an important role in creating a vision and allocating public resources. They also provide strategic leadership, promote networking across administrative units, and create public–private–people partnerships. Utilizers, such as firms and non-governmental associations, produce place-based knowledge and set small-scale objectives, and they pursue the creation of products and services suitable to the area and its residents. Research institutions engage researchers and students in development work, provide innovative methods, and take responsibility for systematic knowledge augmentation. Residents as users produce place-based user experiences, participate in experiments, and empower other citizens through co-creation.

When starting an urban living lab, it is first important to bring multiple actors together and engage them in creating a shared vision. However, collaboration in living laboratories is challenged by power struggles and inadequate cooperation skills, and therefore, actors need to learn to interact with others in the first place (Hakkarainen & Hyysalo, 2013). In particular, the role of residents is vulnerable, because their local knowledge and use of natural language is not compatible with the jargon of experts (Staffans, 2014).

Previous studies raise the question of how urban living labs should be coordinated in order to utilize their full innovation potential, which lies on the boundaries between different groups and actors (Melkas & Harmaakorpi, 2008). Socially oriented urban living labs are based on so-called Mode 2 innovation activity that is organized around a particular application, and innovators need to combine different types of information from scattered sources over lengthy periods (Gibbons et al., 1994; Melkas & Harmaakorpi, 2008.) This raises the question of how Mode 2 innovation activity could be best coordinated. Leminen (2013) distinguishes bottom-up and top-down approaches for coordinating innovation activities, with the former operating at the grassroots levels and focusing on local needs, and latter pursuing centralized and official targets. A bottom-up approach is facilitated rather than managed, whereas a top-down approach is managed rather than facilitated (Leminen et al., 2012). Leminen (2013) further points out that enabler-driven living labs (e.g., driven by city representatives) and user-driven labs (e.g., driven by residents) are characterized by bottom-up coordination, whereas provider-driven labs (e.g., driven by R&D institutions) and utilizer-driven labs (e.g., driven by companies) tend to be top-down coordinated.

Consistent with some previous studies (Lievens et al., 2011; Sauer, 2012) we argue that urban living labs should combine bottom-up and top-down developments. Whereas a bottom-up approach helps to identify needs and unanticipated ideas, a top-down approach is needed to validate ideas and concepts and to provide a formal structure. Urban living labs have usually been led by enablers collecting needs identified within the region through networking, involving a risk that innovation activities remain as information-sharing networks (see Leminen et al., 2012). Although the involvement of enablers gives urban living labs authority and legitimacy, they rarely possess R&D methodologies to transform harvested ideas towards realizing large-scale societal goals. As a result of ineffective collaboration, residents and other actors may become frustrated and give up their participatory efforts (Friedlich et al., 2013).

The aim of the present study is to advance understanding about the early innovation process and its facilitation or management. In-depth studies of microprocesses in living lab development are rare (Hakkarainen & Hyysalo, 2013), and we wish to examine how early innovation process develops within emerging urban living labs activities. For this purpose, we first introduce a model developed by Darsø (2003) based on team processes in an international company. While doing so, we presume that successful innovation processes are basically team processes regardless of the context (Juujärvi & Pesso, 2013b).

The remainders of the article is organized as follows. After discussing the early innovation process, we describe our intervention method for urban living lab development using community workshops. Next, we present our research design and findings. Finally, we provide conclusions.

The Early Innovation Process

Several models of innovation emphasize the importance of the early innovation process for nurturing creativity at the level of interaction. Among the most well-known ones is the model of knowledge creation by Nonaka and Takeuchi (1995) and further developed by Nonaka and Konno (1998). In their model, the innovation process is seen a collective learning spiral that increases knowledge through four arenas *(bas)*. In brief, the phases are described by Nonaka and Konno (1998) as follows:

1. Socialization to originating *ba* involves sharing of tacit knowledge, that is, each individual's mental model, through physical proximity and face-to-face contact, which creates common understanding and mutual trust among group members.

2. Externalization to interacting *ba* means the expression of tacit knowledge and its translation into concepts, and making it understandable to others through dialogue.

3. Combination to cyber *ba* combines new knowledge with existing knowledge into explicit knowledge, transcending the group through different media.

4. Internalization to exercising *ba* converts explicit knowledge into tacit knowledge in practice through experiments or simulations

The model of knowledge creation points out the importance of sharing tacit knowledge and explicates much of what happens in urban living lab activities. Although the model creates general understanding about the innovation process, it seems too theoretical in practical matters of urban living lab management. Therefore, we turn to a model developed by Darsø (2003), who investigated innovation teams of a large international company and identified two distinctive phases for a successful innovation process: a *project* and a *preject*. The project refers to the usual project management with goal definition and limited time; it seeks results, prefers linear progress towards goals, and employs convergent thinking and fast decision making. However, in successful innovation processes, the project is preceded by prolonged goal seeking and the emergence of divergent thinking in an open decision space, where a group of people searches for novel knowledge and probes new possibilities. From the perspective of management, this period – the preject – may seem chaotic, but it is crucial for generating radical innovations. Darsø (2003) emphasizes that the preject needs a different type of management that utilizes diverse leadership roles and functions, as identified previously in the group theory literature (e.g., Johnson & Johnson, 2002). Most importantly, to enhance preject development, one needs to know the critical parameters of the preject, which Darsø has crystallized in the diamond of innovation model: building relationships, developing knowledge, exploring ignorance, innovating new concepts) (Figure 1).

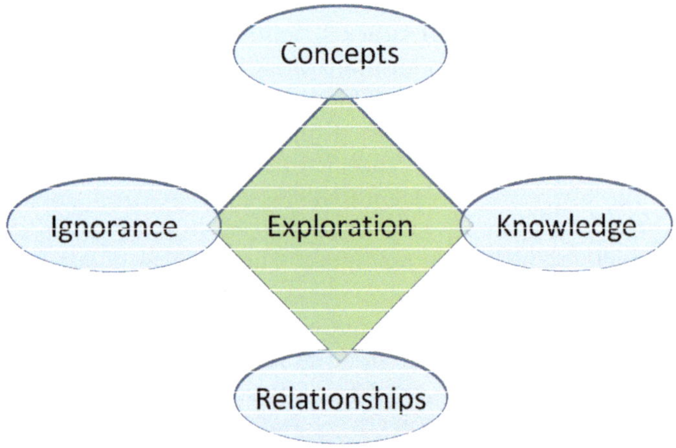

Figure 1. The diamond of innovation model (adapted from Darsø, 2003)

Knowledge in innovation processes is under constant development and has different modes, such as scientific knowledge and personal knowledge that is developed through experience, reflection, and practice. In the innovation literature, personal knowledge is seen mostly as a positive contribution, but Darsø warns that it may include personal beliefs and attitudes that may hinder group development, rather than open up new possibilities. In addition to these modes of knowledge, it is important to acknowledge the role of tacit knowledge that is transformed into explicit concepts in innovation processes (Nonaka & Takeuchi, 1995).

Relationships have a great influence on the quality of the results; mutual trust and honest communication is needed to venture into areas of new possibilities. Relationships can be determined by discussing each participant's expectations and wishes, and their level of ambition in relation to the project. Possibilities to link evolving common goals with personal ones are important, because personal goals motivate participants and enhance commitment (see Bandura, 2001). In the living lab context, it is especially important to identify user motivations that are usually based on personal, rather than professional, interests. Their participation is voluntary, and consequently, strong motivation is needed for long-term engagement.

Ignorance is the most important parameter in the diamond of innovation, because it provokes questions that boost the innovation process. However, participants who reveal their ignorance are susceptible to criticism, which again underscores the need for mutual trust and a supportive atmosphere. Finally, developing new concepts signals the emergence of innovation ideas. Words are often insufficient, and therefore, conceptualization can be advanced through drawings, figures, or 3D models. Whether incremental or radical, concepts are not yet innovations, but can become them through further development (Darsø, 2003).

Darsø (2003) further emphasizes that the poles of the axes in the model are not contradictory but complementary and reinforcing, and they can be worked on at the same time. Knowledge and ignorance can be present simultaneously, and continuous movement between them is important. Similarly, conversation about personal interests may stimulate and expand understanding about concepts, and vice versa.

Thus, the rationale of the present study is based on the assumption that, due to the multidimensional nature of urban problems, innovation processes in urban living labs are at high risk of inadequately defining problems. Therefore, a successful urban living lab initiative may require project management to nurture the early innovation process (i.e., the preject), which can be complemented by project planning, as shown in the case study that follows.

Case Study: Espoo Centre, Finland

This study is part of a three-year participatory action research project (Kemmis &McTaggart, 2000) to examine and enhance residents' participation in urban development and to develop efficient means for residents and stakeholders to collaborate in urban development. The research project included wide context mapping with interview with 32 residents and 64 stakeholders, participation in local development networks, and two main interventions: i) special workshops for residents (Juujärvi & Lund, unpublished) and ii) residents and stakeholders (i.e., community workshops). The focus area is Espoo Centre, a part of the municipal district in the City of Espoo in southern Finland, which consists of the administrative centre of the city and two surrounding neighbourhoods, with a total population of 17,000. The area is characterized by different historical layers in terms of construction of social housing, and waves of migration, mainly refugees, from the 1970s onwards. Cultural diversity in daily life is reflected in a high proportion of immigrants and more than 70 spoken languages. According to social and economic indicators, the area represents the least advantaged suburb in the City of Espoo. The area's strengths include good transportation and services, and access to surrounding natural areas that enable outdoor activities (Hirvonen, 2011; Residents' Welfare in Espoo, 2013). Several academic research projects have pinpointed challenges of the area, and consequently, the city of Espoo has undertaken several projects to improve the environment and launch a regeneration process. In recent decades, non-profit agencies in particular have been eager to start different kinds of development initiatives to improve social cohesion and the wellbeing of citizens.

Community Workshops

The present study was motivated by two main observations within the overall project: i) resident have so far shown low engagement in development endeavours and ii) there has been a lack of systematic collaboration among various stakeholders and developers. The living lab approach was assumed to provide an appropriate innovation platform for a systematic collaboration initiative that would bring together actors who do not necessarily know each other, but who would be expected to exploit each other's resources and expertise. For this purpose, a special method called "community workshops" was designed and implemented. The community workshops method represents an application of the Change Laboratory, which is based on the theory of expansive learning (Engeström & Sannino, 2010) and has been widely applied for promoting innovation and learning within organizations. The Change Laboratory is a formative intervention method where new ideas are developed and put into action in a social process of innovation. Researchers act as interventionists in the process by providing tools for envisioning, designing, and experimenting with novel forms of activities. The rationale behind interventions is to expand participants' understanding about the objects of development work, thereby enabling shared goals and enhancing collaboration. Each workshop has a specific purpose to deepen the innovation process (Virkkunen & Newnham, 2013). Based on our previous pilot (Juujärvi & Pesso, 2012), we expected that the community workshops would combine bottom-up and top-down approaches, because they engage stakeholders and residents in transforming grassroots ideas into new activities and allow enablers to play an active role in shaping the shared vision and boosting activities.

The community workshops included five successive workshops in early 2015 as follows: i) charting the situation; ii) analyzing contradictions and issues with the situation; iii) creating new models; iv) concretizing new models and then experimenting during a period of two months; and v) evaluating the experiments and making decisions about their consolidation. The workshop process yielded four experiments, which each represented new forms of practices in urban development. They represented social innovations, such as co-planning of a local community house, a multi-actor steering group for regional development, a multicultural food festival, and a multi-event square for citizens. Common characteristics in each new practice were that they required developing partnerships and coordinating multiple resources.

The approximately two-hour workshop programme included presentations of pieces of research data, speeches about future lines of development, and various innovation methods. Between 30 and 40 people attended each of the five workshops. The participants were residents, members of resident associations, managers of regeneration projects, city planners, civil servants, representatives of non-profit organizations and local parishes, and managers of shopping malls. The workshops were managed by a consultant qualified to practice the Change Laboratory method in collaboration with four researchers, including the authors, who acted as group facilitators.

Research Design

We employed ethnographic methods of participatory action research (Kemmis & McTaggart, 2000), meaning that we recorded video of all workshop activities, recorded and partly transcribed small-group discussions, and documented the materials. We acted as group facilitators and made observations on interactions and each member's role therein, and we later checked those observations by reviewing the recordings. We acquainted ourselves with the early innovation process and reviewed material from all workshops several times in order to obtain an overall view of the process and re-plan activities for the forthcoming workshop.

It became evident that the third and fourth workshops were the most critical ones. In the third workshop, newly-formed teams started to innovate and plan experiments based on the shared interest (the third workshop); In the fourth workshop, teams finalized and cross-evaluated their plans for experimenting with new ways of collaborating. Therefore, we decided to limit the analysis to these two workshops, which yielded approximately 19 hours of recordings that were partially transcribed.

Qualitative directive content analysis was employed, meaning that data are initially coded with categories derived from existing theories then are complemented with themes emerging from the data. The ultimate purpose is to validate or extend a conceptual framework or theory (Boyatzis, 1998; Zhang & Wildemuth, 2000). This analysis was guided by the following research questions:

1. *How are the features of the preject manifested in workshop activities?*

2. *How can the process of the preject be enhanced by workshop interventions?*

The analysis proceeded as follows. The first author listened to recordings and wrote down observations that were cross-checked by the second author and compared with earlier observations made by the group facilitators. Then, the written observations were coded into categories derived from the components of the preject (i.e., building relationships, developing knowledge, exploring ignorance, innovating new concepts). The content of each category was compared with the theory, and some sub-categories providing evidence of new knowledge were added. Finally, the categories were written as descriptions presented in the findings. With regard to the second research question (How can the process of the preject be enhanced by workshop interventions?), we decided to describe the interventions at the third workshop because they reveal the dynamics between them and the innovation of new concepts.

Workshop Interventions

The objective of the third workshop was to shape goals for near-future development and start to plan experiments based on the previous analysis of contradictions and conflicts in current practices that have hampered development actions in the area (Virkkunen & Newnham, 2013). The participants were guided to choose a group with a pre-determined theme that emerged from the previous workshops and that would match their interests (e.g., multicultural integration, common premises, coordination of urban development, "wild card"). Following the guidelines of the Change Laboratory method, three sets of stimuli were given at different points of the workshop to provoke innovative thinking: i) a synopsis of contradictions in current development presented by a principal researcher, ii) a speech about future urban development given by a director of the city planning department, and iii) a shared group reflection on future possibilities by five volunteering participants. The purpose of these interventions was to enhance common understanding on goal setting and to help participants shape their roles in planned experiments, but the retrospective analysis also revealed unintended positive effects. The most powerful one was the director's speech, which triggered creating new concepts among the participants.

The director told about future lines of local urban planning up to 2030, based on an envisioned zoning scheme of the area. City planning would be focused on laying foundations for a physical environment of high quality, which in turn shall enhance residents' wellbeing and sense of community, for example, by creating meeting places. However, he also emphasized that urban planning procedures do not involve means that would directly address to the problems identified by the participants in the current workshops. Social and cultural aspects are not sufficiently taken into account in the zoning process, and there is a lack of multi-professional cooperation due to the rigid boundaries of administrative units. He admitted his ignorance about how to proceed with these deficiencies and invited the audience to give him some advice and ideas.

The speech was followed by reflections in pairs and a lively plenary discussion, in which participants pointed out critical aspects lacking in city planning: the plan did not cover aspects of social and cultural development, and more specifically, it did not provide any means to prevent further segregation of immigrant groups. The recordings in small-group reflections revealed that the speech had triggered innovative new concepts to overcome limitations of the current city planning. The concept was later explicated in the plenary discussion as social zoning, as illustrated by the following participant comments:

> *"I have never thought before that, in zoning, there are no marks for social things. It is a weird idea, an interesting idea, it fascinates me."*
>
> *"Is community-building a solution for involving social and cultural development in the zoning process? But what is a name for this process? Is it a zoning scheme? How could social aspects be marked on the scheme in some way? How do you put them on it?"*
>
> *"We need social and cultural strategies in zoning, but what is the word for this?"*
>
> *"We can elaborate what it [social zoning] could be. Now it is hidden between the lines of scheme markings."*

After the interventions, the groups were instructed to start planning an experiment for a new way of stakeholder collaboration within the following two months. The participants were encouraged to change their group choice if it no longer matched their interests. The planning continued in the successive workshop, and the evolving plans were cross-evaluated at several points.

The Preject's Group Process

All established teams succeeded in generating a solid plan for their experiments, which represented new co-operational initiatives, including a regional development committee, the co-design of a local community house, a citizen square, and a multicultural food festival. Each planning process involved the elements of the early innovation process: building relationships and sharing knowledge preceding exploring ignorance and innovating new concepts complemented with refinements, as shown in Figure 2. The order of the elements is logical rather than chronological; the teams changed their focus back and forth during the process. In addition, even though all elements could be discerned from the flow of discussion, they were not equally balanced across the teams.

Building relationships took place through the whole process. In the beginning, group members introduced themselves and gave some background information, but later on, they started to ask about and reveal their personal motivations and interests, leading to in-depth discussions. Participants also built friendships and working alliances across the teams. Multiple discussions as a part of the Change Laboratory procedure encouraged them to informally reveal to each other their personal interests, motivations, and feelings.

The important step in building relationships was the emergence of *actor roles,* which took place after the teams started to plan their innovation projects on the later rounds of the cycle (see Figure 2). In particular, city representatives as enablers had a distinguished role in creating a vision and allocating resources. The representatives of the firms and non-governmental organizations (NGOs) were eager to pursue new services and operational models, and residents showed nuanced knowledge of local conditions and empowered other citizens to participate in forthcoming experiments, which is consistent with previous findings (Juujärvi & Pesso, 2013a). These roles, however, emerged as a result of the planning process and were situation-contingent rather than based on participants' acknowledged positions or expertise (Nyström et al., 2014). One team member could play multiple roles while being a key driver in the process. Some participants adopted the role of provider/developer, which is strongly encouraged by the Change Laboratory methods. The teams also missed some roles in beginning, typically utilizer or enabler, to accomplish intended actions, and started to look for them outside.

Sharing knowledge was critically tied to building relationships, because participants got to know each other by exchanging knowledge. The exchange of tacit knowledge included sharing personal ways of thinking and emotional outbursts revealing values and attitudes, and its exchange was present through the process. Delivering *tacit knowledge* was supported by physical proximity in small groups and intensive working periods, and it enabled building a highly positive, lighthearted with plenty of joking and laughing (Nonaka & Takeuchi, 1995). Knowledge included participants' specific *local knowledge* that was especially useful when defining problems. The *expert knowledge* of team members was exploited when the plans were realized; during the ideation process, it was largely ignored.

Scrutinizing limitations of current practices is a starting point for development actions in the Change Laboratory method, and therefore, exploring one's own ignorance grows a collective effort to explore the limits of shared knowledge. Ignorance was explored to varying degrees among teams and it was largely induced by the interventions, especially the director's speech, which revealed his own ignorance, as described above. When reflecting on ignorance, participants also hinted about their own skills and expertise, and therefore, this element was renamed as *exploring limits of expertise* (see Figure 2). This in turn prompted *innovative ideas,* which were further elaborated into *new concepts.* The teams struggled to find an appropriate name for their future experiments, because the familiar ones did not align with the core intent of their innovation, as illustrated by the following comments from participants:

> *"Even though the regional welfare group we used to have in past has been ceased, it should be something like this. But it must have a different name. But unlike it, this group must have responsibility, duties, and resources; it cannot be any sort of discussion or coffee drinking club."*

> *"Based on its tasks, it ought to be a regional development group. It is an awfully dull name, but it is what this all is about."*

> *"We are not satisfied with this name. It must be much cooler, more attractive. Let's put it in quotation marks."*

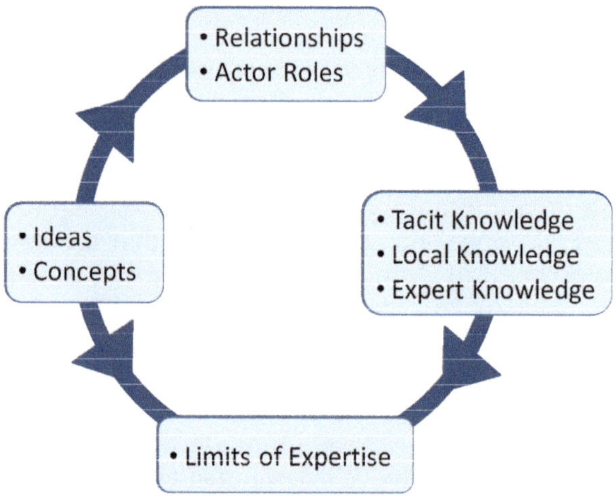

Figure 2. Group process of a preject

The complexity of the concept was related to the length of the joint group process. In two teams, most members had joined together in the previous workshops, and had already shared ideas to some extent. These teams had lively, enriching discussions. In contrast, two other teams had several member changes between the workshops. They spent a lot time exchanging knowledge and getting to know each other and were forced into premature decisions due to time limits in the previous workshop, which caused them to spend time reshaping ideas in the successive workshop.

Conclusion

This article elaborated the early innovation process in an urban living lab initiative and resulted in several new practices in urban development. They represented social innovations created through collaboration between participants representing diverse living lab roles. It became evident that all living lab roles were needed to realize innovation intents. The strength of socially oriented urban living labs situated in a certain geographical region lies in the dissemination of local knowledge along emerging social networks (Melkas & Harmaakorpi, 2008). The bottom-up approach is a suitable starting point for problem definition and brainstorming, but it must be adjusted by a top-down approach that provides information about official visions, goals, plans, and procedures. The top-down approach can be empowered by civil servants and politicians who can be equal participants or otherwise engaged in living lab activities. Our findings suggest that urban living labs provide a promising approach for neighbourhood renewal, which has been dominated by centralized top-down planning and urgently needs tools for citizen involvement (Pennen & Bortel, 2015). Urban living labs could work as an intermediary bringing self-organizing groups and city developers together to co-create urban space (Horelli et al., 2015). However, this potential can be lost if urban living labs are poorly managed. In conclusion, three key lessons can be taken from the study:

1. *Successful urban living lab activities require sufficient time dedicated to early innovation process.* Urban living labs are usually established to solve complex problems for which several unsuccessful attempts have already been made. Urban living labs provide an opportunity to bring together stakeholders with diverse knowledge and experience, and to collaborate in tackling those problems. In order to actualize innovation potential, sufficient time should be dedicated to building relationships, exchanging knowledge, and establishing shared goals. Even in low-threshold settings, the participation of residents seems to require some citizen skills and relevant basic knowledge, and consequently, additional support and encouragement are needed (Veeckman & Graaf, 2015). Even though not all participants would engage in development projects as a result of the bottom-up process, building alliances across different boundaries is a valuable result from the perspective of networking.

2. *Innovations result from successful team processes.* Early innovation process took place in teams that enable relationships to be built and relevant knowledge to be constructed, as well as the exploration of ignorance and concepts (Darsø, 2003; Darsø & Høyrup, 2011). The present findings suggest that building relationships and sharing knowledge precede the exploration of ignorance and the innovation of new ideas and concepts. Iterating cycles form a progressive spiral, leading to the exploration of actor roles and the limits of expertise, and to the development of more nuanced concepts. There must be sufficient trust and confidence before members dare to reveal their ignorance and to express unconventional ideas, as well as to confront the opinions of others. Members also need to understand and share an evolving innovation concept in order to commit themselves to development projects.

3. *Prejects can be managed.* The present study provides some insights into preject management. First, the alternation of plenary discussions and working in groups created tension between general goals and the interests of participants, and it cross-fertilized innovative thinking among participants. The plenary discussions helped spread ideas across the teams and encouraged the sharing of feedback and additional resources. Second, special attention should be given to protecting and nurturing group processes, for example, using group facilitators. Under time pressure, teams tend to make premature decisions leading to rather conventional projects with limited commitment; therefore, the team should process their idea until something truly new to them emerges and they become emotionally engaged. However, group processes are difficult to maintain across the workshops due to fluctuations in participation. It could be advisable to build teams around key people who are strongly motivated. Third, of crucial importance is a constructive atmosphere that encourages discussions of ignorance and welcoming questions and criticism. Innovations can further be enhanced

through specific interventions aimed at exploring ignorance and the limits of expertise. This can be simply done by asking questions, but more sophisticated tools are available (e.g., Virkkunen & Newnham, 2013). With these conditions, an urban living lab can provide a forum for creative collaboration and problem-solving in community and urban development.

Acknowledgements

An earlier version of this article was presented at the XXVI International Society for Professional Innovation Management (ISPIM) Conference – Shaping the Frontiers of Innovation Management, Budapest, Hungary, June 14–17, 2015.

About the Authors

Soile Juujärvi is a Principal Lecturer at the Laurea University of Applied Sciences and an Adjunct Professor in Social Psychology at the University of Helsinki in Finland. She holds a Doctor of Social Science degree from the University of Helsinki. Her research interests include moral and ethics education, and innovation processes in living labs. She worked as a principal researcher in the project Caring and Sharing Networks (2013–2015), which aims to enhance citizen participation and stakeholder collaboration in the city of Espoo, southern Finland.

Virpi Lund is a Senior Lecturer in Social Services at the Laurea University of Applied Sciences in Finland. Her research interest is residents' agency and learning through participation in urban development. She worked as a researcher in the project Caring and Sharing Networks funded by Developmental Programme for Residential Areas and the Finnish Ministry of the Environment. She holds a Master of Education degree from the University of Helsinki in Finland.

References

Almirall, E., & Wareham, J. 2008. Living Labs and Open Innovation: Roles and Applicability. *The Electronic Journal for Virtual Organizations and Networks,* 10: 21–26.

Bandura, A. 2001. Social Cognitive Theory: An Agentic Perspective. *Annual Review of Psychology,* 52: 1–26.
http://dx.doi.org/10.1146/annurev.psych.52.1.1

Baynes, T. M. 2009. Complexity in Urban Development and Management: Historical Overview and Opportunities. *Journal of Industrial Ecology,* 13: 214–227.
http://dx.doi.org/10.1111/j.1530-9290.2009.00123.x

Boyatzis, R.D. 1998. *Transforming Qualitative Information.* Thematic Analysis and Code Development. London: Sage.

Darsø, L. 2003. *Findes der en formel for innovation?* [Is there a Formula for Innovation?] Copenhagen: Børsen Forum.

Darsø, L., & Høyrup, S. 2011. Developing a Framework for Innovation and Learning in the Workplace. In H. Melkas, and V. Harmaakorpi (Eds.), *Practice-based Innovation: Insights, Applications and Policy Implications:* 135–154. Berlin: Springer.

Engeström, Y., & Sannino, A.-L. 2010. Studies of Expansive Learning: Foundations, Findings and Future Challenges. *Educational Research Review,* 5(1): 1–24.
http://dx.doi.org/10.1016/j.edurev.2009.12.002

Franz, Y., Tausz, K., & Thiel, S.-K. 2015. *A Comparison of Different Living Lab Concepts.* Paper presented at The XXVI ISPIM Conference – Shaping the Frontiers of Innovation Management, Budapest, Hungary, June 14–17, 2015.

Friedrich, P., Karlsson, A., & Federley, M. 2013. *Report 2.1. Boundary Conditions for Successful Urban Living Labs.* The Hague, The Netherlands: SubUrbanLab.
http://jpi-urbaneurope.eu/wp-content/uploads/2014/01/Report-Boundary-conditions-for-successful-Urban-Living-Labs.pdf

Gibbons, M., Limoges, C., Nowotny, H., Schwarzman, S., Scott, P., & Trow, M. 1994. *The New Production of Knowledge.* London: Sage.

Hakkarainen, L., & Hyysalo, S. 2013. How Do We Keep the Living Laboratory Alive? Learning and Conflicts in Living Lab Collaboration. *Technology Innovation Management Review,* 3(12): 16–22.
http://timreview.ca/article/749

Horelli, L., Saad-Suhonen, J., Wallin, S. & Botero, A. 2015. When Self-Organization Intersects with Urban Planning: Two Cases from Helsinki. *Planning Practice and Research,* 3(13): 286–302.
http://dx.doi.org/10.1080/02697459.2015.1052941

Hirvonen, J. 2011. *Suvela tilastoissa ja asukkaiden kokemana.* [Suvela in statistics and residents' experiences] Aalto University Publications TIEDE + TEKNOLOGIA 17/2011. Helsinki, Finland: Aalto University.

Johnson, D. W., and Johnson, F. P. 2002. *Joining Together: Group Theory and Group Skills.* Boston: Allyn and Bacon.

Juujärvi, S. & Pesso, K. 2012. *Ihmiset tekevät Suurpellon: Koulii-hankkeen arviointitutkimus* [Evaluation study on the Koulii project] Laurea Publications 11. Vantaa, Finland: Laurea University of Applied Sciences.

Juujärvi, S. & Lund, V. Unpublished. *Engaging Residents: Insights from Participatory Action Research into Urban Living Labbing.* A manuscript submitted for publication.

Juujärvi, S. & Pesso, K. 2013a. Actor Roles in an Urban Living Lab: What Can We Learn from Suurpelto, Finland? *Technology Innovation Management Review,* 3(11): 22–27. http://timreview.ca/article/742

Juujärvi, S. & Pesso, K. 2013b. Teams as a Tool for Regional Innovations in Professional Education. *Interdisciplinary Studies Journal,* 3(3): 19–29.

Kemmis, S., & McTaggart, R. 2000. Participatory Action Research. In N. Denzin, & Y. Lincoln (Eds.), *Handbook of Qualitative Research:* 567–614. Thousand Oaks, CA: Sage.

Leminen, S. 2013. Coordination and Participation in Living Lab Networks. *Technology Innovation Management Review,* 3(11): 5–14. http://timreview.ca/article/740

Leminen, S. & Westerlund, M. 2015. Cities as Labs: Towards Collaborative Innovation in cities. In P. Lappalainen, M. Markkula, & H. Kune (Eds), *Orchestrating Regional Innovation Ecosystems. Espoo Innovation Garden:* 167–175. Espoo, Finland: Aalto University.

Leminen, S., Westerlund, M., & Nyström, A.-G. 2012. Living Labs as Open-Innovation Networks. *Technology Innovation Management Review,* 2(9): 6–11. http://timreview.ca/article/602

Lievens, B., Schaffers, H., Turkama, P. Ståhlbröst A., & Ballon, P. 2011. Cross Border Living Labs Networks to Support SMEs Accessing New Markets. In P. Cunningham and M. Cunningham (Eds), *eChallenges e-2011 Conference Proceedings.* Dublin: IIMC International Information Management Corporation.

Melkas, H., & Harmaakorpi, V. 2008. Data, Information and Knowledge in Regional Innovation Networks. *European Journal of Innovation Management,* 11(1): 103–124. http://dx.doi.org/10.1108/14601060810845240

Nonaka, I. & Konno, N. 1998. The Concept of "Ba": Building a Foundation for Knowledge Creation. *California Management Review,* 40(3): 40–54. http://dx.doi.org/10.2307/41165942

Nonaka, I. & Takeuchi, H. 1995. *The Knowledge Creating Company.* London: Oxford University Press.

Nyström, A.-G., Leminen, S., Westerlund, M., & Kortelainen, M. 2014. Actor Roles and Role Patterns Influencing Innovations in Living Labs. *Industrial Marketing Management,* 43: 483–485. http://dx.doi.org/10.1016/j.indmarman.2013.12.016

Pennen van der, T., & Bortel van, G. 2015. Exemplary Urban Practitioners in Neighbourhood Renewal: Survival of the Fittest... and the Fitting. *VOLUNTAS: International Journal of Voluntary and Nonprofit Organizations,* June 2015: 1–20. http://dx.doi.org/10.1007/s11266-015-9600-4

Residents' Welfare in Espoo. 2013. *Espoolaisten hyvinvoinnin tila 2013.* Espoo, Finland: City of Espoo. Accessed October, 20, 2015: http://www.espoo.fi/fiFI/Espoon_kaupunki/Tietoa_Espoosta/Tilastot_ja_tutkimukset/Hyvinvointi/Espoolaisten_hyvinvoinnin_tila_2013(39045)

Sauer, S. 2012. Do Smart Cities Produce Smart Entrepreneurs? *Journal of Theoretical and Applied Electronic Commerce Research,* 7(3): 63–73. http://dx.doi.org/10.4067/S0718-18762012000300007.

Staffans, A. 2004. *Vaikuttavat asukkaat: Vuorovaikutus ja paikallinen tieto kaupunkisuunnittelun haasteina.* [Influencing Residents: Interaction and Local Knowledge as Challenges of Urban Planning]. Espoo: Helsinki University of Technology.

Veeckman, C., & van der Graaf, S. 2015. The City as Living Laboratory: Empowering Citizens with the Citadel Toolkit. *Technology Innovation Management Review,* 5(3): 6–17. http://timreview.ca/article/877

Virkkunen, J., & Newnham, D. S. 2013. *The Change Laboratory: A Tool for Collaborative Development of Work and Education.* Rotterdam: Sense Publishers.

Wallin, S. 2013. Urban Complexity Challenging Urban Planning. In L. Horelli (Ed.), New Approaches to Urban Planning. *Insights from Participatory Communities:* 23–41. Aalto University Publication Series Aalto-ST 10/2013. Helsinki: Aalto University, 2013.

Zhang, Y., & Wildemuth, B. M. 2009. Qualitative Analysis of Content. In B. Wildemuth (Ed.), *Applications of Social Research Methods to Questions in Information and Library Science:* 308–319. Westport, CT: Libraries Unlimited.

Keywords: Change Laboratory; innovation management, innovation process; preject; urban living labs

3

Developing Tacit Knowledge of Complex Systems: The Value of Early Empirical Inquiry in Healthcare Design

Chantal Trudel, Sue Cobb, Kathryn Momtahan, Janet Brintnell, and Ann Mitchell

> *To know that you do not know is the best.*
> *To think you know when you do not is a disease.*
> *Recognizing this disease as a disease is to be free of it.*
>
> Lao-Tzu (6th–5th century BC)
> Ancient Chinese philosopher

Infection prevention and control has been the subject of much study in medical and epidemiological research and a variety of best practice guidelines have been developed to support healthcare workers and related stakeholders. Yet, despite the availability of information, managing healthcare-associated infections remains a challenge because the relevant explicit knowledge is not being adequately developed and mobilized as tacit knowledge for use "on the front lines". Some researchers have called for a human factors perspective to help address challenges in designing for infection prevention and control, but relatively few studies have been conducted to date. Researchers also suggest that empirical inquiry is needed to better inform the design process, and particularly the design of complex systems where attention to detailed processes and interactions can support the success of an intervention. A human factors approach can help designers develop a deeper understanding of work processes, technology considerations, as well as physiological, psychological, cultural, and organizational factors. The need is particularly pressing in low-resource healthcare environments where funds, time, and human resources may be scarce and strategic design decisions based on evidence are needed to support meaningful and effective changes. With this in mind, a human factors study was conducted in an existing neonatal intensive care unit to identify the influence of product and environment design on infection prevention and control and to inform recommendations for improvement. In this case study, we illustrate how the application of an empirical, methodical approach can help design professionals and stakeholders develop tacit knowledge of complex systems – knowledge that can be used to better inform design priorities, the design process, decision making, and the allocation of resources to help maximize improvements.

Introduction

Every year, more than 200,000 patients in Canada acquire an infection while receiving healthcare and more than 8,000 die as a result (Public Health Agency of Canada, 2013). Infections can arise from the entry and multiplication of a microorganism in a person's tissue (Public Health Ontario, 2012a, 2012b). As healthcare workers perform tasks, they circulate among patients, objects, and surfaces, thereby creating opportunities for transmitting microorganisms and possibly contributing to the spread of hospital-acquired infections. Newborns hospitalized in neonatal intensive care units are particularly vulnerable because of inherent risk factors such as low birth weight, underlying illness, undeveloped immune systems, and greater skin permeability. Other risk factors include poor staff-to-patient ratios, crowded environments, and exposure to invasive devices (Public Health Ontario, 2012a, 2012b).

These challenges point to the need for more care in the study of design for infection prevention and control. Best practice guidelines and protocols do exist to prevent and control infections; the key challenge is developing and mobilizing the tacit knowledge required to realize the intended improvements.

Currently, there are two basic protocols in infection prevention: i) routine practice, which is a fundamental requirement used on "all patients at all times in all healthcare settings" (Public Health Agency of Canada, 2012) and ii) additional precautions, which include a more rigorous process required when risk assessments performed on a patient suggest extra barriers are needed to mitigate the risk of infection transmission. Four "moments" of hand hygiene (Figure 1) are recommended in routine practice to break the chain of infection transmission. But in neonatal care, five moments are recommended, with an additional hand hygiene step required prior to entering the neonate environment (Public Health Ontario, 2012a). Despite the prevalence of guidelines for infection prevention and control and new technologies focused on improving survival rates and outcomes, infectious complications in neonates is challenging for healthcare institutions (Pessoa-Silva et al., 2007). For example, a survey of 997 pediatric patients across 19 Canadian hospitals revealed that 80 children had a combined total of 91 healthcare acquired infections (or 9.1 % of patients surveyed) (Gravel et al., 2007). The study also showed that the highest prevalence of healthcare associated infections was in the neonate age category. This group was 1.5 times as likely to have a healthcare associated infection than all other groups combined.

Some common reasons given for poor compliance with infection prevention measures include: poorly located and insufficient quantities of sinks and hand sanitizers (e.g., Graham, 1990; Muto et al., 2000; Pittet, 2000); poorly located supplies (e.g., Hendrich, 2003); lack of knowledge of protocols or disagreement with protocols (e.g., Pittet, 2001); perceptions that the risk of transmitting infections is low; perceptions that protocols interfere with staff–patient relations; perceptions that patient needs take priority over hand hygiene (e.g., Pittet, 2001); and issues such as understaffing, forgetfulness, insufficient time, and high workload (e.g., Archibald et al., 1997; Pittet, 2001; Ulrich et al., 2004). This brief list illustrates the broad range of considerations in infection prevention and control, and researchers are suggesting that multimodal, multidisciplinary, and systemic approaches are needed to adequately address the scope of issues (Alvarado, 2012; Pessoa-Silva et al., 2007; Pittet, 2001).

Here, we present the results of a study into the challenges of infection prevention and control experienced by a neonatal intensive care unit (NICU) as a case study

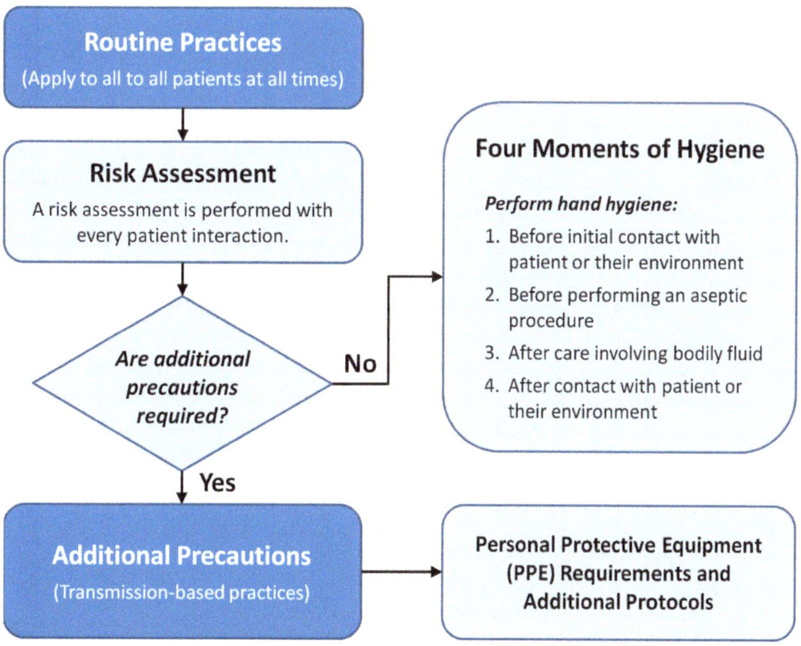

Figure 1. The "four moments" of hand hygiene in routine practice

of the broader issue of knowledge mobilization in complex systems. We discuss the rationale for the study design and briefly outline the methods and the principal outcomes that were developed from applying the approach. The two main outcomes from the study are: i) a framework illustrating what healthcare workers are experiencing in infection prevention and control and how this relates to design and ii) the NICU IPAC [Infection Prevention and Control] Design Exploration Guide, which categorized issues and opportunities for future study and design development.

Another, less obvious outcome that we discuss is the benefits associated with the inquiry process itself in facilitating a deeper knowledge of the issues – knowledge that led to the development of the NICU IPAC Design Exploration Guide. The process required a thorough and systematic approach to data collection and analysis, which we propose helped improve our understanding of the issues healthcare workers are experiencing in infection prevention and control. We provide evidence that suggests human-centred design can deepen our knowledge of complex work and support a more informed path for designing in the field.

The Existing Context and the Need for Pre-Design Investigation

The context of this study, a Level III neonatal intensive care unit (NICU), was renovated prior to the introduction of standards for NICU design, which recommend single patient rooms, appropriate space per infant, and dedicated hand-wash sinks for each patient (White, 2006, 2007). Apart from these basic considerations, stakeholders explained that "adhering to "the five moments" of hand hygiene was difficult due to the nature of their work in relation to the design of products and their environment. As a result, the unit adopted the "four moments" model and zoned the unit into the hospital environment and the patient environment (Figure 2). In this model, healthcare workers were required to perform hand hygiene before contact with each zone, bring only sterile supplies into the patient environment, and disinfect items moving between zones in order to break the potential path of pathogen transmission. Despite reports of high hand hygiene compliance (approximately 90%) during the time of the study, the unit wanted to improve measures to support infection prevention.

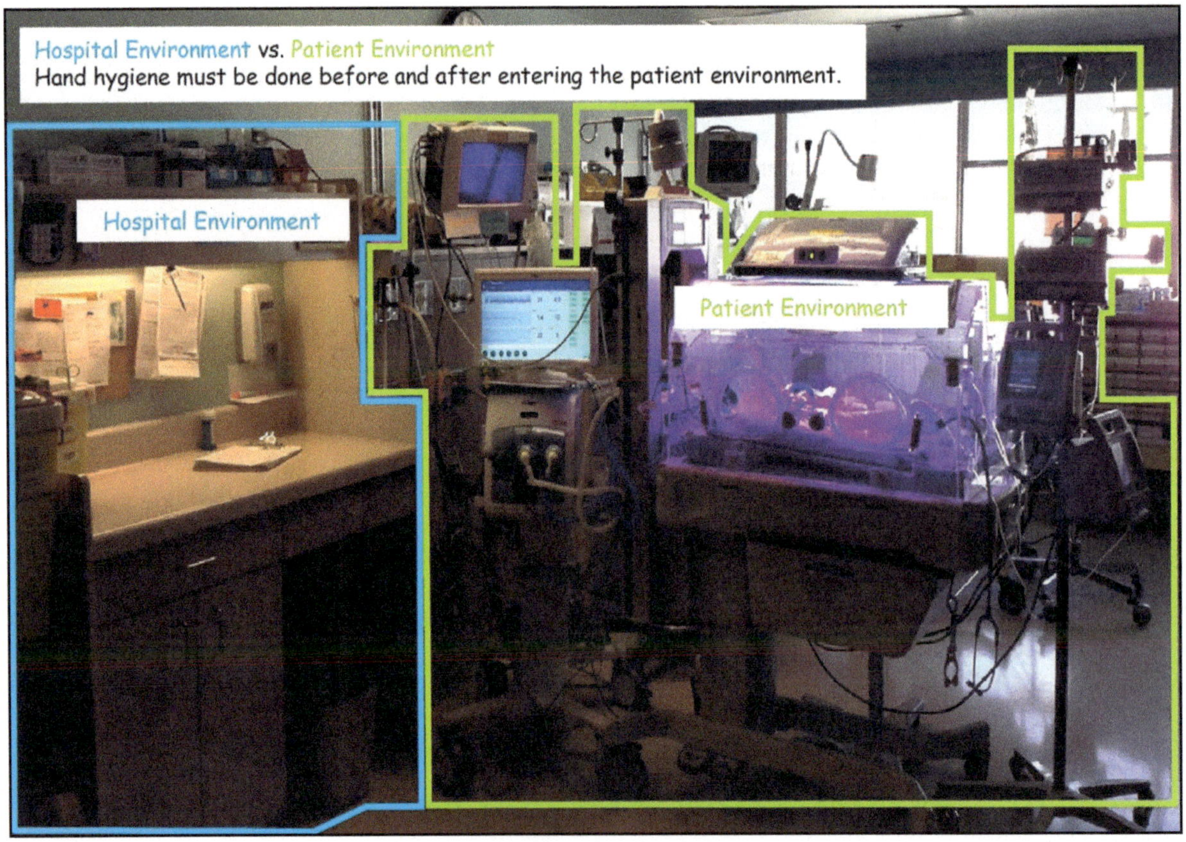

Figure 2. Zoning of the patient environment and the hospital environment in the neonatal intensive care unit

Stakeholders suggested their main challenges in infection prevention and control involved overstocking of bedside supplies, shared equipment, and workflow relationships to products and the environment. The researcher, who had past experience in healthcare design, asked if supply carts had been considered to replace the bedside counter, given that some were capable of being decontaminated, could move with the patient if needed, and might provide opportunities for mounting and organizing supplies and accessories. This suggestion was met with hesitation by stakeholders, who felt this would not resolve their challenges. However, they were interested in studying the issue in greater depth and using the knowledge developed in the study to inform future design initiatives. Considering the health and safety and financial risks involved in developing an effective strategy for improvement, the value of conducting empirical inquiry early on – indeed prior to – developing specific design interventions was clear.

Support for Empirical Inquiry in the Design of Complex Systems

There is growing interest in using empirical inquiry prior to design to help inform development (Brehmer et al., 2014). Friedman (2003) has voiced concerns with designers taking on increasingly complex projects, stating that failures and issues related to design are commonly due to a "lack of method and absence of systematic and comprehensive understanding". In discussing architecture, Remjin (2006) highlights several problems with top-down design approaches, including that they rely too heavily on: the objectives and goals set by management; the past experience of the architect, which may not necessarily fit the project needs; and generalized programmatic requirements, which may not recognize the difficulties and intricacies of complex work. Remjin (2006) notes that, in "a complex work situation it is plausible that practice differs from the expected situation by architect and perception of management" and that this can lead to designs that do not support work processes.

A human factors/ergonomics approach in design focuses on minimizing the effects of constraints, complementing the strengths and abilities of end users, and not forcing people to adapt to undesirable conditions (Chartered Institute of Ergonomics and Human Factors, 2016). These principles are supported by analyzing existing conditions as well as possibilities for new states or situations (Remijn, 2006). The experiences and perspectives of front-line workers, as well as goals set by management, are incorporated to develop a "systems approach" to generating proposals. A systems perspective may also help identify factors that can improve a situation outside the traditional domain of design (e.g., the impact of education and training) or outside a particular field of design (e.g., the impact of product versus architectural design).

The Study Design

Data collection methods from human-centred design frameworks (see ISO 2015,; Maguire, 2001) can help foster a greater understanding of front-line workers, their tasks and objectives, and the greater work context (Rogers et al., 2012). Data collection methods for human-centred design, such as naturalistic observation, encourage participant feedback during observations, which can help people — particularly in complex work — explain what they do (Rogers et al., 2012).

With regards to analyzing data, thematic analysis respects the natural context of the phenomenon by integrating the views and experiences of participants to support an in-depth understanding of their situation (Vaismoradi et al., 2013). The process involves familiarizing oneself with the data, generating codes, searching for themes among codes, and defining themes to produce a framework (Braun & Clarke, 2006) that can move analysis beyond describing individual experiences (Guest et al., 2012) to theorizing why certain behaviours are prevalent. Coding actions keeps the codes tied to activities, reducing the tendency "to make conceptual leaps" from the data (Charmaz, 2014) and helps support the validity of the interpretation (Saldaña, 2009). The resulting framework serves as an organizing principle for illustrating and disclosing the interpretation of the data to the researcher and translating knowledge to a wider audience (Attride-Stirling, 2001).

These methods were selected to identify breaches in infection prevention, behavioural patterns and perceptions related to observed breaches, and the potential influence of design on infection prevention practice. The methods were chosen to help generate a framework that illustrates the issues staff are experiencing, to verify this understanding with stakeholders, and to guide recommendations for future study or design development.

Approach

The study took place in a Level III NICU primarily at the patient bedside, where the majority of patient contact occurs. Field notes, photos, and sketches were used to document the site, comments from stakeholder meetings, observations, and participation of 81 healthcare workers (e.g., nurses, respiratory therapists, housekeepers, supply workers, physicians, technicians). Fifty hours of observations in 4-hour intervals were conducted over the course of twelve shifts on weekdays and weekends, distributed equally over a 24-hour work cycle, covering the beginning, middle, and end of shifts. Observing a large pool of healthcare workers for 50 hours during a 24-hour work cycle produced a rich dataset.

Thematic analysis took place throughout the collection of information. The researcher assigned action codes to the data and wrote memos that captured reflections throughout the study about what was being observed but also on the study process itself. The process helped "chunk" (or categorize) the data into themes, which led to a final framework outlining the main issues healthcare workers are experiencing in infection prevention and control and a recommendations guide to assist the unit with future developments.

Findings

The study revealed that healthcare workers lack a shared mental model of infection prevention and that design is failing to support a shared, functional model. In some cases, the model was functional, with some healthcare workers taking extra precautions in infection-prevention practice beyond what would be required. Other healthcare workers exhibited hazy or faulty models, which resulted in lapses or breaches in infection-prevention practice. The study also showed that the design failed to provide appropriate or clear cues, space requirements, and other necessary attributes to support the understanding and practice of infection prevention and control. This core theme was illustrated in a larger framework of understanding and helped in the development of a recommendations guide to support future initiatives for improvement. For more detailed information on the overall study findings and framework, please refer to Trudel and colleagues (2016a, 2016b). Apart from these findings, with a particular focus on knowledge mobilization and study design, we asked:

- What evidence is there to suggest that the methods themselves fostered a deeper understanding of the experience of healthcare workers in infection-prevention practice?
- Can this approach help inform and strengthen strategies for design development?

Evidence for the value of the approach

The use of multiple data-collection methods and media led to a rich dataset of the environment, products, equipment, supplies, work processes, participant behaviour, and participant comments. Spending 50 hours on the unit over a 24-hour work cycle allowed the researcher to observe processes related to infection prevention and control that occur only at specific points in the work cycle (e.g., blood work, rounds). The data and insights from observing such processes may be missed in design approaches that rely predominantly on collecting information from user-group meetings held outside the unit or during regular work hours. Participant comments and observations demonstrate that the methods helped clarify misinterpretations or correct assumptions held by the researcher (Box 1). Observing front-line staff on site helped them discuss the complexities of their work and infection prevention and control in relation to the existing design, allowing them to physically illustrate the issues and suggest possibilities for improvement.

Field notes and codes were reviewed during and after the observation to produce a thematic framework illustrating the relationship between design and infection prevention and control. This process helped the researcher "chunk" codes into categories (Table 1), identify dominant categories, identify relationships between categories, and develop a high-level, systemic picture of the experience with infection prevention and control within the unit. Recurring words from the dataset were also assessed. Whereas the frequency of some words could perhaps be foreseen (e.g., isolette, supplies, bedside counter), others were not readily anticipated by the researcher (e.g., drawers, chart, walking), suggesting more detailed investigation may be needed in these areas.

The researcher's memos (Table 2) demonstrate that using empirical and systematic methods prior to design may help deepen our understanding of the experience of healthcare workers in infection prevention and control and may inform recommendations for further research and design development. The memos illustrate:

- missing details or micro-interactions relevant to infection prevention and control, which required verification in subsequent observations

Box 1. Examples of observation scenarios and participant quotations

The scenarios below illustrate the rich detail that came from the participation of healthcare workers in the study and how methods from human-centred design such as naturalistic observation and stakeholder meetings can help clarify unknowns or misinterpretations about the context within which infection prevention and control is practiced.

Scenario 1: Routine Checks

Researcher (speaking to Nurse A): "Can you explain what you're doing when you silence the alarms?"

Nurse A: "The oxygen saturation levels usually drop when an infant comes close to the end of a feeding since the stomach expands and compresses the diaphragm and lungs. When this happens the physiological monitor alarm rings off warning that oxygen saturation levels are low. The nurses then adjust the siPAP or biPAP to adjust the oxygen supply to increase oxygen saturation. With the IV pumps, the alarm might go after telling me the infusion is complete or the rate we've set it at is complete."

Researcher: (watching Nurse B set up feeding supplies in the corner of the work surface closest to the bedside): "Why do you set them up there?"

Nurse B: "To avoid going in the drawers."

Researcher (speaking to Nurse B): "Why have you left the stethoscope on the cot instead of putting it back on the blender?"

Nurse B: "I'll be using it again and it reminds me to disinfect it before putting it back."

Scenario 2: Feeding

Researcher: I move to room 2 to observe. Nurse D is feeding the infant in one of the parent chairs. Nurse E is burping the infant within the isolette, her arms and body are in an awkward posture. I missed where Nurse E's infant was fed. Nurse D and Nurse E are each taking care of two infants. After burping the infant, Nurse E is trying to set up a feeding syringe to give the remainder of the feed by gravity in the incubator.

Researcher (speaking to Nurse E): "Can you explain what you're setting up?"

Nurse E: "When infants are not at risk for regurgitating and are capable of handling a gravity feed we will set up this way. But if the feed amount is really large we will put it in the syringe pump and set the flow rate to suit the infant's status."

Researcher: Nurse E struggles a bit with setting up the gravity feed trying to use the ventilator tubing holder but this is not working well: it keeps dropping out.

Nurse E: "The tube is too rigid and the holder's not the right size."

Scenario 3: The Family Chair

Researcher (speaking to Nurse F): "I'm confused about the family chair because from my understanding it's part of the hospital environment yet it's just dawned on me that infants are pulled out and fed in these chairs. Are the chairs disinfected between use, what about the arms in particular or the top of the chair?"

Nurse F: (looking a little crestfallen by my question): "But they need to be held." (understandably a bit defensive).

Researcher: "I understand and know how important it is to hold the infant, but I'm trying to understand the chair in relation to infection prevention."

The conversation is informal. No one else is in the room and I am sitting beside her in one the family chairs as she sits in another. Nurse F looks up in the air, pausing (to reflect?) as she continues to feed the infant.

Nurse F: "I never thought about it, but yes feeding in the chair, that could be a problem. I touch the arm, I stroke the infant's head: that's a breach. I've always thought about it as the seat being the issue but I never thought about the arms. We had a mum once who was a carrier of MRSA [infection]. Even though she doesn't have it, she can spread it to the infants. The mum would be sitting in the chair, she'd get up and I'd wipe it down every time she used it. But I never really thought about the chair otherwise. I would question the chair... my hands are going to go to that arm and then go to another chair. I try to kick things out of the way, but it makes so much noise... I try to do a lot with my feet."

Table 1. Examples of early themes from process coding

Example Themes	Process Codes	
Handling the baby with physical barriers (an emotional contradiction?)	• holding the baby in chair • feeding the baby • wearing protection • putting on gown • making a pouch • reaching the counter from chair	• carrying baby outside of bay • moving gloved hand between isolette & supply drawers • going into isolette • sitting in the chair
Chart intricately tied to caring for the baby	• touching the chart • looking at the chart • writing in chart • leaning over chart	• charting at the counter • charting in the chair • not having room
Trying to carve out a "safe space"	• finding space to work • using hands to define • creating perimeters to remind • creating perimeters to define area • creating a barrier • separating supplies	• separating us from surfaces • managing equipment and controls • wanting things closer • moving back and forth • touching equipment

- considerations in infection prevention and control that needed clarification by participants to deepen the understanding of processes and constraints
- aspects of the context that were striking, not previously known, or not understood
- sensitivities around infection prevention and control and discussing it with participants
- benefits of the method and strategies to improve the method
- thoughts on future research and design development

Discussion

Early empirical and systematic inquiry and accountability in design

As noted earlier, the researcher had initially suggested perhaps replacing the existing work counter with a supply cart to help with issues around infection practice, a suggestion which was met with some hesitation by stakeholders. At the time, this exchange did not seem significant, but it gained importance during the study. Applying a qualitative, multi-method approach focused primarily on the experiences and perceptions of front-line workers led to a better understanding of this hesitation and the complex challenges healthcare workers had been contending with in infection prevention and control.

It is not clear whether a traditional "top-down" design approach as described by Remijn (2006) would have yielded the same depth in data collected or the level of analysis and recommendations that resulted from the approach. The changes that occurred in the researcher's own level of understanding of the healthcare workers' experiences of infection prevention and control, evidenced by field notes, memos, and the process of thematic analysis, suggest that early empirical and systematic inquiry can help support design development and perhaps improve accountability in design.

Challenges and opportunities in the method

The main advantage of conducting empirical inquiry prior to design is to help deepen our understanding of end users and their workspace in order to inform design requirements for subsequent development. Developing a framework that structures the main issues end users are experiencing in their workplace can help designers assess whether their understanding of the issues is comprehensive and resonates with that of stakeholders. The results of this study, the framework and the NICU IPAC [Infection Prevention and Control] Design Exploration and have been shared with the leadership team and the organization is working towards a major renovation of the unit to address these challenges. Requirements may change as the design progresses and new information is discovered. A framework that outlines the main issues healthcare workers are experiencing and a design guideline, can

Table 2. Examples of researcher memos

Topic	Example Memos
Identifying lapses in observation - missed interactions relevant to IPAC	• I'm finding it difficult to see the interactions at times so feel like I am missing breaches that are occurring. I also find it difficult to watch multiple people at the same time and feel like I see partial tasks. Have to think about this and how to improve the observation process. • Nurses put supplies in the corner of the counter. They Virox it when they go on shift and sometimes throughout the shift, perhaps before a procedure. The aspect I find confusing is that the chart is often on the counter and is considered part of the hospital environment. I have not seen it wiped down and it touches the counter where the supplies are often set up. Sometimes the supplies are set up on sterile packages that have been opened but the outside of the packages have touched the inside of the drawers which don't seem to get cleaned regularly, don't get cleaned between baby transfers or discharges, so effectively the outside of the package is dirty?
Identifying lack of clarity in what constitutes a breach or best practice in IPAC	• Is the isolette cover 'dirty'? It gets touched after hand hygiene then nurses go into the isolette. Are they contaminating the bed? • Talk to nurses about chair being 'dirty' and feeding infants .
Awareness and questions emerging from observations	• There's no Purell on the left side of the baby in their infant bay. So staff working on that side have to come around to the right to Purell or use the Purell in the adjacent infant bay (if a bay is located there). There is no Purell on the chair which is also POC [point-of-care]. • 'Corners' are little carved out spaces for the nurses. They want a clean corner and sterile corner to manage the transition of supplies. • Being caught off guard or unprepared justifies going into the drawers. Wanting to complete tasks justifies going into drawers. • There may be confusion on when it makes sense to use gloves and hand sanitizing practice before and after glove use. Too many rules and exceptions? Can steps be distilled?
Thoughts on sensitivities around IPAC	• I think the hardest thing for me is telling people I have observed a breach. Participants seem to shut down when I tell them and feel nervous around me. I hope it does not impact further interaction I have with them or a sense of trust.
Thoughts on method	• I feel the coding is working as an observation tool to better focus observations because as I code, I question very precise interactions which prompts me to look even closer at seemingly simple actions. The process coding forces you to distill things to the string of actions, interactions. When you see them together in a suite, it generates questions, like I missed that, what happened there? Why did she do that? Why is it done that way? Is there a problem with these codes being side by side? The codes allow you to see things side by side without the visual break of time and space and thus allow you to see "distance" interactions better. The coding is helping me visualize a path a pathogen could potentially take from surface to surface.
Thoughts on future explorations to investigate, test, evaluate design requirements	• Discussed with one nurse central backup supply within room itself and reduce supplies at the counter to eliminate waste and encourage discarding supplies after baby is discharged. Challenge will be space and rearranging space.

help assess the introduction of new information and refinement of requirements.

Additional measures could have increased the depth of the analysis. More detailed and comprehensive data related to specific work domains and data from unavailable stakeholders may have helped fill gaps in "the work system approach" to understanding infection prevention (Alvarado, 2013). Other challenges included the fast pace of work and vigilant observation of micro-interactions that may be associated with breaches in infection prevention and control. But transcribing and process coding field notes after an observation session helped reveal micro-interactions or steps in work processes that had been missed. Memos were also helpful in catching such lapses and informing subsequent observation strategies.

The use of naturalistic observation may be criticized for encouraging "appropriate" behaviour. But, because the approach was non-punitive, the identity of front-line workers was protected, and the study emphasized the importance of front-line participation, this may have in fact helped support natural behaviour. The study time frame limited the extent and type of methods that could be implemented, but future work would ideally include:

- participatory development and assessment of mockups to draw out more information and insight on infection prevention issues and possibilities for improvement

- post-occupancy evaluation of design proposals in simulations or lower acuity areas supported by statistical analysis

Is the method sustainable?
Implementing this type of approach may be met with resistance, because it goes beyond a traditional design scope. At the same time, the term "evidenced-based design", which involves basing designs “on credible research to achieve the best possible outcomes” (The Centre for Health Design, 2016) is becoming increasingly required in healthcare design. Organizations should be able to demonstrate that they have resources and strategies in place to truly support a design process that is based in evidence. The use of systematic, empirical, and multi-method approaches can help support evidence-based practice and accountability over processes that depend solely on the past experience of designers, goals set by management, or generalized programs (Remjin, 2003), or approaches that rely predominantly on unsystematic or singular techniques to understand the issues.

Some may argue this level of inquiry compromises "the bottom line". However, discovering critical issues early can inform design requirements, improve the focus and consensus-building ability of the client-design team, and provide a better measure for evaluation. Further, without implementing such a strategy, how can an organization truly know if it will negatively impact the bottom line without the evidence to support this assumption? Organizations may also consider partnering with academic institutions to pilot and measure such initiatives to mitigate against the risks of developing and integrating "research systems" into an organization’s culture and strategic plan.

Conclusion

This study demonstrates how early empirical and systematic inquiry using methods from human-centred design and thematic analysis led to a framework of understanding and a recommendations guide to help stakeholders improve their understanding and practice of infection prevention. Specifically, the study provides evidence illustrating how our own understanding of designing for healthcare was clarified, broadened, and changed through the application of scientific methods in design. In applying such approaches, we can help professional designers gain the necessary knowledge to make better decisions in the design of complex systems, and at a broader level, sensitize a new generation of designers to the importance of rigour in knowledge acquisition, transfer, application, and assessment.

About the Authors

Chantal Trudel is an Assistant Professor at Carleton University's School of Industrial Design in Ottawa, Canada. Chantal draws on her background in industrial design (B.I.D, Carleton University), human factors and ergonomics (MSc, University of Nottingham) and 10 years professional experience in commercial and healthcare design to support her research. Chantal is interested in human-centred and participatory design methods to improve our understanding and design of peoples' experiences within complex systems.

Sue Cobb is Associate Professor and Head of the Human Factors Research Group in the Faculty of Engineering at the University of Nottingham in the United Kingdom. Sue has been working in this group for 30 years. Over this period she has worked on many projects involving analysis of user requirements and user feedback or opinion on a variety of issues in contexts including industrial workplaces, educational settings, and community environments. Her specialist interest is in multi-disciplinary research and the use of user-centred and participatory design methods in a variety of contexts including special education and healthcare.

Kathryn Momtahan has 10 years of critical care nursing experience and a PhD in Experimental Psychology, focusing on healthcare human factors. Dr. Momtahan's human factors professional experience includes several years working in the hi-tech sector and fourteen years in a research-lead capacity in a hospital environment. She has published in human factors, healthcare, and engineering journals and holds several adjunct professor appointments in various faculties and programs including engineering, psychology, nursing, business, and human–computer interaction.

Ann Mitchell is the Director of Obstetrics, Gynecology and Newborn Care at The Ottawa Hospital. Ann has 28 years of experience in NICU and maternal newborn care in a variety of clinical, regional, and administrative roles. Her current focus is on staff and patient safety, engaging families, and improving the patient experience.

Janet Brintnell is the Corporate Clinical Manager of the Neonatal Intensive Care Unit and Special Care Nursery at The Ottawa Hospital. Her nursing career spans 32 years of which 26 have been focused in the Neonatal Intensive Care Unit. Her experience ranges from a clinical bedside capacity to her current role with an administrative focus while constantly maintaining a critical focus on the delivery of family-centered care and best practices.

References

Alvarado, C. J. 2012. Human Factors and Ergonomics in Infection Prevention. In P. Carayon (Ed.), *Handbook of Human Factors and Ergonomics in Healthcare and Patient Safety* (2nd Ed.): 793–802. Boca Raton, FL: CRC Press.

Archibald, L. K., Manning, M. L., Bell, L. M., Banerjee, S., & Jarvis, W. R. 1997. Patient Density, Nurse-To-Patient Ratio and Nosocomial Infection Risk in a Pediatric Cardiac Intensive Care Unit. *The Pediatric Infectious Disease Journal,* 16(11): 1045–1048. http://dx.doi.org/10.1097/00006454-199711000-00008

Attride-Stirling, J. 2001. Thematic Networks: An Analytic Tool for Qualitative Research. *Qualitative Research,* 1(3): 385–405. http://dx.doi.org/10.1177/146879410100100307

Braun, V., & Clarke, V. 2006. Using Thematic Analysis in Psychology. *Qualitative Research in Psychology,* 3(2): 77–101. http://dx.doi.org/10.1191/1478088706qp063oa

Brehmer, M., Carpendale, S., Lee, B., & Tory, M. 2014. Pre-Design Empiricism for Information Visualization: Scenarios, Methods, and Challenges. In H. Lam, P. Isenberg, T. Isenberg, & F. M. Sedlmair (Eds.), *Proceedings of the ACM Workshop on beyond Time and Errors: Novel Evaluation Methods for Visualization (BELIV)*: 147–151. New York: ACM. http://dx.doi.org/10.1145/2669557.2669564

The Centre for Health Design. 2016. EDAC: Evidence-Based Design Accreditation and Certification. *The Centre for Health Design.* Accessed June 4th, 2016: https://www.healthdesign.org/certification-outreach/edac/about

Charmaz, K. 2014. *Constructing Grounded Theory* (2nd ed.). London, England: Sage Publications Ltd.

Chartered Institute of Ergonomics and Human Factors. 2016. What Is Ergonomics? *Chartered Institute of Ergonomics and Human Factors.* Accessed August 9th, 2016: http://www.ergonomics.org.uk/what-is-ergonomics/

Friedman, K. 2003. Theory Construction in Design Research: Criteria: Approaches, and Methods. *Design Studies,* 24(6): 507–522. http://dx.doi.org/10.1016/S0142-694X(03)00039-5

Graham, M. 1990. Frequency and Duration of Handwashing in an Intensive Care Unit. *American Journal of Infection Control,* 18(2): 77–81. http://dx.doi.org/10.1016/0196-6553(90)90085-7

Gravel, D., Matlow, A., Ofner-Agostini, M., Loeb, M., Johnston, L., Bryce, E., Sample, M-L., Roth, V. R., Goldman, C., Taylor, G., & The Canadian Nosocomial Infection Surveillance Program. 2007. A Point Prevalence Survey of Health Care–Associated Infections in Pediatric Populations in Major Canadian Acute Care Hospitals. *American Journal of Infection Control,* 35(3): 157–162. http://dx.doi.org/10.1016/j.ajic.2006.06.006

Guest, G., MacQueen, K. M., & Namey, E. E. 2012. *Applied Thematic Analysis.* Los Angeles, CA: Sage Publications.

Hendrich, A. 2003. *Optimizing Physical Space for Improved Outcomes: Satisfaction and the Bottom Line. Optimizing the Physical Space for Improved Outcomes, Satisfaction, and the Bottom Line.* Mini-Course Sponsored by the Institute for Healthcare Improvement and the Center for Health Design, Atlanta, GA.

ISO. 2015. *ISO 9241-210:2010 Ergonomics of Human-System Interaction. Part 210: Human-Centred Design for Interactive Systems.* International Organization for Standardization (ISO). Switzerland: ISO Copyright Office. Accessed August 11th, 2016: http://www.iso.org/Iso/Home/Store/catalogue_ics/catalogue_detail_ics.Htm?Csnumber=52075

Maguire, M. 2001. Methods to Support Human-Centred Design. *International Journal of Human-Computer Studies,* 55(4): 587–634. http://dx.doi.org/10.1006/Ijhc.2001.0503

Muto, C. A., Sistrom, M. G., & Farr, B. M. 2000. Hand Hygiene Rates Unaffected by Installation of Dispensers of a Rapidly Acting Hand Antiseptic. *American Journal of Infection Control,* 28(3): 273–276. http://dx.doi.org/10.1067/Mic.2000.103242

Public Health Agency of Canada. 2012. *Routine Practices and Additional Precautions for Preventing the Transmission of Infection in Healthcare Settings.* Ottawa: Public Health Agency of Canada. http://publications.gc.ca/collections//collection_2013/aspc-phac/HP40-83-2013-eng.pdf

Public Health Agency of Canada. 2013. *The Chief Public Health Officer's Report on the State of Public Health in Canada, 2013: Infectious Disease – The Never-Ending Threat.* Ottawa: Public Health Agency of Canada. http://www.phac-aspc.gc.ca/cphorsphc-respcacsp/2013/infections-eng.php

Public Health Ontario. 2012a. *Best Practices for Infection Prevention and Control in Perinatology: In All Health Care Settings That Provide Obstetrical and Newborn Care.* Toronto: Public Health Ontario, Provincial Infectious Diseases Advisory Committee (PIDAC). https://www.publichealthontario.ca/en/erepository/IPC%20in%20Perinatology_ENGLISH_Final_2012-05-25%5B1%5D.pdf

Public Health Ontario. 2012b. *Best Practices for Infection Prevention and Control Programs in Ontario: In All Health Care Settings* (3rd ed). Toronto: Public Health Ontario, Provincial Infectious Diseases Advisory Committee (PIDAC). http://www.publichealthontario.ca/en/eRepository/BP_IPAC_Ontario_HCSettings_2012.pdf

Pessoa-Silva, C. L., Hugonnet, S., Pfister, R., Touveneau, S. Dharan, S., Posfay-Barbe, K., & Pittet, D. 2007. Reduction of Health Care-Associated Infection Risk in Neonates by Successful Hand Hygiene Promotion. *Pediatrics,* 120(2): e382–e390. http://dx.doi.org/10.1542/Peds.2006-3712

Pittet, D. 2000. Improving Compliance with Hand Hygiene in Hospitals. *Infection Control Hospital Epidemiology,* 21(6): 381–386. http://dx.doi.org/10.1086/501777

Pittet, D. 2001. Improving Adherence to Hand Hygiene Practice: A Multidisciplinary Approach. *Emerging Infectious Diseases,* 7(2): 234–240. http://dx.doi.org/10.3201/eid0702.700234

Remijn, S. L. M. 2006. Integrating Ergonomics into the Architectural Design Processes: Tools for User Participation in Hospital Design. In *Proceedings of IEA 2006 16th World Congress on Ergonomics: Meeting Diversity in Ergonomics,* Maastricht, Netherlands: 10–14.

Rogers, Y., Sharp, H., & Preece, J. 2012. *Interaction Design – Beyond Human Computer Interaction* (3rd ed.). West Sussex, England: John Wiley & Sons Ltd.

Saldaña, J. 2009. *The Coding Manual for Qualitative Researchers.* Los Angeles, CA: Sage Publications.

Trudel, C., Cobb, S., Momtahan, K., Brintnell, J., & Mitchell, A. 2016a. *A Human Factors Approach to Understanding and Designing for Infection Prevention and Control in a Neonatal Intensive Care Unit.* Paper presented at Ergonomics & Human Factors 2016, Daventry, United Kingdom, Chartered Institute of Ergonomics & Human Factors. http://programme.exordo.com/ehf2016/Delegates/Presentation/48/

Trudel, C., Cobb, S., Momtahan, K., Brintnell, J., & Mitchell, A. 2016b. Disconnects in Design and Infection Prevention and Control – How the Design of Products and the Environment in Neonatal Intensive Care May Be Undermining Infection Prevention Practice. *Proceedings of the International Symposium on Human Factors and Ergonomics in Health Care June 2016,* 5(1): 44–49.

Ulrich, R., Quan, X., Zimring, C., Joseph, A., & Choudhary, R. 2004. *The Role of the Physical Environment in the Hospital of the 21st Century: A Once-In-A-Lifetime Opportunity.* Concord, CA: The Center for Health Design.

Vaismoradi, M., Turunen H., & Bondas, T. 2013. Content Analysis and Thematic Analysis: Implications for Conducting a Qualitative Descriptive Study. *Nursing & Health Sciences,* 15(3): 398–405. http://dx.doi.org/10.1111/nhs.12048

White, R. D. 2006. Recommended Standards for the Newborn ICU. *Journal of Perinatology,* 26: S2–S18. http://dx.doi.org/10.1038/sj.jp.7211587

White, R. D. 2007. Recommended Standards for the Newborn ICU. *Journal of Perinatology,* 27(Suppl 2): S4–S19. http://dx.doi.org/10.1038/sj.jp.7211837

Keywords: healthcare design, human factors, ergonomics, design research, knowledge mobilization, infection prevention and control, neonatal intensive care unit

4

Mobilizing Knowledge: The Evidence Gap for Assistive Devices

Edward D. Lemaire

> *The temptation to form premature theories upon insufficient data is the bane of our profession.*
>
> Sherlock Holmes
> Fictional private detective
> In *The Valley of Fear*
> by Arthur Conan Doyle (1859–1930)

Knowledge mobilization can be hindered in healthcare technology settings where the pace of change outpaces the ability to perform high-quality research methodologies that provide timely knowledge to enable informed prescription and technology application to the end user. Although well-controlled research with appropriate sample sizes is needed, this approach must be balanced with other evidence sources to address the knowledge immediacy requirements. Using carbon-fibre ankle–foot orthoses (i.e., lower-limb braces that improve stability, alignment, and foot-to-ground placement) as a case study, various sources of assistive device evidence were explored for their contribution to the continuum of knowledge in this area. A basic level of knowledge exists, but the quality is insufficient to inform the physical rehabilitation community on selecting from the almost 70 different devices on the market and the expected clinical outcomes for a target population. A combination of enhanced single-participant reports should be considered as an important part of the knowledge continuum and essential for knowledge immediacy. This approach must also be expanded to national and multinational database initiatives that provide a better base from which to extract knowledge on assistive device performance and mobilize this knowledge to provide optimal care for people with physical disabilities.

Introduction

Knowledge mobilization relies on evidence of sufficient quality to enable appropriate decision making. Although many established sectors produce well-controlled research with sufficient sample sizes to guide decision making confidently, the knowledge mobilization process is challenged in sectors where the pace of change outpaces the ability to complete high-quality research methodologies. For example, technology-sector research involving products and humans can take much longer to complete that the annual or faster product-revision cycles. Therefore, the outcomes from a large multi-site, randomized control study may not apply to a new product that replaces the tested technology. Although well-controlled research with appropriate sample sizes is needed, this approach must be balanced with other evidence sources to address the knowledge-immediacy requirements. The assistive device sector is an interesting group for exploring this human–technology knowledge mobilization issue.

Assistive devices that improve mobility for people with disabilities cover a range of technologies, from wheelchairs to prosthetics and orthotics to robotic exoskeletons. Although the target populations and core technologies vary between these mobility assistive devices, a common element is the rapid pace of innovation in relatively small markets (Baljko & Hamidi, 2014). Modern assistive devices are on annual or biannual revision cycles, often with faster revision cycles for devices controlled by high-end microprocessors. While this pace of innovation should be encouraged to provide the best technology for people with mobility disabilities, constant change presents challenges for evidence-based practice and knowledge mobilization.

In the field of physical rehabilitation, several challenges have been identified for knowledge development and mobilization (Johnston et al., 2009). These challenges include:

- the complexity of interactions between people with a disability and their environment (WHO, 2016)
- the need to evaluate people as they participate in their daily lives
- small sample sizes
- the difficulty or impossibility of implementing blinding or placebos
- ethical issues for control groups who do not receive treatment
- integrating assistive technology evaluation with other research methods and design factors
- the lack of funding for large effectiveness studies
- the difficulty of addressing issues between large social systems and physical, economic, and environmental factors

These challenges are directly related to knowledge mobilization for assistive devices. In fact, poor consideration of assistive technology use in rehabilitation research can be a confounding factor when interpreting research results (Rust & Smith, 2005).

To positively affect practice, the appropriate evidence with the appropriate quality must exist. For mobility assistive devices, appropriate evidence may be in forms other than peer reviewed academic publications. Appropriate assistive device research for knowledge mobilization can be classified as technical, biomechanical, and clinical. An aspect that is often overlooked on the technical category are standards-based evaluations that may be conducted by the manufacturer, independent laboratories, or government laboratories. Although countries may implement additional test procedures, the base for assistive device technical evaluations are the International Standards Organization (ISO, 2016; Rust & Smith, 2005) and Rehabilitation Engineering Society and Assistive Technology Society for North America (RESNA, 2016). Standard tests typically cover structural capacity (load tests, etc.), device function (wheelchair stability, etc.), and other factors such as flammability.

The study by Gebrosky and colleagues (2013) demonstrated the usefulness of these standard tests by performing the ANSI/RESNA standard tests on a series of lightweight wheelchairs. Most of the wheelchairs tested did not meet durability standards and a recommendation was made for stronger regulations and testing by independent and certified facilities. Ensuring that standard tests are appropriately administered is essential, but access to these test results is also lacking given that companies, test facilities, and regulators are not required to share test outcomes. Even if the intent to share this information existed, a reference and indexing source that can easily be used by clinicians is not available. For knowledge mobilization, systems such as clinical trial registries (such as ClinicalTrials.gov) are needed to provide an access gateway to the standard evaluation outcomes for assistive devices. This would enable direct access by the healthcare community and provide a base for knowledge synthesis by researchers and other stakeholders.

Assistive devices are typically categorized under Class 1 or Class 2 for regulatory approval. In Canada, Class 1 devices have the lowest medical risk and do not require a Medical Device License. Therefore, much of the background evidence is not provided through the regulatory process. Even with regulatory information, independent research evidence to guide clinical decision making is usually lacking when new or modified devices appear in the market.

From a biomechanical perspective, a modest amount of literature exists for the effect on movement and body structures from assistive device use. A search using the Scopus database with keywords "wheelchair biomechanics" returned 953 results, and 151 results since 2012. A search using "(prosthetics or orthotics) and biomechanics : excluding implants (surgical), Animal, Arthroplasty, Bone Cements, Biomedical Engineering – Surgical Implants" returned 2,689 publications, with 406 results since 2012. In comparison, a search using "physical rehabilitation and biomechanics" returned 55,042 publications. Most of the biomechanics literature involves motion laboratory studies, with the majority of locomotor research on level ground. However, the emergence of wearable sensor systems and room-scale virtual reality systems (Sinitski et al., 2015) are providing research tools to generate device-function evidence that better reflects movement environments encountered in daily living.

Clinical research on assistive technology includes questionnaire-based tools for evaluating user perceptions of using an assistive device for daily living and function-based tasks. On a base level, classification tools can be used to facilitate collection of data related to assistive devices and disability and to define categories related to disability and assistive technology (e.g., function, body structures, activities, participation, environmental factors) (Jutai et al., 2005; Lenker & Paquet, 2003). Other tools measure psycho-social aspects of assistive technology use (Jutai & Day, 2002; Scherer et al., 2011), user satisfaction (Demers et al., 2002; Samuelsson & Wressle, 2008), factors for assistive technology selection (Bernd et al., 2009), and device specific tools (Condie et al., 2006; Miller & Swanson, 2009; Wright, 2009).

Ideally, a continuum of evidence can be amassed that covers the need for immediate information on new technologies, broader research that quantifies how a device compares with similar assistive devices in the market or literature, and high-quality evidence for assistive device concepts that exist throughout a device class. To explore this continuum concept, a case study involving carbon-fibre ankle–foot orthoses is presented here.

Evaluating the Current State of Carbon-Fibre Ankle–Foot Orthoses

A carbon-fibre ankle–foot orthosis (CF-AFO) is a lower limb brace that controls movement of the foot relative to the shank for people with lower limb weakness, paralysis, or excessive tone/spasm. Carbon-fibre materials produce a device that is thin and light, stiff, or capable of storing and returning energy when walking. A market survey conducted by the investigator returned 68 distinct CF-AFO products from 13 manufacturers (i.e., devices with distinct designs or functions, not including different sizes). Most CF-AFOs consist of a carbon-fibre foot and shank section that inserts into the shoe and a strap at the proximal end to anchor the orthosis to the shank (Figure 1). The energy storage design consists of separate foot and shank sections that both connect to a centre energy storing part (rods or rectangular section) that stores energy after foot strike and then releases stored energy at toe off. An expiring patent on this technology should lead to many more devices entering the market place in the next few years.

The first level of evidence for device function and unique contributions is patents. The investigator identified 25 relevant patents with a Google Patent search using the keywords "carbon fibre ankle foot orthosis AFO". This information provides claims and design concepts for many of the commercial devices. Peer review is from patent office experts. Although this level of evidence is insufficient to guide clinical practice, it does provide information to help understand the objectives, concepts, and novel contributions that differ between devices.

From a regulatory perspective, CF-AFOs are Class 1 medical devices and therefore the manufacturer is not required to provide evidence on device safety or function. Although regional regulatory requirements may provide a level of evidence, this information cannot be expected for this class of assistive device. In general, when assembling knowledge for mobilization, regulatory testing results are typically not considered, but could be a useful source of information that would be available as a technology enters the market.

From the literature, searches of Scopus and Google Scholar databases with the keywords "ankle foot orthosis afo carbon fibre" produced 29 relevant articles published between 2006 and 2016. Of these, 9 were related to AFO mechanical testing, 15 using biomechanical and physiological analysis, and 4 using clinical tests.

For mechanical testing, a series of experimental protocols were used to apply loads and measure the resulting AFO movement. These methods evaluated AFO stiffness, engineering modelling and structural analysis, and repeated loading to determine how the device could fail. However, the lack of a standard measurement method prohibits valid comparisons between studies and meta-analysis. The orthosis–limb interac-

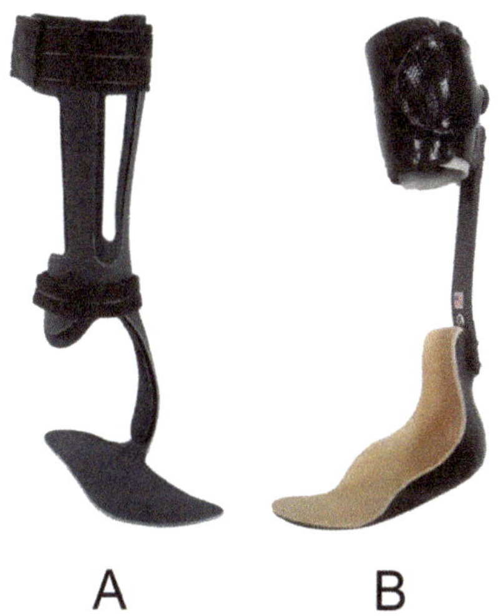

Figure 1. One piece prefabricated ankle–foot orthosis (A: Ossur AFO Dynamic) and custom ankle–foot orthosis with carbon-fibre strut (B: Fabtech Posterior Dynamic Element).

tion is also a consistent deficiency. Orthoses are designed to intimately interact with the foot and lower limb; therefore, the limb is part of the structural system and therefore must be considered when testing device performance. Most studies did not consider the limb and, of the three studies that tested the AFO with a limb surrogate, a rigid plaster limb shape with a single axis joint was employed. For carbon-fibre AFO with multiple axes of possible movement, a solid single-axis foot–leg model is insufficient for replicating AFO movement under daily activity loads. Future research should develop a more appropriate foot–leg surrogate to enable mechanical testing so that the results can be more directly translated to the real world. For knowledge mobilization, this example applies broadly to knowledge obtained through simulation and modelling. Insufficient or inappropriate identification of the person within the system being evaluated could lead to poor evidence that would not be easily identified by the knowledge user following knowledge translation.

Biomechanical testing can be grouped by design: custom AFO with a posterior strut that stores and releases energy during gait, or lighter one-piece carbon-fibre AFO that is typically prefabricated. 3D quantitative motion analysis demonstrated that devices with a posterior strut typically stored and returned more energy, thereby helping the person walk more naturally and with less effort, and fine control of strut stiffness may not be required for prescription but can affect comfort and range of movement. Although the biomechanical evidence is adequate, most papers originated from one study at one laboratory and many of the other studies did not provide appropriate comparative data. Biomechanical research on the one-piece carbon-fibre designs included different populations in each study, such as able-bodied, Charcot-Marie-Tooth, polio, multiple sclerosis, peroneal nerve injury, and stroke. Given that each population had different walking characteristics and requirements, comparisons across populations become difficult. Analyses also vary, from full 3D motion analysis to simple stride analysis to physiological energy expenditure. Although the available biomechanical information provides a basis for understanding how various CF-AFOs perform during walking, the level of evidence remains insufficient to support clinical decisions when choosing between the many designs available in the marketplace.

Biomechanical laboratory-based analyses have difficulty meeting the immediacy requirement for knowledge mobilization, where barriers for providing timely outcomes include costs and time for completing the study (fitting devices for each person), laboratory access, and recruitment fatigue from a regional participant pool that are approached for multiple evaluation studies. Economically, funding these studies may be difficult for small manufacturers where profits from individual products are much less than other health sectors, such as pharmaceuticals. Motion-laboratory analyses may be better positioned for assessing broader classes of devices and first occurrence of disruptive technologies.

Methodologically, many biomechanical studies compared AFO gait to walking without a device. Given that the literature has previously established that any appropriately prescribed AFO improves gait over walking without the orthosis (Tyson et al., 2013), this comparison does not inform the clinical community as to the device's relevance in the marketplace. This is in contradiction to the typical placebo methodology where the placebo is a harmless intervention prescribed for the psychological benefit of the participant. For knowledge mobilization, comparisons between the assistive technology and the current standard of practice for the target population are required to ensure that a clinical decision maker can use the translated knowledge to recommend the most appropriate device for the patient.

Clinical testing also varied between the four studies. Across all studies, different combinations of 14 tests were employed to evaluate balance, muscle strength, functional status, walking and movement capacity, mobility capacity, quality of life, and disability. As with the biomechanical studies, different populations participated in each study, and each population had different movement characteristics. These clinical tests provided functional performance measures that helped to complete the knowledge base for CF-AFO; however, the quantity of information is insufficient for making evidence-based decisions for CF-AFO prescription and fitting. The advantage of these clinical tests are relatively quick implementation (minutes), portability, quick reporting, and standardized protocols. However, these tests only provide a high-level assessment of the parameter (i.e., we could learn that a person walks faster with a new assistive device, but not why they walk faster).

As with much of the assistive technology literature, participants were not well described in reference to how their disability could affect CF-AFO performance. Detailed research participant descriptions are required to enable the reader to assess the study and make decisions about the relevance of outcomes to a specific clinical population (e.g., deciding whether a study is relevant to a clinician's practice). Similar guidelines to the "ISPO Recommendation for Defining Participants in Prosthetics Research" (ISPO, 2016) would benefit knowledge mobilization for CF-AFO. Also, CF-AFOs are often not well described, which is essential for study replication and applying the results in practice. This is especially relevant for custom manufactured orthoses. With many academic publishers being able to include additional documents online, detailed orthotic device descriptions should be included with publications in this area. For prefabricated devices, the model and date are essential because a CF-AFO brand name may differ annually as device characteristics evolve at the manufacturer level. For knowledge mobilization involving technology, insufficient description of the tested technology occurs often. For example, the technology's model number may be provided but the software/firmware version that is essential to the technology's performance is typically not provided. A consistent effort from multiple stakeholders would be required to affect change and solve these reporting deficiencies.

Conclusion

As demonstrated by the CF-AFO case study, it is apparent that typical medical research methods are insufficient to generate timely evidence to guide assistive device prescription and consumer/clinician decision making and enable effective knowledge mobilization (Johnston et al., 2009). Immediacy is a core requirement that is contrary to the time required to complete controlled research with a sufficient sample size. Overall, a series of recommendations can be considered.

First, as outlined by Ottenbacher and Hinderer (2001), single-participant studies should be considered for evidence-based practice when quick reports on new devices or practices are needed. The lack of generalizability from single-participant studies can be partially addressed by adding additional replications of the measured treatment effect, systematic replication of the program or treatment over a wider range of situations, and meta-analysis. However, consistent data collection and reporting in the field are needed to provide the appropriate information.

Second, beyond single-participant studies, the assistive device field requires a multisite, multinational clinical database approach to support data assembly, knowledge extraction, and knowledge mobilization (Fuhrer, 2001). Secure international data networks are now available for collecting de-identified information on the user, assistive device characteristics, and outcome measures (e.g., SwedeAmp quality registry [Emilsson et al., 2015]). By engaging multinational clinical, non-government, and government organizations, the infrastructure to participate efficiently in a global registry can be created, thereby providing timely data on device outcomes that can be synthesized computationally to reveal patterns, trends, and associations related to assistive-technology utilization (i.e., Big Data approaches). The obstacles to achieve such an objective include data-security requirements across jurisdictions, engagement at the clinician level to collect high-quality standardized outcome measures, engagement at the local administration level to enable de-identified data sharing, and partnerships to enable timely and relevant analysis and reporting from the system to the appropriate recipients (i.e., knowledge mobilization). Differences in clinical practices between sites is also a potential problem because the local therapy methods may influence outcomes beyond factors attributed to the assistive device.

Third, in addition to clinician-generated information, assistive-device consumers may also participate by sharing their wearable mobility device information (smart watch, smart phone, etc.) or stored data from their intelligent assistive devices (microprocessor controlled prosthesis, powered wheelchair, etc.). Industry must be engaged by multiple stakeholders (including end users, healthcare providers, professional associations, and privacy regulators) to make the transition from only using device sensor data for device control and product development to making this device-user-centric data an integral part of the evidence base.

The research process cannot keep pace with innovation, so evidence to guide prescription is lacking; however, the continuous knowledge generation era is upon us and the assistive technology field is well positioned to take advantage and bring immediacy to knowledge mobilization. For assistive devices, partnerships and action are required to "mobilize" knowledge mobilization.

About the Author

Edward Lemaire, PhD, is actively involved with research on technologies that improve mobility for people with physical disabilities. He is a Clinical Researcher at The Ottawa Hospital Research Institute's Centre for Rehabilitation Research and Development; Professor at the University of Ottawa Faculty of Medicine; and Adjunct Professor in Human Kinetics, Mechanical Engineering, and Systems Design Engineering. He is also active with the International Society of Prosthetics and Orthotics, as a board member and international congress scientific chairman. Dr. Lemaire's research has resulted in over 350 published papers and presentations that include intelligent prosthetics and orthotics, biomechanical walking analysis in 3D virtual environments, smartphone approaches to improve decision making, and eHealth technology to enhance access to education and rehabilitation services.

References

Baljko, M., & Hamidi, F. 2014. Knowledge Co-Creation and Assistive Technology. *Scholarly and Research Communication,* 5(3).

Bernd, T., Van Der Pijl, D., & De Witte, L. P. 2009. Existing Models and Instruments for the Selection of Assistive Technology in Rehabilitation Practice. *Scandinavian Journal of Occupational Therapy,* 16(3): 146–158. http://dx.doi.org/10.1080/11038120802449362

Condie, E., Scott, H., & Treweek, S. 2006. Lower Limb Prosthetic Outcome Measures: A Review of the Literature 1995 to 2005. *Journal of Prosthetics and Orthotics,* 18(6): 13–45.

Demers, L., Weiss-Lambrou, R., & Ska, B. 2002. The Quebec User Evaluation of Satisfaction with Assistive Technology (QUEST 2.0): An Overview and Recent Progress. *Technology and Disability,* 14(3): 101–105.

Emilsson, L., Lindahl, B., KöSter, M., Lambe, M., & Ludvigsson, J. F. 2015. Review of 103 Swedish Healthcare Quality Registries. *Journal of Internal Medicine,* 277(1): 94–136. http://dx.doi.org/10.1111/joim.12303

Fuhrer, M. J. 2001. Assistive Technology Outcomes Research: Challenges Met and Yet Unmet. *American Journal of Physical Medicine & Rehabilitation,* 80(7): 528–535.

Gebrosky, B., Pearlman, J., Cooper, R. A., Cooper, R., & Kelleher, A. 2013. Evaluation of Lightweight Wheelchairs Using ANSI/RESNA Testing Standards. *Journal of Rehabilitation Research and Development,* 50(10): 1373–1390. http://dx.doi.org/10.1682/JRRD.2012.08.0155

ISO. 2016. *ISO Standard 11.180: Aids for Disabled or Handicapped Persons.* Geneva: International Organization for Standardization (ISO). http://www.iso.org/Iso/Home/Store/catalogue_ics/catalogue_ics_browse.Htm?ICS1=11&ICS2=180

ISPO. 2016. Defining Participants in Prosthetics Research. *International Society for Prosthetics and Orthotics.* Accessed August 11, 2016: http://www.ispoint.org/Defining-Participants-Prosthetics-Research

Johnston, M. V., Vanderheiden, G. C., Farkas, M. D., Rogers, E. S., Summers, J. A., & Westbrook, J. D. 2009. *The Challenge of Evidence in Disability and Rehabilitation Research and Practice: A Position Paper.* Austin, TX: The National Center for the Dissemination of Disability Research (NCDDR).

Jutai, J., & Day, H. 2002. Psychosocial Impact of Assistive Devices Scale (PIADS). *Technology and Disability,* 14(3): 107–111.

Jutai, J. W., Fuhrer, M. J., Demers, L., Scherer, M. J., & DeRuyter, F. 2005. Toward a Taxonomy of Assistive Technology Device Outcomes. *American Journal of Physical Medicine & Rehabilitation,* 84(4): 294–302.

Lenker, J. A., & Paquet, v. L. 2003. A Review of Conceptual Models for Assistive Technology Outcomes Research and Practice. *Assistive Technology,* 15(1): 1–15. http://dx.doi.org/10.1080/10400435.2003.10131885

Miller, L. A., & Swanson, S. 2009. Summary and Recommendations of the Academy's State of the Science Conference on Upper Limb Prosthetic Outcome Measures. *Journal of Prosthetics and Orthotics,* 21(9): 83–89.

Ottenbacher, K. J., & Hinderer, S. R. 2001. Evidence-Based Practice: Methods to Evaluate Individual Patient Improvement. *American Journal of Physical Medicine & Rehabilitation,* 80(10): 786–796.

RESNA. 2016. AT Standards. *Rehabilitation Engineering & Assistive Technology Society of North America (RESNA).* Accessed May 11, 2016: http://www.resna.org/At-Standards

Rust, K. L., & Smith, R. O. 2005. Assistive Technology in the Measurement of Rehabilitation and Health Outcomes: A Review and Analysis of Instruments. *American Journal of Physical Medicine & Rehabilitation,* 84(10): 780–793.

Samuelsson, K., & Wressle, E. 2008. User Satisfaction with Mobility Assistive Devices: An Important Element in the Rehabilitation Process. *Disability and Rehabilitation,* 30(7): 551–558. http://dx.doi.org/10.1080/09638280701355777

Scherer, M. J., Craddock, G., & Mackeogh, T. 2011. The Relationship of Personal Factors and Subjective Well-Being to the Use of Assistive Technology Devices. *Disability and Rehabilitation,* 33(10): 811–817. http://dx.doi.org/10.3109/09638288.2010.511418

Sinitski, E. H., Lemaire, E. D., & Baddour, N. 2015. Evaluation of Motion Platform Embedded with Dual Belt Treadmill Instrumented with Two Force Plates. *Journal of Rehabilitation Research and Development,* 52(2): 221–234. http://dx.doi.org/10.1682/JRRD.2013.11.0244

Tyson, S. F., Sadeghi-Demneh, E., & Nester, C. J. 2013. A Systematic Review and Meta-Analysis of the Effect of an Ankle-Foot Orthosis on Gait Biomechanics after Stroke. *Clinical Rehabilitation,* 27(10): 879–891.
http://dx.doi.org/10.1177/0269215513486497

WHO. 2016. International Classification of Functioning, Disability and Health (ICF). *World Health Organization (WHO).* Accessed May 17, 2016.
http://www.who.int/classifications/icf/en/

Wright, V. 2009. Prosthetic Outcome Measures for Use with Upper Limb Amputees: A Systematic Review of the Peer-Reviewed Literature, 1970 to 2009. *Journal of Prosthetics and Orthotics,* 21(9): 3–63.

Keywords: knowledge mobilization, assistive devices, orthosis

5

Towards Third-Generation Living Lab Networks in Cities

Seppo Leminen, Mervi Rajahonka, and Mika Westerlund

> *All the evolution we know of proceeds from the vague to the definite.*
>
> Charles Sanders Peirce (1839–1914)
> Philosopher, logician, mathematician, and scientist

Many cities engage in diverse experimentation, innovation, and development activities with a broad variety of environments and stakeholders to the benefit of citizens, companies, municipalities, and other organizations. Hence, this article discusses such engagement in terms of next-generation living lab networks in the city context. In so doing, the study contributes to the discussion on living labs by introducing a framework of collaborative innovation networks in cities and suggesting a typology of third-generation living labs. Our framework is characterized by diverse platforms and participation approaches, resulting in four distinctive modes of collaborative innovation networks where the city is: i) a provider, ii) a neighbourhood participator, iii) a catalyst, or iv) a rapid experimenter. The typology is based on an analysis of 118 interviews with participants in six Finnish cities and reveals various ways to organize innovation activities in the city context. In particular, cities can benefit from innovation networks by simultaneously exploiting multiple platforms such as living labs for innovation. We conclude by discussing implications to theory and practice, and suggesting directions for future research.

Introduction

Living labs are increasingly accepted as a prominent form of open innovation (e.g., Bergvall-Kåreborn et al., 2015; Brankaert et al., 2015; Guimont & Lapointe, 2016; Hakkarainen & Hyysalo, 2016). The roots of the concept may be traced back to Knight (1749), who referred to "living laboratory" as the elements and conditions of a body and an environment of an experiment. More recent studies apply living labs in heterogeneous fields and suggest that this phenomenon provides ample research opportunities (cf. Leminen, 2015). Following the definition of Westerlund and Leminen (2011), the present study views living labs as: *"physical regions or virtual realities, or interaction spaces, in which stakeholders form public–private–people partnerships (4Ps) of companies, public agencies, universities, users, and other stakeholders, all collaborating for creation, prototyping, validating, and testing of new technologies, services, products, and systems in real-life contexts."*

Although the literature on living lab is rich with various concepts, methodologies, research streams, and tools (Dutilleul et al., 2010; Følstad, 2008; Leminen & Westerlund, 2016, 2017), studies increasingly document the plurality of living labs using different conceptualizations (e.g., Leminen et al., 2012; Rits et al., 2015; Savelkoul & Peutz, 2017; Schuurman et al., 2016; Ståhlbröst & Lassinantti, 2015). Among them, Leminen and colleagues (2012) classify living labs as user-, enabler-, utilizer-, or provider-driven. Moreover, the outcomes of innovation activities are linked with the characteristics of the living lab, its driving party, and the selected strategy – and the living lab's structure is that of an open innovation network (Leminen & Westerlund, 2013; Leminen et al., 2016; Steen & van Bueren, 2017; Veeckman et al., 2013). Similar to the notion of open innovation networks (Jarvenpaa & Wernick, 2012), living labs typically comprise different stakeholders, such as suppliers, customers and users, competitors, research units of universities, and other institutions and organizations, all of whom brings their interests to the collaboration and innovation.

Living labs also may be classified by stakeholder roles (Leminen et al., 2014; Leminen, Turunen, & Westerlund, 2015; Nyström et al., 2015). A city or an urban environment as well as involved stakeholders and their roles are encompassed in many recent living lab studies (e.g., Juujärvi & Lund, 2016; Steen & van Bueren, 2017). Previous research is unified in that cities have a crucial role to support plurality of innovation activities in the urban context (e.g., Leminen & Westerlund., 2015; Markkula & Kune, 2015; Tukiainen et al., 2015; Tukiainen & Sutinen, 2015). Given that various types and modes of collaborative innovations are flourishing in the city context (Sutinen et al., 2016), cities have drawn increasing attention from both innovation scholars and practitioners. Experimentation, innovation, and development activities in cities include a variety of modes of collaborative innovation, including hackathons, innovation labs, innovative purchasing, open spaces, participatory budgeting, makerspaces, fablabs, co-working places, and innovation spaces (e.g., Bogers et al., 2017; Hyysalo et al., 2014, 2016; Kohtala & Hyysalo, 2015, Schuurman & Tõnurist, 2016).

Acknowledging the categorization of living labs phenomenon by Leminen (2015) – in other words, viewing living labs as a context, a method, and a conceptualization – the present study contributes to this perspective and labels the variety of collaborative innovation as "third-generation living lab networks". The first of generation living labs focused on the landscape(s) of living labs as real-life environments intertwined with users and stakeholder activities. The second generation of living labs considered methods and methodologies as a part of innovation activities in the real-life environment. The third-generation living labs portray different modes of collaborative innovation, where different stakeholders and particularly users have crucial roles in innovation on platforms. Following Habib, Westerlund, and Leminen (2015), the present study defines third-generation living labs as: *"platforms with shared resources, which organize their stakeholders into a collaboration network(s), that relies on representative governance, participation, open-standards, and diverse activities and methods to gather, create, communicate, and deliver new knowledge, validated solutions, professional development, and social impact in real-life contexts."*

Numerous studies document innovation activities in the smart city context (e.g., Khomsi, 2016; Ojasalo & Kauppinen, 2016; Ojasalo & Tähtinen, 2016), where various types of collaborative innovations and platforms have emerged in practice and that have been discussed in the scholarly literature (Bollier, 2016; Raunio et al., 2016; Walravens & Ballon, 2013).

Among the many definitions of "platforms" provided in the literature, Raunio and colleagues (2016) propose that a platform refers to *"any operating environment, technology, system, product or service, whose development has been systematically opened up to outside developers, and whose key aims are the benefit produced by the platform's users to each other and the network effect brought by participation."* The platform-based operating method is a key to digitalized participatory urban development, which significantly increases the innovation impact and participatory nature of development (Raunio et al., 2016). One of the key concepts used in this connection is "innovation platform", which requires that a city can shift its mindset from government to governance so that its focus will shift to the development and realization of development goals instead of regulation and enforcement of decisions. In other words, the city should adopt the role of coordinator rather than executor. Similarly, cities have begun to see their citizens as co-designers, co-producers, and co-learners (Bollier, 2016), suggesting that citizens move away from being subjects to being active participants in innovation (Leminen et al., 2014). Simultaneously, cities increasingly rely on expertise and resources on different communities (Anttiroiko, 2010). Moreover, platform orientation arises from profound social changes in cities (Raunio et al., 2016). Taken together, prior research has suggested the importance and role of the city as an enabler of innovation, yet studies on living labs are sparse on various roles that cities can adopt. In particular, there is a need for research on the implications of next-generation living lab networks in the city context. Hence, through this study, we aim to understand collaborative innovation networks in cities, herein referred to as "third-generation living lab networks". Accordingly, we pose the following research questions:

- *What are collaborative innovation networks and their roles in cities?*

- *How can cities exploit such collaborative innovation networks?*

The article is organized as follows. First, we review previous literature to create a framework of collaborative innovation in cities. Then, we describe our research design and research process. Thereafter, we describe the key findings regarding collaborative innovation in cities through six cases resulting in four types of novel third-generation living lab networks. Finally, we discuss the theoretical and managerial implications and provide directions for future research on third-generation living labs.

Towards Third-Generation Living Labs

We propose a framework on collaborative innovation based on two dimensions arising from previous literature on living labs and cities. The dimensions are i) "platform" (Anttitroiko, 2016; Bollier, 2016; Ojasalo & Tähtinen, 2016; Raunio et al., 2016; Walravens & Ballon, 2013) and ii) "participation approach" (Hossain, 2016; Leminen, 2013; Leminen & Westerlund, 2015, 2017; Steen & van Bueren, 2017). The framework demonstrates the differences between collaborative innovation networks in the city context. The platform dimension distinguishes between the city and the neighbourhood, building on the notion that cities or their parts are increasingly documented as *platforms* (Anttiroiko, 2016). A neighbourhood or a suburb could also refer to a smaller entity or unit within a city, such as a school, a hospital, a community house, or a geographical area such as a park.

As to platform as the first dimension, living labs are generally viewed as platforms for innovation (Almirall & Wareham, 2008; Anttiroiko, 2016; Dell´Era & Landoni, 2014; Habib et al., 2015). Ojasalo and Tähtinen (2016) argue that, in the context of cities, the owner of the innovation platform is usually a city, and the platform functions as an innovation vehicle between the city and external actors. Walravens and Ballon (2013) study platform business models for smart cities (in particular, business models of mobile service offerings of cities). The authors put forward a "public business model grid", where they have a dimension of public value, spanning from direct to indirect public value, and a dimension of governmental involvement, spanning from limited to strong. Raunio and colleagues (2016) propose that, through platforms, citizens become an active part of public service development and the city's role changes from being a service provider to a facilitator of innovative services. The authors conclude that platform thinking has also been viewed as the next development stage of conventional cluster policy, suggesting a re-organization of innovation collaboration in the city community. Furthermore, Raunio and colleagues (2016) make a "simplistic but practical division" between platforms, by categorizing them into i) intermediary platforms that create value by conveying the products or services of others (e.g., Uber, Alibaba, eBay); ii) development platforms or platform ecosystems that produce value by co-creating products and services with other companies (e.g., Microsoft, Intel, SAP); and iii) integrated platforms that function as intermediaries but also have a large external developer network (e.g., Google, Facebook, Apple, Amazon) (Gawer, 2009; Evans & Gawer, 2016; Thomas et al., 2014).

The platform owner (usually a city, a higher education institute, or a development company) facilitates, or organizes the facilitation of, the activities and defines the goal(s) of the platform. Platforms can be rather permanent physical or digital environments or less permanent environments, such as pop-up events, co-creation competitions, and hackathons. Anttiroiko (2016) documents participatory innovation platforms of three case cities, and states that, given that the city government facilitates these platforms and that they are integrated with the official planning system and local development policy, they resemble enabler-driven living labs. Furthermore, Anttiroiko (2016) observes three points of business–citizen interaction, namely open data, public services, and urban development. All of Anttiroiko's (2016) case cities support open data and knowledge sharing, focus on the development of public services with platforms within the smart city framework, and utilize innovation platforms in neighbourhood revitalization. He also highlights that citizens are, in most cases, customers or users, but they sometimes play the role of empowered residents or citizens whose needs push the design of local services. Thus, living labs can either span over the whole city (i.e., the "city as a platform") or focus on a specific neighbourhood.

Leminen (2013) classifies living labs into four types based on their coordination approach (i.e., bottom-up versus top-down) and participation approach (exhalation-dominated versus inhalation-dominated). He argues that a top-down approach is led or coordinated in accordance with centralized and official targets, whereas a bottom-up approach focuses on local needs and operates at the grassroots level. Whereas the inhalation-dominated innovation approach aims at fulfilling the needs of the driving party of the living lab, the exhalation-dominated innovation approach aims at fulfilling the requirements of other stakeholders. Leminen (2013) proposes to encourage parties to share their knowledge, expertise, and resources with the open innovation network. The exhalation-dominated approach engages stakeholders in collective action in the open innovation network to fulfill the needs of the others (Leminen, 2013).

The second dimension of our framework, "participation approach", depicts the innovation approach either as exhalation-dominated or inhalation-dominated. In this respect, Steen and van Bueren (2017) operationalized a definition of urban living labs, which was used to assess 90 sustainable urban innovation projects in the city of Amsterdam. They summarized the characteristics of living labs as four elements: aims, activities, participants, and context. Living labs are aimed at innovation and formal learning, and activities of living labs include development, co-creation, and iteration. Specifically, urban living labs aim at urban sustainability. Participants are public and private actors, users and knowledge institutes, and all the involved stakeholders have decision-making power. The context of the living lab is that of a real-life, and in many urban living labs, this means a territory or a space-bound place. Notably, Steen and van Bueren (2017) argue that most of the projects that label themselves as living labs do not include all the defining elements of a living lab.

To summarize, our conceptual framework captures the characteristics of collaborative innovation in the city context. Using the bipolar dimensions of platform and participation approach as principal axes in the framework, we can distinguish between four different modes of collaborative innovation networks in cities. We anticipate that the two-dimensional framework, as shown in Figure 1, can help us to identify existing collaborative innovations in cities, and a further analysis of the dimensions enables us to capture differences and similarities between the models.

Research Design

We chose collaborative innovation networks, particularly living labs in cities, by exploring their innovation processes and contexts in order to contribute to the discussion on open innovation networks. The study applies a qualitative, multiple case study approach (Yin, 1989) by analyzing a unique data set encompassing 118 interviews in six cities in Finland. The selected case cities are at the forefront of development of collaborative innovation networks, and they represent a broad variety of collaborative innovation, such as living labs, hackathons, innovative purchasing, participatory budgeting, open spaces, makerspaces, fablabs, co-working places, innovation spaces, and so forth. These various modes or types met the suggested criteria of collaborative innovation networks in cities, where one specific form, a living lab, is associated with a real-life environment, multiple stakeholders, and the pivotal role of users (Almirall & Wareham, 2011; Bergvall-Kåreborn & Ståhlbröst, 2009; Leminen, 2013, 2015; Leminen et al., 2014; Leminen, Nyström, & Westerlund, 2015). As suggested by Jensen and Rogers (2001), we organized the cases as snapshot studies, meaning that the cases represented the diversity of innovation activities driven by different actors in networks (Leminen et al., 2012). In addition, we utilized secondary data consisting of websites, bulletins,

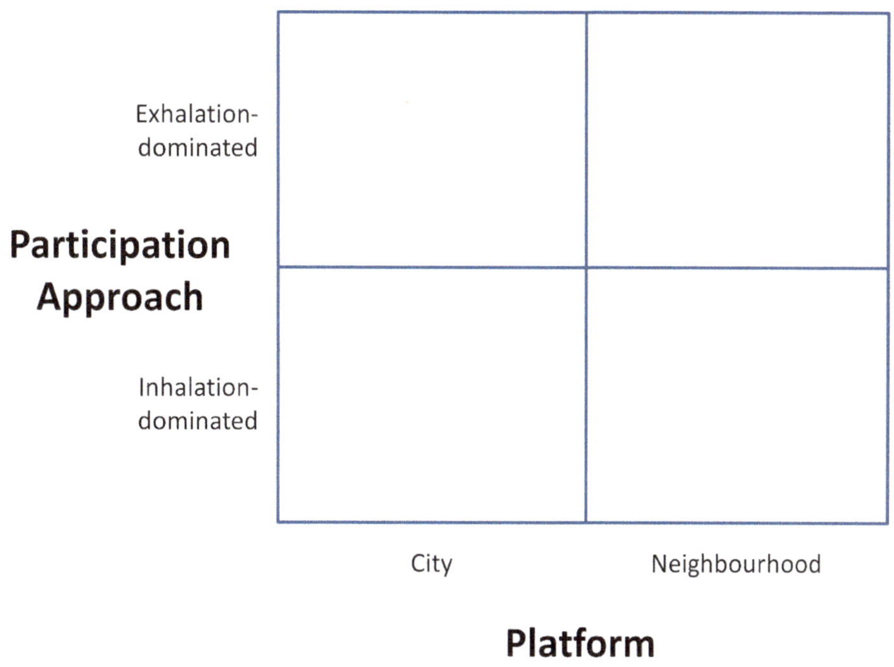

Figure 1. A conceptual framework for collaborative innovation networks in cities

magazines, and reports to gain further understanding of some of the collaborative innovation networks or to resolve arising issues or inconsistencies in the interviews.

Data collection
We collected all of the interview data in 2017. We audio recorded and transcribed all face-to-face meetings and meetings by phone, and followed an interview guide when collecting information from various themes of collaboration innovations (Patton, 1990), and we had the informants verify the findings. Understanding different modes of collaborative innovations in cities and the roles of platform(s) and the gained benefits for different stakeholders in such collaborative innovation networks exemplify the themes of the semi-structured and open-ended questions. Our informants comprised various stakeholders representing different modes of collaborative innovation networks, especially living labs. The selected informants were interviewed because they have in-depth knowledge and first-hand experience of collaborative innovation in cities. The informants included CEOs, civil servants, directors, managers, professors, researchers, project coordinators, technical specialist, and citizens (users as innovators). The names of organizations and the identities of informants are withheld to maintain confidentiality.

Data analysis
An overview of the data analysis and the phases of the study is presented in Table 1. We organized the empirical data according to the informant, the date of interview, the type of informant, and the case. Then, we followed a multi-staged data analysis process consisting of open coding, focused coding, identification of innovation processes, and theorizing the codes. The main unit of analysis was the collaborative innovation: stakeholder activities and the characteristics of collaboration innovation networks. The original transcribed interviews were analyzed and coded by the researchers. We searched the words associated with activities, innovation processes, contexts, methods, methodologies, platforms, stakeholders, and tools using a content analysis technique. For instance, we coded stakeholders as utilizers, enablers, providers, or users to identify the characteristics of third-generation living labs. So doing, we followed the examples of Roberts (1997) and Neuendorf (2002) to understand the cases by coding and content analysis. We first coded the original, word-by-word transcribed empirical material independently and then compared, discussed and agreed on the results.

In the second phase, the first round of coding resulted in describing and identifying participation approaches

Table 1. Data analysis process

Data Analysis Phases	Task	Outcome
1. Open coding	• Organize dataset • Identify collaborative innovation networks and informants	Overview of cities, collaborative innovation networks, informant, type of informant, and time of interview
2. Focused coding #1	• Identify and briefly describe participation approaches and platforms in cities	Overview of innovation activities resulted in identifying participation approaches and platforms in cities
3. Focused coding #2	• Analyze innovation through the identified innovation activities and participation approaches as well as contexts and platforms • Compare data to theory	Detection of previously identified innovation activities and participation approaches (Leminen, 2013; Nyström et al., 2014) as well as contexts and platforms (Anttiroiko, 2016)
4. Theorizing the codes	• Synthesize phases 1 to 3: analyze identified modes in collaborative innovation networks • Identify theoretical implications • Identify managerial implications	Classification of the participation approaches and platforms resulting in four archetypes of collaborative innovation networks in cities (Figure 2)

and platforms in cities. We then analyzed the four archetypes of collaborative innovation networks by categorizing participation approaches and platforms in six cities (Figure 2). We anticipate that the four archetypes of collaborative innovation networks in cities are our key findings.

Findings and Discussion

In this study, we analyze and classify the variety of collaborative innovation activities in six Finnish cities. So doing, we establish a framework based on platform and participation approach, which puts forward four diverse archetypes, or modes, of collaborative innovation in the city context, which are illustrated in Figure 2:

A. The city as a provider
B. The city as a neighbourhood participator
C. The city as a catalyst
D. The city as a rapid experimenter

A. The city as a provider
The mode of *city as a provider* (lower-left corner of Figure 2) represents an inhalation-dominated participation approach where improvements are done to the city's own service provisioning, and an entire city is seen as a platform. In brief, this mode refers to exposing the service provisioning of a city to others in order to improve its services and processes for citizens.

Improvements to services and processes are undertaken with a broad variety of stakeholders such as companies and research institutes by providing expertise for a city. Activities are often initiated by the city's strategic aims to pursue predefined improvements for its services. The city endeavours to create points where stakeholders can anchor their activities to the city's operations, facilities, areas, and routes, and to gather information, test, co-create, and validate products, services, and systems. The city has specific city-wide targets, and it spells out how companies and other actors need to act with it, and what kinds of benefits they can receive. When the city exposes its processes to others, a network or an ecosystem forms around the city that organizes activities to streamline and develop the city's service provisioning.

An innovation platform produces ideas, solutions, and knowledge for making public services and their production more efficient, while the city acts as a utilizer of the results. The city scales processes by providing guidebooks while companies and research institutes gather information, test, develop, and co-create products, services, and systems. Companies may also be utilizers benefiting from the results of innovation activities in their product and service development processes. Rather than being active actors, users are essentially treated as "lab rats" for testing products, services, and systems. Therefore, this mode does not make use of the full expertise and potential of citizens. The innovation

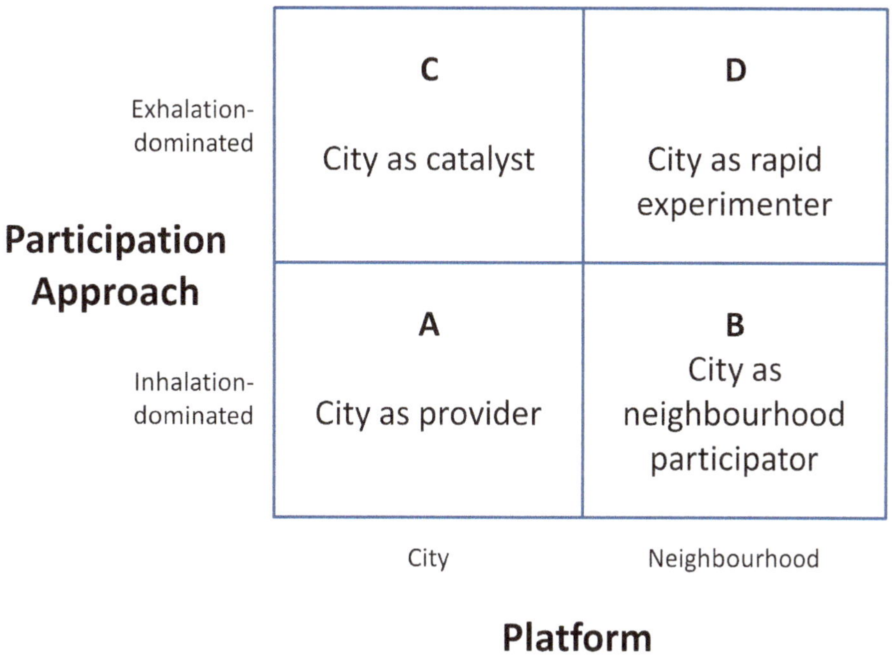

Figure 2. Collaborative innovation modes in cities

mechanism of this mode assumes a city exposes its activities to companies and research institutes that collect information from processes of the city and improve the city´s service provisioning.

B. The city as a neighbourhood participator
The mode of *city as a neighbourhood participator* (lower-right corner of Figure 2) represents an inhalation-dominated participation approach where the neighbourhood is seen as a platform. The mode refers to improving the neighbourhoods of citizens and their living conditions by local, grassroot activities initiated by the citizens. Similarly to the previous mode, innovation activities are initiated by and aimed at improving the conditions of the driving party. Specifically, citizens lead and benefit from innovation activities in this mode. In other words, a city engages itself in the collaborative innovation process, participates in activities, and supports citizen activities rather than attempting to steer the innovation activities.

A network or an ecosystem forms around a neighbourhood/community that organizes action to solve citizens' needs and aims to bring benefits for its citizens. Success is based on the activity and enthusiasm of citizens, and activities in this mode require patience from the city, not vast resources. The implemented operations are often small and quick, and easy to accomplish by the city. Examples of social innovations in our data included gardening activities initiated by the citizens in a neighbourhood, a village fête in the neighbourhood, and a digital bulletin board installed in stairwells – all of them jointly developed with the citizens. Another example: a residential area was isolated and there were hardly any services, but citizens, a local grocery store, and the developer of the residential area jointly ideated a drop-off location where the grocery store delivers online food purchases for pick-up by customers. Later, this resulted in the establishment of a specific e-grocery.

In this mode, the city is an enabler by participating in and supporting innovation activities in neighbourhoods. Scalability into citywide solutions is not as important as in the previous mode. However, the platform is the source of ideas and needs, which are the cultivated and developed into commercialized products, start-up companies, or social innovations. In contrast to the previous mode, where citizens act as lab rats, citizens here lead innovation activities or are participants in innovation activities with other stakeholders and can be perceived as co-creators or creative consumers (Leminen et al., 2014, 2015). The innovation mechanism of this mode assumes that the city not only initiates, participates in, and supports activities, but also collects the best ideas for further development.

C. The city as a catalyst
The mode of *city as a catalyst* (upper-left corner of Figure 2) represents an exhalation-dominated participation approach where the entire city is a platform. The city boosts the development of companies and increases value of their operations by combining other aims and connecting other actors to the service provisioning in a city region. The main objective of the city is not to develop more efficient services for itself but to enhance and nourish business ecosystem(s) through living labs, when no companies take a role in order to boost and cultivate new networks and ecosystems in the city.

A network or an ecosystem forms around the city`s own service provisioning, where living labs have a built-in role in the city's operations and service production. The city is a catalyst that opens up its service production and processes. The platform can be physical, virtual, or hybrid, and it consists of processes and procedures of the city, such as city planning and land use, wellbeing and healthcare, and the educational system. Although the city opens up the service production and data resources, it becomes a development platform for companies to develop, experiment, test, and validate products, services, and systems.

Because living labs and their activities are intertwined with the catalyst's service production and processes, they generate diverse value for the stakeholders. Put differently, by combining the conventional service provisioning of the city, as well as its processes, the city pursues benefits that are difficult to obtain otherwise. For example, residential area planning can be arranged with an innovative conveyancing competition, where construction companies and others compete on ideas that they implement, and they seek to identify potential companies interested in jointly building and experimenting with new types of houses and housing solutions such as zero-energy homes. Such operations catalyze development and stimulate adoption and creation of new solutions and services in the building industry.

In addition, co-operation between various sectors increases, and cross-pollination and learning take place between different sectors. The long-term benefits for the city are realized through activated business life in the city and through the success of companies developing their products and services, as the mode assumes prolonged development in the city region. Further, the scaling mechanism of the mode postulates learning and understanding of the principles of living labs by sharing and transferring knowledge regarding experiences of innovation activities between humans rather than formulating knowledge in manuals as the activities of living labs are at different maturity levels.

A city acts as an enabler by enhancing and nourishing a business ecosystem. That is, the innovation mechanism of this mode assumes opening the city's service provisioning and boosting business ecosystem(s) in the city. Companies and research institutes test, develop, and co-create their products, services, and systems. Although the roles of platforms are twofold, they enhance development, experimentation, testing, validation of companies' services, products, and systems, and act as showrooms for companies' activities and their outputs. Similar to the first mode, users act as mere "lab rats" for testing products, services, and systems; thus, the mode does not benefit from the full potential of citizens.

D. The city as a rapid experimenter

The mode of *city as a rapid experimenter* (upper-right corner of Figure 2) represents an exhalation-dominated participation approach, where the platform is a neighbourhood, unit, or a specific theme or activity rather than the entire city. This mode refers to accomplishing trials of new products, services, and systems by companies to gather experience and knowledge, to learn fast from such experiments, and to accelerate their product and service development processes and growth. The mode assumes development in predetermined thematic areas or neighbourhoods through rapid experiments that the city supports with a modest financial or non-financial stake in publicly funded projects. The city arranges competitions of rapid experimentation dedicated to certain predefined thematic fields, activities, or areas such as energy efficient solutions, smart mobility, health, and other solutions for smart cities. In other words, the city supports the growth of small companies and the business ecosystem by enabling rapid experiments.

The city has no specific short-term targets but can realize long-term benefits as it initiates a business network or an ecosystem around the needs that will be solved through rapid development. The development process will provide benefits to other stakeholders, bring new solutions for the city or citizens, and develop the platform or its processes. The benefits of rapid experimentation increase, at least indirectly, and include flexibility, learning, and knowledge transfer. The developed solutions can be scalable to other contexts, but the scaling is conducted by the involved companies.

The benefits of the participating companies are twofold. First, the companies can gather information, test, develop, and co-create their products, services, and systems. Second, they may look for references for their products and services in cities. Users' or citizens' specific roles may vary, and they may act as "lab rats" for testing products, services, and systems; yet, their full potential and expertise as a part of innovation activities may be involved. This mode assumes learning from trial and error; such flexibility can be achieved by bringing in new actors and developing limited and rapidly implementable solutions for topical problems in real environments. Table 2 presents characteristics of collaborative innovation in different types of third-generation living labs.

To summarize, our findings indicate that cities may simultaneously use several collaborative innovation modes and that innovation can adopt different modes at the same time. Furthermore, because the needs of cities are often versatile, various modes of platforms (virtual, physical, or hybrid) and operational models (ranging from everyday basics to complex collaborative innovation networks) are increasingly used. Consequently, a city must possess capabilities to simultaneously handle the variety of forms. A city has to be able to develop its basic services that it has legal obligations to provide to its citizens, and to activate companies, act efficiently and innovatively, and at the same time conduct small experiments benefiting its citizens and companies. In contrast to prior studies on living lab networks (e.g., Leminen et al., 2012; Leminen et al., 2014, 2015, 2016; Nyström et al., 2014; Steen & van Bueren, 2017), which documented innovation activities through the importance of users in various innovation networks characterized by openness in cities, we argue that cities play a pivotal role by enabling innovation activities using different mechanisms to boost innovation with different parties.

Conclusion

This study classified the variety of collaborative innovation activities in six cities in Finland. The study identified two essential dimensions in previous literatures on living labs and cities, namely platforms and participation approaches, in order to propose a framework that demonstrates differences of current and potential collaborative innovation networks in cities. The study aimed to understand the plurality and variety of collaborative innovation networks in cities, referred to as third-generation living lab networks. Particularly, the study attempted to take a step towards research that would review implications of the third-generation of living labs in cities. Therefore, this study not only illuminates four collaborative innovation modes but also contributes to the growing literatures of open innovation networks and living labs by describing the ways living lab networks are exploited in the city context.

Table 2. Characteristics of collaborative innovation in different types of third-generation living labs

	Types of Third-Generation Living Labs			
Characteristic	**A. The city as a provider**	**B. The city as a neighbourhood participator**	**C. The city as a catalyst**	**D. The city as a rapid experimenter**
Definition	Expose service provisioning of a city to improve its services and processes for citizens with a broad variety of providers providing expertise for a city	Improve neighbourhood or living conditions of citizens by local, grassroot activities initiated by citizen(s)	Boost development of companies and increase value of operations by combining other aims and connecting other actors to service provisioning in a city region	Accomplish trials of new products, services, and systems by companies to gather experience and knowledge and learn fast, and to accelerate companies' development processes and growth
Strategy	Strategic R&D activity with preset objectives, where city utilizes the outcomes to streamline its service provisioning	Problem solving by collaborative accomplishments with neighbourhood	Enhancing or nourishing city's business ecosystem(s)	Learning and operations development through rapid experimentations
Organization	Network/ecosystem forms around a city that organizes actions to streamline its service provisioning	Network/ecosystem forms around a neighbourhood/ community, organizing action to solve citizens' needs and bring value for citizens	Network/ecosystem forms around a city`s own service provisioning or its data storages	Network/ecosystem initiated by needs to be solved by rapid development
Participation approach	Inhalation-dominated (city – city)	Inhalation-dominated (citizens – citizens)	Exhalation-dominated (city – companies)	Exhalation-dominated (city – companies)
Platform	Entire city and its spaces, places, areas, processes, and routes	Neighbourhood, suburb of a city, or theme	Entire city and its spaces, places, areas, processes, and routes	Neighbourhood, suburb of a city, or theme
Role of platform	Provide new ideas, solutions and knowledge Boost service provisioning and make production of public services more efficient	Ideas and needs are cultivated and developed towards commercialized products, startups, or social innovations	Enhance development, experimentation, testing, validation of companies' services, products, and systems A showroom for companies' products and services	Bring new solutions for the city or citizens, and develop the platform or its processes

Table 2. ***(continued)*** Characteristics of collaborative innovation in different types of third-generation living labs

	Types of Third-Generation Living Labs			
Characteristic	**A. The city as a provider**	**B. The city as a neighbourhood participator**	**C. The city as a catalyst**	**D. The city as a rapid experimenter**
Enabler (city)	*Expose, align, and improve* Initiate innovation activities and create anchorage points	*Initiate, participate, and support* Participate in and support innovation activities	*Open, activate, and boost* Enhance or nourish business ecosystem(s)	*Learn from trial and error* Promote rapid prototyping with small incentives
Utilizer (city)	Benefit directly and scale up results of streamlined service provisioning	Harvest ideas and needs from innovation activities and use them in development activities	Bring new solutions for citizens and develop the platform or its processes	Bring new solutions for citizens and develop the platform or its processes
Utilizer (companies)	Benefit from the results of innovation activities	Harvest ideas, needs, and results from innovation activities and use them in development activities	Benefit from results of innovation activities Platform as a showroom for products and services	Learn through rapid experimentations Look for references
Provider (companies and research institutes)	Gather information, test, develop, and co-create products, services, and systems	Gather information, test, develop, and co-create products, services, and systems	Gather information, test, develop, and co-create products, services, and systems Develop the platform	Gather information, test, develop, and co-create products, services, and systems Develop the platform
User/citizen	"Lab rats"	Leader or equal participants	"Lab rats"	"Lab rats" or equal participants
Innovation mechanism	Parallel skiing (alignment)	Participate, do not activate Collect the best ideas	Added value is produced for others by adding different processes and activities to city's own operations	Flexibility achieved by bringing in new actors and developing limited and rapidly implementable solutions
Outcome	New ideas, knowledge, and solutions for streamlined service provisioning	Social innovations and solutions offering pleasure and benefits for citizens Commercialized products or companies (e.g., digital notice boards, food halls)	New products and services Showroom for companies' activities Business ecosystem growth, new jobs	New solutions for a city and its citizens Showroom for companies' activities Business ecosystem growth, new jobs
Lifespan	Short, medium, or long	Long	Long	Short

Theoretical implications

There are three theoretical contributions that the study highlights in regard to the discussions on collaborative innovation and open innovation networks particularly in the area of living labs. First, the paper suggested a *new conceptual framework* for revealing collaborative innovation networks in the city context: the third-generation of living lab networks (Figure 1). Second, the framework distinguished *four archetypes of collaborative innovation* through third-generation living lab networks based on their participation approach and the platform: A. the city as a provider, B. the city as a neighbourhood participator, C. the city as a catalyst, and D. the city as a rapid experimenter. Third, the study proposed that *cities reinforce long-term participation and engagement of stakeholders,* suggesting various benefits to all stakeholders. Each of these contributions is described as follows:

1. *Conceptual framework:* Mulder (2012) argues that the existing living labs fail to benefit from their full potential, because they rely too much on traditional user-centric lab methodologies, forgetting the "living part" that makes a living lab an exceptional methodology. The framework suggested in the present study illuminates various types of collaborative innovation. The dimensions of the framework include the platform (in terms of "city" versus "neighbourhood") and the participation approach (in terms of "inhalation-dominated" versus "exhalation-dominated"). Whereas the former dimension is grounded on exploiting different platforms in cities, the latter is grounded on the assumption on the participation approach.

2. *Four archetypes of third-generation living labs:* The conceptual framework distinguishes four archetypes of third-generation living labs based on the participation approach and the platform. The *city as a provider* assumes that an entire city is viewed as a platform, and its service provisioning is exposed to other stakeholders in order to improve and make services and their processes more efficient, as well as to provide expertise for the city. The mode of the *city as a neighbourhood participator* refers to improving neighbourhood of citizens or their living conditions by local, grassroot innovations by citizens, where the platform is a neighbourhood or a suburb of city, and such innovation activities are conducted for the benefits of citizens themselves. The *city as a catalyst* refers to a mode where the city boosts the development of companies and increases value of its own operations by combining other aims and connecting other actors to its service provisioning in the entire city region. Finally, the *city as a rapid experimenter* considers a part of city (e.g., a neighbourhood) as a platform, where it attempts to learn fast from the rapid experiments and to accelerate companies' service and product development processes.

3. *Cities reinforce long-term participation and engagement of stakeholders:* The extant literature proposes many benefits from engaging multiple stakeholders and particularly users in organization's innovation activities (e.g., Leminen & Westerlund, 2012; Leminen, 2015). Although the benefits are widely acknowledged, Hannukainen and colleagues (2017) note that user-oriented innovation activities may not be rooted in part of an organization's innovation and development activities even though the organizations are excited by such modes and find them useful. One explanation for this might be that many company-driven living lab targets are, by nature, short term; for instance, the goal may be to solve a company's instant needs in their innovation activities (Leminen et al., 2012). Our study proposes that cities increasingly reinforce the long-term participation and engagement of users, citizens, and other stakeholders particularly in the city as provider and city as catalyst modes because cities' innovation and development activities are increasingly coupled into their service-provisioning. Therefore, if a city succeeds in aligning its modes in collaborative innovation networks with its long-term mission and goal, and in building appropriate anchorage points for other stakeholders, the structure can become a long-lasting part of the city's innovation system.

Managerial implications

From the managerial perspective, the study contributes a framework, or tool, to identify and categorize collaborative innovation networks in cities. The framework and the identified characteristics of the modes with regards to collaborative innovation networks portray different stakeholders and their activities and benefits. We described four different types of third-generation living lab networks based on their participation approach and platform whose interests dominate the network's operation. By identifying each mode in collaborative innovation networks in cities, managers may link their own innovation and development processes as a part of the city's activities. In other words, cities may provide many benefits for managers when cities are seen as platforms, source(s) of data, and sources of needs by the citizens and the city. More specifically, managers may learn that cities have a key role in boosting companies' own innovation and development activities, ranging from testing and validating their products to co-creating and developing new ones. Particularly companies should prepare for a revision of their roles and activities corresponding to identified collaborative innovation networks in cities.

Limitations and future research
All studies have their limitations. First, the present study put forward a matrix where we selected participation approach as one dimension while excluding the dimension of coordination approach presented by Leminen (2013). The coordination approach could be included in the matrix in future research. We were not able to include all the different stakeholders in the studied collaborative innovation networks, the third-generation networks, because of the limited resources. However, we believe that our data set is sufficiently rich and covers multiple types of informants and diverse collaborative innovation networks in six cities. Yet, the limitation may affect the results on modes in collaborative innovative networks in the city context. We share the view that living labs are coupled into the contexts (Leminen, 2015), and further research is needed for different modes of collaborative innovation networks. For example, new platforms enable citizens to participate and engage in development and innovation activities in cities, and it is crucial to understand the mutual interests and mechanisms of open and collaborative innovation activities. Therefore, we propose more research on how different stakeholders should be motivated in order to be engaged in the development and innovation processes in collaborative innovation networks, and on what actions are necessary to keep stakeholders engaged. Further, we propose the importance of studying the relations of different collaborative and open innovation networks. Also, we suggest a need for additional research on the characteristics in open and collaborative innovation networks. Therefore, we call for further analyses of specific cases, eventually including how different stakeholders employ collaborative and open innovation networks in cities. Are there relations (or correlations) between different types of cities, collaborative innovation networks, and the position of the informants? To conclude, we call for more research on collaborative innovation networks, the third-generation networks.

Acknowledgements

The authors gratefully appreciate the anonymous blind reviewer and the Editor-in-Chief, Chris McPhee for their valuable comments, suggestions, and guidelines to improve this article. Special thanks are also extended to Director for Services Development Päivi Sutinen and all of the interview subjects who kindly shared their knowledge and experience as stakeholders in living labs and other forms of collaborative innovations in cities. The authors express their gratitude to City of Espoo and the 6-aika program for funding the development of the article as a part of the Living Lab 3.0 project.

About the Authors

Seppo Leminen is a Principal Lecturer at the Laurea University of Applied Sciences in Espoo, Finland, and he serves as an Adjunct Professor of Business Development at Aalto University in Helsinki, Finland, and an Adjunct Research Professor at Carleton University in Ottawa, Canada. He holds a doctoral degree in Marketing from the Hanken School of Economics and a doctoral degree in Industrial Engineering and Management from the School of Science at Aalto University. His research and consulting interests include living labs, open innovation, innovation ecosystems, robotics, the Internet of Things (IoT), as well as management models in high-tech and service-intensive industries. Results from his research have been reported in *Industrial Marketing Management*, the *Journal of Engineering and Technology Management*, the *Journal of Business & Industrial Marketing*, *Management Decision*, the *International Journal of Technology Management*, the *International Journal of Technology Marketing*, the *International Journal of Product Development*, and the *Technology Innovation Management Review*, among many others.

Continued on next page...

About the Authors *(continued)*

Mika Westerlund, DSc (Econ), is an Associate Professor of Technology Innovation Management at Carleton University in Ottawa, Canada. He previously held positions as a Postdoctoral Scholar in the Haas School of Business at the University of California Berkeley and in the School of Economics at Aalto University in Helsinki, Finland. Mika earned his doctoral degree in Marketing from the Helsinki School of Economics in Finland. His current research interests include open and user innovation, the Internet of Things, business strategy, and management models in high-tech and service-intensive industries.

Mervi Rajahonka, DSc (Econ), works as an RDI Advisor at the Small Business Center (SBC), currently a part of the South-Eastern Finland University of Applied Sciences XAMK, Finland, and as an Adjunct Research Professor at Carleton University in Ottawa, Canada. She has been working at the SBC for about 10 years, participating in numerous EU-funded projects. She earned her doctoral degree in Logistics from the Department of Information and Service Economy at Aalto University School of Business in Helsinki, Finland. She also holds a Master's degree in Technology from the Helsinki University of Technology and a Master's degree in Law from the University of Helsinki. Her research interests include sustainable logistics and supply chain management, high-tech and service business models, service modularity, and service innovations. Her research has been published in a number of journals in the areas of logistics, services, and operations management.

References

Almirall, E., & Wareham, J. 2008. The Role of Living Labs in Open Innovation, In J. Schumacher & V.-P. Niitamo (Eds.), *European Living Labs – A New Approach for Human Centric Regional Innovation:* 48–165. Berlin: Wissenschaftlicher Verlag Berlin.

Almirall, E., & Wareham, J. 2011. Living Labs: Arbiters of Mid- and Ground-Level Innovation. *Technology Analysis & Strategic Management,* 23(1): 87–102.

Anttiroiko, A.-V. 2010. *Luova kaupunkikehittäminen. Kaupunkikonseptit innovatiivisen kaupunkikehittämisen apuna* (Creative Urban Development – City Concepts as Support Tools for Innovative Urban Development). Tampere, Finland: Research Unit for Urban and Regional Development Studies, Sente publications 32/2010, University of Tampere.

Anttiroiko, A.-V. 2016. City-as-a-Platform: The Rise of Participatory Innovation Platforms in Finnish Cities. *Sustainability,* 8(9): 2–31.

Ballon, P., Glidden, J., Kranas, P., Menychtas, A., Ruston, S., & Van der Graaf, S. 2011. Is there a Need for a Cloud Platform for European Smart Cities? In P. Cunningham & M. Cunningham (Eds.), *Proceedings of the eChallenges e-2011 Conference.* Dublin: International Information Management Corporation (IIMC).

Bergvall-Kåreborn, B., Ihlström Eriksson, C., & Ståhlbröst, A. 2015. Places and Spaces within Living Labs. *Technology Innovation Management Review,* 5(12): 37–47. http://timreview.ca/article/951

Bergvall-Kåreborn, B., & Ståhlbröst, A. 2009. Living Lab: An Open and Citizen-Centric Approach for Innovation. *International Journal of Innovation and Regional Development,* 1(4): 356–370. https://doi.org/10.1504/IJIRD.2009.022727

Bogers, M., Zobel, A.-K., Afuah, A., Almirall, E., Brunswicker, S., Dahlander, L., Frederiksen, L., Gawer, A., Gruber, M., Haefliger, S., Hagedoorn, J., Hilgers, D., Laursen, K., Magnusson, M. G., Majchrzak, A., McCarthy, I. P., Moeslein, K. M., Nambisan, S., Piller, F. T., Radziwon, A., Rossi-Lamastra, C., Sims, J., & Ter Wal, A. L. J. 2017. The Open Innovation Research Landscape: Established Perspectives and Emerging Themes Across Different Levels of Analysis. *Industry and Innovation,* 24(1): 8–40. http://doi.org/10.1080/13662716.2016.1240068

Bollier, D. 2016. *The City as Platform: How Digital Networks Are Changing Urban Life and Governance.* Washington, DC: The Aspen Institute.

Brankaert, R., den Ouden, E., & Brombacher, A. 2015. Innovate Dementia: The Development of a Living Lab Protocol to Evaluate Interventions in Context. *Info,* 17(4): 40–52. http://dx.doi.org/10.1108/info-01-2015-0010

Dell'Era, C., & Landoni, P. 2014. Living Lab: A Methodology between User-Centred Design and Participatory Design. *Creativity and Innovation Management,* 23(2): 137–154. http://dx.doi.org/10.1111/caim.12061

Dutilleul, B., Birrer, F. A. J., & Mensink, W. 2010. Unpacking European Living Labs: Analysing Innovation's Social Dimensions. *Central European Journal of Public Policy,* 4(1): 60–85.

Guimont, D., & Lapointe, D. 2016. Empowering Local Tourism Providers to Innovate through a Living Lab Process: Does Scale Matter? *Technology Innovation Management Review,* 6(11): 18–25. http://timreview.ca/article/1031

Hakkarainen, L., & Hyysalo, S. 2016. The Evolution of Intermediary Activities: Broadening the Concept of Facilitation in Living Labs. *Technology Innovation Management Review,* 6(1): 45–58. http://timreview.ca/article/960

Hannukainen, P., Mäkinen, S., & Hyysalo, S. 2017. Organisational Adoption of the Lead User Method: A Follow-Up Study on Intentions Versus Actions. *International Journal of Business Excellence,* 12(4): 508–536. https://doi.org/10.1504/IJBEX.2017.085016

Habib, C., Westerlund, M., & Leminen, S. 2015. *Living Labs as Innovation Platforms.* Paper presented at the 6th ENoLL Living Labs Summer School, August 25–28, 2015, Istanbul, Turkey.

Hossain, M. 2016. Grassroots Innovation: A Systematic Review of Two Decades of Research. *Journal of Cleaner Production,* 137: 973–981. https://doi.org/10.1016/j.jclepro.2016.07.140

Hyysalo, S., Kohtala, C., Helminen, P., Mäkinen, S., Miettinen, V., & Muurinen, L. 2014. Collaborative Futuring with and by Makers. CoDesign: *International Journal of CoCreation in Design and the Arts,* 10(3-4): 209–228.
https://doi.org/10.1080/15710882.2014.983937

Hyysalo, S., Repo, P., Timonen, P. Hakkarainen, L., & Heiskanen, E. 2016. Diversity and Change of User Driven Innovation Modes in Companies. *International Journal of Innovation Management,* 20(2): 1650023.
https://doi.org/10.1142/S1363919616500237

Jarvenpaa, S. L., & Wernick, A. 2012. Open Innovation Networks: The Evolution of Bureaucratic Control. In A. Bøllingtoft, L. Donaldson, G. P. Huber, D. D. Håkonsson, & C. C. Snow (Eds.), *Collaborative Communities of Firms: Purpose, Process, and Design:* 9–33.

Juujärvi, S., & Lund, V. 2016. Enhancing Early Innovation in an Urban Living Lab: Lessons from Espoo, Finland. *Technology Innovation Management Review,* 6(1): 17–26.
http://timreview.ca/article/957

Jensen, J. L., & Rodgers, R. 2001. Cumulating the Intellectual Gold of Case Study Research. *Public Administration Review,* 61(2): 235–260.
http://dx.doi.org/10.1111/0033-3352.00025

Knight, T. 1749. *Reflections upon Catholicons, or Universal Medicines.* London: Printed for T. Osborne in Gray's-Inn.

Khomsi, M. R. 2016. The Smart City Ecosystem as an Innovation Model: Lessons from Montreal. *Technology Innovation Management Review,* 6(11): 26–31.
http://timreview.ca/article/1032

Kohtala, C., & Hyysalo, S. 2015. Anticipated Environmental Sustainability of Personal Fabrication. *Journal of Cleaner Production,* (99): 333–344.
https://doi.org/10.1016/j.jclepro.2015.02.093

Leminen, S. 2013. Coordination and Participation in Living Lab Networks. *Technology Innovation Management Review,* 3(11): 5–14.
http://timreview.ca/article/740

Leminen. S. 2015. *Living Labs as Open Innovation Networks – Networks, Roles and Innovation Outcomes.* Doctoral dissertation. Helsinki, Finland: Aalto University.

Leminen, S., Nyström, A.-G., & Westerlund, M. 2015. A Typology of Creative Consumers in Living Labs. *Journal of Engineering and Technology Management,* 37: 6–20.
http://dx.doi.org/10.1016/j.jengtecman.2015.08.008

Leminen, S., Turunen, T., & Westerlund, M. 2015. The Grey Areas Between Open and Closed in Innovation Networks. *Technology Innovation Management Review,* 5(12): 6–18.
http://timreview.ca/article/948

Leminen, S., & Westerlund, M. 2012. Towards Innovation in Living Labs Network. *International Journal of Product Development,* 17(1/2): 43–59.
http://dx.doi.org/10.1504/IJPD.2012.051161

Leminen, S., & Westerlund, M. 2013. Incremental and Radical Service Innovation in Living Labs. In B. Christiansen, S. Yildiz, & E. Yildiz, E (Eds.), *Transcultural Marketing for Incremental & Radical Innovation:* 281–295. Hershey, PA: Information Science Reference

Leminen, S., & Westerlund, M. 2015. Cities as Labs: Towards Collaborative Innovation in Cities. In P. Lappalainen, M. Markkula, & H. Kune, H. (Eds.), *Orchestrating Regional Innovation Ecosystems Espoo Innovation Garden:* 167–175. Espoo: Otavan Kirjapaino Oy.

Leminen, S., & Westerlund, M. 2016. A Framework for Understanding the Different Research Avenues of Living Labs. *International Journal of Technology Marketing,* 11(4): 399–420.
https://doi.org/10.1504/IJTMKT.2016.079731

Leminen, S., & Westerlund, M. 2017. Categorization of Innovation Tools in Living Labs. *Technology Innovation Management Review,* 7(1): 15–25.
http://timreview.ca/article/1046

Leminen, S., Westerlund, M., & Nyström, A.-G. 2012. Living Labs as Open-Innovation Networks. *Technology Innovation Management Review,* 2(9): 6–11.
http://timreview.ca/article/602

Leminen, S., Westerlund, M., & Nyström, A.-G. 2014. On Becoming Creative Consumers – User Roles in Living Labs Networks. *International Journal of Technology Marketing,* 9(1): 33–52.
http://dx.doi.org/10.1504/IJTMKT.2014.058082

Leminen, S., Westerlund, M., Nyström, A.-G., & Kortelainen, M. 2016. The Effect of Network Structure on Radical Innovation in Living Labs. *Journal of Business & Industrial Marketing,* 31(6): 743–757.
http://dx.doi.org/10.1108/JBIM-10-2012-0179

Leminen, S., Westerlund, M., Sánchez, L., & Serra, A. 2014b. Users as Content Creator, Aggregator, and Distributor at Citilab Living Lab. In R. DeFillippi & P. Wikstrom (Eds.), *Business Innovation and Disruption in Film, Video and Photography:* 247–266. Cheltenham, UK: Edward Elgar.

Markkula, M., & Kune, H. 2015. Making Smart Regions Smarter: Smart Specialization and the Role of Universities in Regional Innovation Ecosystems. *Technology Innovation Management Review,* 5(10): 7–15.
http://timreview.ca/article/932

Mulder, I. 2012. Living Labbing the Rotterdam Way: Co-Creation as an Enabler for Urban Innovation. *Technology Innovation Management Review,* 2(9): 39–43.
https://timreview.ca/article/607

Neuendorf, K. A. 2002. *The Content Analysis Guidebook.* Thousand Oaks, CA: Sage.

Nyström, A.-G., Leminen, S., Westerlund, M., & Kortelainen, M. 2014. Actor Roles and Role Patterns Influencing Innovation in Living Labs. *Industrial Marketing Management,* 43(3): 483–495.
http://dx.doi.org/10.1016/j.indmarman.2013.12.016

Ojasalo, J., & Kauppinen, H. 2016. Collaborative Innovation with External Actors: An Empirical Study on Open Innovation Platforms in Smart Cities. *Technology Innovation Management Review,* 6(12): 49–60.
https://timreview.ca/article/1041

Ojasalo, J., & Tähtinen, L. 2016. Integrating Open Innovation Platforms in Public Sector Decision Making: Empirical Results from Smart City Research. *Technology Innovation Management Review,* 6(12): 38–48.
https://timreview.ca/article/1040

Patton, M. 1990. *Qualitative Evaluation and Research Methods* (2nd ed.). Newbury Park, CA: Sage.

Raunio, M., Nordling, N., Ketola, T., Saarinen, J. P., & Heinikangas, A. 2016. *Open Innovation Platforms: Handbook for Developers.* Tampere, Finland: 6Aika.

Rits, O., Schuurman, D., & Ballon, P. 2015. Exploring the Benefits of Integrating Business Model Research within Living Lab Projects. *Technology Innovation Management Review,* 5(12): 19–27. http://timreview.ca/article/949

Roberts, C.W. (Ed.). 1997. *Text Analysis for the Social Sciences: Methods for Drawing Statistical Inferences from Texts and Transcripts.* Mahwah, NJ: Lawrence Erlbaum.

Savelkoul, L., & Peutz, M. 2016. Needsfinding in Living Labs: A Structured Research Approach. *Technology Innovation Management Review,* 7(2): 42–51. http://timreview.ca/article/1057

Schuurman, D., De Marez, L., & Ballon, P. 2016. The Impact of Living Lab Methodology on Open Innovation Contributions and Outcomes. *Technology Innovation Management Review,* 6(1): 7–16. http://timreview.ca/article/956

Schuurman, D., & Tõnurist, P. 2016. Innovation in the Public Sector: Exploring the Characteristics and Potential of Living Labs and Innovation Labs. *Technology Innovation Management Review,* 7(1): 7–14. http://timreview.ca/article/1045

Steen, K., & van Bueren, E. 2017. The Defining Characteristics of Urban Living Labs. *Technology Innovation Management Review,* 7(7): 21–33. http://timreview.ca/article/1088

Ståhlbröst, A., & Lassinantti, J. 2015. Leveraging Living Lab Innovation Processes through Crowdsourcing. *Technology Innovation Management Review,* 5(12): 28–36. http://timreview.ca/article/950

Sutinen P., Erkkilä K., Wollstén P., Hagman K., Hirvikoski T., & Äyväri A. 2016. *KYKY Living Lab Handbook for Co-Creation by Schools and Companies.* Espoo, Finland: City of Espoo.

Tukiainen, T., Leminen, S., & Westerlund, M. 2015. Cities as Collaborative Innovation Platforms. *Technology Innovation Management Review,* 5(10): 16–23. http://timreview.ca/article/933

Tukiainen, T., & Sutinen, P. 2015. Cities as Open Innovation Platforms for Business Ecosystems. In P. Lappalainen, M. Markkula, & H. Kune (Eds), *Orchestrating Regional Innovation Ecosystems: Espoo Innovation Garden:* 313–322. Espoo: Aalto University in cooperation with Laurea University of Applied Sciences and Built Environment Innovations RYM Ltd.

Veeckman, C., Schuurman, D., Leminen, S., & Westerlund, M. 2013. Linking Living Lab Characteristics and Their Outcomes: Towards a Conceptual Framework. *Technology Innovation Management Review,* 3(12): 6–15. http://timreview.ca/article/748

Walravens, N., & Ballon, P. 2013. Platform Business Models for Smart Cities: From Control and Value to Governance and Public Value. *IEEE Communications Magazine,* 51(6): 72–79. https://doi.org/10.1109/MCOM.2013.6525598

Westerlund, M., & Leminen, S. 2011. Managing the Challenges of Becoming an Open Innovation Company: Experiences from Living Labs. *Technology Innovation Management Review,* 1(1): 9–25. http://timreview.ca/article/489

Keywords: living lab, third-generation, open innovation, networks, innovation, co-creation, city, smart city, collaborative innovation

6

Action Research as a Framework to Evaluate the Operations of a Living Lab

Sara Logghe and Dimitri Schuurman

> *"Seeking objective truth, the modern worldview makes no connection between knowledge and power. This positivist worldview has outlived its usefulness: as Habermas announced, 'modernism is dead'."*
>
> Peter Reason and Hilary Bradbury
> In *Handbook of Action Research* (2001)

In this article, we propose an action research approach to capture and act upon the delights and frustrations of panel members who participate in living lab research in order to optimize the operations of the living lab itself. We used this approach to test the effectiveness of action research in providing guidelines to practitioners to evaluate and design effective and sustainable user involvement processes in living labs. We conducted a focused literature review and an in-depth case study of both the integration of a researcher within the community and the implementation of an action research project within an existing living lab. This living lab is regarded as both a forerunner and a best-practice example in Europe. Based on our findings, we recommend co-creating the "operations" of a living lab with the users themselves following a combined action research and living lab approach.

Introduction

The fast pace of technological change and globalization – and the associated increase in access to knowledge – have enabled a growing number of users to engage in the innovation process. Companies, too, have sought out user contributions to their (new) products and services (Bogers et al., 2010). In addition, research has indicated that different kinds of users have different kinds of needs. People have a higher willingness to pay for a product or service that perfectly satisfies their personal needs (Franke & Piller, 2004). To develop these customized products or services, it is possible to let users adapt products themselves and thus become part of the innovation process itself (Franke & von Hippel, 2003).

Active end-user contribution is one of the building blocks of innovation processes in living labs. Living labs are public–private partnerships established to foster user-driven innovation and are supported by the European Commission through policy measures (Schuurman, 2015). Living lab research consists of user co-creation and experimentation of innovations in real-life contexts (Eriksson et al., 2005). The underlying idea is that people's ideas, experiences, and knowledge, as well as their daily needs for support from products, services, or applications, should be the starting point in innovation (Bergvall-Kåreborn & Ståhlbröst, 2009). The living lab approach is also a form of open innovation, because technology is developed and tested in a physical or virtual real-life context with multiple innovation stakeholders, and end users are important informants and co-creators during this process (Kusiak & Tang, 2006).

The efficiency of a living lab is based on the creative power of user communities. For this reason, it is important for the research activities and living lab operations to be aligned with the expectations of the participating users. Motivated users, willing to participate in research and co-creation activities, are essential for the functioning of a living lab, given that the underlying philosophy is that people's ideas, experiences, and knowledge, as well as their daily needs and wants, should be the starting point in innovation (Bergvall-Kåreborn & Ståhlbröst, 2009). However, in terms of the three layers within living lab activities (Schuurman, 2015; Schuurman et al., 2016), the input from users is collected at the micro level (living lab user involvement activities), but the panel management and strategy towards user involvement should be taken into account at the macro level of the living lab – meaning its internal organizational processes for coping with different living lab projects.

Although this active user involvement is regarded as essential, few studies have focused on the motivation, attrition, and behaviour of users in Living Labs (Baccarne et al., 2013; Logghe et al., 2014; Ståhlbröst & Bergvall-Kåreborn, 2011). Moreover, we are not aware of any literature that reports end-user involvement in the design of user activities and operations in living labs, although this might seem to be a logic step for living lab practitioners who regard active user involvement as cornerstone of the living lab philosophy. Therefore, within this study, users were empowered to participate in the design of living lab activities and operations – at the macro level (organization), rather than at the micro level (user activities). We propose action research as a method to iteratively capture and implement this feedback because it emerges over time in an evolutionary process, as individuals develop skills of inquiry, and as communities of inquiry develop within communities of practice. This process not only leads to new practical knowledge, but also to new abilities to create knowledge (Reason & Bradbury, 2001).

Therefore, as a first part of this article, we examined the literature on action research for frameworks that could guide this process of user involvement. Within a participatory action research process, "communities of inquiry and action evolve and address questions and issues that are significant for those who participate as co-researchers" (Reason & Bradbury, 2008). Contrary to other research methods, action research does not emphasize disinterested researchers and reproducibility of findings. Among others, Ståhlbröst (2008) has already used action research as a methodology within a living lab environment to involve users early and throughout the whole development process, including the design of new information technology systems based on these users' needs. What is not dealt with in the literature is how this research approach can be used to construct a framework for user involvement and participation in the construction and optimization of living lab operations.

This aspect is addressed in the second part of this article; following our review of the literature on action research, we put the selected frameworks to the test in a single case study carried out at imec.livinglabs (imec-int.com/en/livinglabs; previously iMinds Living Labs). The first step of our implementation of action research was to gain knowledge about the current situation, meaning we sought to identify the basis for the organization's desire to change or alter its behaviour (Baskerville & Pries-Heje, 1999). After six years of living lab research at imec.livinglabs, many of the operation processes have changed. Following Ståhlbröst (2008), we started our action research approach by describing a main research theme within imec.livinglabs: to get to know the delights and frustrations of our panel members regarding their participation in living lab projects. In order to answer this research question, a user researcher from imec.livinglabs became a panel member within specific living lab projects. Thus, our case study uses action research as a method to involve panel members in the organization processes of a living lab. We conclude the article by drawing main conclusions and recommendations on the use of action research in living labs.

Action Research in Living Labs

Lewin (1946) first described action research as "comparative research on the conditions and effects of various forms of social action and research leading to social action" that uses "a spiral of steps, each of which is composed of a circle of planning, action, and fact-finding about the result of the action". Action research has become an established research method that is often used in social sciences, but it is now used to not only to build knowledge on a certain topic but also to bring about changes to the topic. In order to stimulate these changes, several authors suggest that the researcher should become part of the user panel so that the obtained knowledge can be immediately applied (Baskerville, 1999). Furthermore, Checkland and Holwell (1998) distinguish three main phases of the process of action research: i) the researcher enters a real-world situation, ii) actions begin, and iii) the researcher leaves the situation and reflects on it in order to find a variety of lessons learned (Checkland & Holwell, 1998; Rönnerman, 2004). Because of its foundation in practical action and its aim to solve an immediate problem while informing theory, action research is seen to produce highly relevant results (Ståhlbröst, 2008).

Action research starts with a practical problem owned by a certain group of people. The aim of this methodology is to find a solution for this problem, but also to develop theoretical knowledge for the wider research community (Chiasson et al., 2009). It requires understanding the context of the field and bringing changes to a current situation in order to solve certain problems through collaboration (Donnelly & O'Keeffe, 2013). Therefore, action research – and especially participatory action research – is also associated with the tradition of citizen science (Hand, 2010), wherein "the crowd" participates in scientific data collection and processing. Next to solving the specific practical problem, this research approach is mostly used to facilitate the understanding of complex human processes, rather than constructing universal social laws (Baskerville, 1999). It is particularly relevant when trying to "solve an identified class of problems" and producing guidelines for best practice (Sein et al., 2011). Reason and Bradbury (2001) argue that the characteristics of action research lead to a more valid research output because the practical and theoretical outcomes of the research process are grounded in the perspective and interests of those immediately concerned (in this study, the living lab panel members), and they are not filtered through an outside researcher's preconceptions and interests – a process that normally characterizes the positivist research approach.

The action research methodology and the living lab approach both appear to be user-centric research approaches, although the former has been used from the starting point (Lewis, 1946) in social research contexts and the latter has been used in technology innovation contexts. Compared to living lab research, action research fails to sufficiently empower users for co-creation in open development environments. As mentioned, action research has already been used in living lab research (Ståhlbröst, 2008), but mainly as a method on a meso (project) or micro (user activities) level, to gather insights during a living lab *project* by means of the different *user activities.* In this study, we used the framework of Ståhlbröst (2008) as a starting point for our research on the satisfaction and motivation of our panel members regarding their participation in living lab projects in general, which is part of the general operations of the living lab *organization,* or the macro level.

Previous research has shown that intrinsic motivation is very important for users to remain part of a living lab community (Baccarne et al., 2013; Ståhlbröst & Bergvall-Kåreborn, 2011). Not only are users empowered by living labs (Veeckman et al., 2013), living labs are dependent on the involvement and motivation of their participating users. It is not easy for living lab researchers to motivate possible end users to take part in their research activities (Logghe et al., 2014) and to retain them. Action research "seeks to bring together action and reflection, theory and practice, in participation with others, in the pursuit of practical solutions to issues of pressing concern to people, and more generally the flourishing of individual persons and their communities" (Reason & Bradbury, 2001). Thus, we decided to use action research as a framework to co-create policies and guidelines for long-term user involvement in living labs, and to validate and implement them through a case study of a particular living lab.

However, despite the advantages described above, action research is not without criticism in the literature. In our literature review, we identified three main areas of criticism of action research:

1. Because of the fact that a researcher becomes part of the study, a more personal view can become dominant during observations and deductions (Donnelly & O'Keeffe, 2013; Baskerville & Pries-Heje, 1999; Baskerville & Wood-Harper, 1996).

2. There is a lack of a common theoretical description of action research, which results in various approaches to action research (Chiasson et al., 2009; Donnelly and O'Keeffe, 2013).

3. The results of an action research setup are very specific, which may cause action research to simply result in more action research (Baskerville & Wood-Harper, 1996; Donnelly & O'Keeffe, 2013).

During our action research process, we took these valid criticisms into account and tried to establish reliability and validity for action research as a useful framework for researching living lab operations. Therefore, we tried to take into account each criticism as follows:

1. In order to avoid injecting our personal views, we asked questions in different ways. In this way, we were able to detect certain needs that were mentioned after asking various questions.

2. We compared different descriptions and implementations of action research in order to deduct one consistent theoretical framework for our use cases.

3. We selected multiple projects to implement action research in order to detect needs on a meso level rather than on a micro level. In this way, we tried to avoid to specific insights in order to gather more general results.

Case Study

Case description, methodology, and overall process

The case focused on three living lab projects conducted at imec.livinglabs (imec-int.com/en/livinglabs; previously iMinds Living Labs), the living lab division of the research institute imec. Based in Ghent, Belgium, the former iMinds research institute became a separate business unit of imec following their merger in October 2016. The organization's experience conducting living labs dates back to 2009, with a particular emphasis on projects for startups and SMEs. imec.livinglabs has its own group of users (about 18,000 panel members or users) who are invited to participate in its living lab projects. For every living lab project, these panel members are invited to become part of a project community. In this way, they belong to innovation communities hosted by a neutral actor, in this case imec.livinglabs. These panel members cooperate as private participants during their spare time. But, according to Ståhlbröst and Bergvall-Kåreborn (2011), it is not sufficient to merely implement an innovation community in an organization to make a user innovation approach successful: it is also important to take into account what motivates the panel members to be part of this community.

In order to understand these motivations from an action research point of view, the main author of this article (an imec.livinglabs researcher) became a panel member from March 2015 until April 2016 and took part in different research steps in three living lab projects:

1. De Kopploeg (8 months): a living lab project focusing on online privacy issues

2. We Run (2 months): a living lab project about an application for runners

3. Spott (7 months): a living lab project about a new way to buy products with your smartphone you see during a TV show or TV commercial

These living lab projects were selected because they lasted for several months (so we were able to gather enough data) and consisted of multiple research steps (so we were able to ask the same questions in different ways). These three projects were also the first projects within imec.livinglabs whereby we focused on community building as a considerable aspect of a living lab. This approach fits with the aspects of action research of doing research in communities and emphasizing participation and action. Every community was inquired about in a collective way. The researcher became a member of the panel member community in order to experience the main motivations and thresholds for people to become and remain a panel member in a more intense way. The imec.livinglabs panel members of these three living lab projects were invited to give feedback on the general operation of imec.livinglabs by means of a survey (299 participants) and two co-creation session (12 and 8 participants, respectively. This allowed us to compare the experience of the researcher relative to the other imec.livinglabs panel members.

The goal of the survey (n=299) was to evaluate the experience of the panel members with the De Kopploeg living lab project. This survey resulted in insights on how to manage expectations from the panel members regarding a living lab project, on how to keep a panel engaged throughout the research track (short term vs. long term), and on how to keep the most active members engaged or involved during a research project. This was also the first living lab project where imec.livinglabs used a Facebook group to stimulate a community feeling between the panel members who joined this project. In the Facebook group, imec.livinglabs was able to collect reactions on certain statements from the survey, or the panel members asked both practical and substantive questions in this group and answered each other's questions faster than imec.livinglab coworkers were able to do so. imec.livinglabs was able to apply the main insights on the use of a Facebook group to the two other living lab projects under study.

In order to reflect on the involvement of researchers and panel members (Baskerville & Pries-Heje, 1999; Checkland & Holwell, 1998), we organized a survey and invited experienced panel members from each imec.livinglabs persona type (Logghe et al., 2015) to co-creation sessions to create solutions for the current frustrations of our panel members. Each co-creation session included the creation of a mock-up of the most interesting solutions. These mock-ups were validated by means of a specific validation survey and were handed over to the imec.livinglabs Marketing and Communication team who are currently constructing the community platform to be used over all living lab projects.

In December 2016, a final session was organized to ask our panel members for feedback regarding the platform in order to give our panel members the possibility to give an indication about what features or aspects should be modified before the final launch of the community platform. Based on these final remarks, a static version of the platform is being created and will be launched in March 2017. By inviting experienced panel members to these co-creation sessions, they became part of the reflection process on the operations of imec.livinglabs. In this way, learning occurs as an ongoing process in our research while reflecting on the method and projects as a whole (Ståhlbröst, 2008).

Findings

As a researcher who becomes a panel member, it is important to separate the application of knowledge to a project from the derivation of knowledge from each case (Ståhlbröst, 2008). As an example, we used the outcomes from "De Kopploeg", where we saw that a blog did not add value and that a Facebook group was consulted frequently, to plan the community strategy for the "We Run" and "Spott" projects. We also tested and implemented other measures such as a more personal communication approach (including names and photos of the panel managers in the communication), a higher "fun factor" in the incentives (personal challenges were used as "prize questions") and a faster, clearer feedback loop (creating infographics that summarize the user contributions from the research activity). We were able to capture this input during the project because the researcher took part in actions (research steps) in the situation (research projects) and acted upon it on an iterative basis for the subsequent projects. Also, through being a panel member and having conversations with other panel members, we found out that the imec.livinglabs panel has multiple unfulfilled needs and wants:

- They want more detailed information about the initiator of our living lab projects.
- They want more detailed results than they receive now.
- They need an overview of all the calls for research participation. They want to be able to look back at a finished project and ask: What was it about? What research steps were organized? In which research steps did I participate?

During these reflective co-creation sessions, we found the following underlying needs and wants:

1. Panel members want to define their role in the innovation process more explicitly.
2. They expect that the innovation will be implemented according to their inputs and feedback.
3. They accept that they are co-creating and experimenting with innovations that are not yet finished.

Based on these needs and wants, the panel members were asked to think about ways to address them. First, they indicated that it would be useful to have the opportunity to consult an online platform with a project flow for each living lab research project indicating in which research step users are needed and what will happen with their input. Second, they asked for more concrete follow-up information. Nowadays, panel members receive an infographic via email with the main conclusions of a research activity, but they want to receive a more detailed report with extensive findings. They want to consult this report online and not via email. Also, they want to receive more information about the final product when the research project is finished, including (where possible) a link to the app store or a newsletter of the company who created the innovation. Eventually, these sessions resulted in mock-ups of an imec.livinglabs community platform for panel members, which would provide the required information. At the end of the co-creation sessions, the participating panel members told us they felt appreciated because they were involved in a feedback moment regarding the living lab operations.

Discussion and Conclusion

In our study, we found it useful to have a researcher becoming a panel member and to directly gather feedback from panel members using more traditional methods (i.e., a survey and co-creation sessions). By combining living lab activities with an action research methodology, we were able to quickly gathering issues and frustrations on the one hand and rapidly co-create and implement practical solutions on the other. The added value of a researcher being part of the panel was the fact that the researcher was able to elucidate the results and put the insights into perspective. In this way, the action research methodology shed light on the differences between what people say and what people do (van Merrienboer, 2015).

Moreover, by means of a case study, we illustrated that the end users themselves can also be part of this action research process, which dealt with the question of how to improve the operations of a living lab. We facilitated the reflection process of panel members by adding a researcher as participant, which ensured that the researcher was completely socialized in the role of test user and attained a deep level of understanding of the current needs and wants of panel members while also lowering the threshold for end users to reflect and give feedback freely. Moreover, this approach also allowed for quick testing and evaluation of new approaches and solutions with panel members. In this way, imec.livinglabs decided to stick to a Facebook group to keep the panel members aware of the living lab projects.

In other words, adopting action research as a framework to evaluate and improve the operations of our living lab yielded positive results.

A major take-away was the fact that the action research approach and the living lab methodology strengthened and reinforced each other: the action research allowed us to uncover needs and wants with regards to user involvement in living lab projects and made it easier to experiment with new ways of involvement, which in turn enhanced the quality of the living lab research itself as the panel members felt empowered and involved in the living lab activities.

However, action research also has some caveats: as an action researcher, it is tempting to try to "act" like one of the regular panel members (Baskerville & Pries-Heje, 1999; Baskerville & Wood-Harper, 1996; Donnelly & O'Keeffe, 2013). The action researcher must stay more or less impartial in this situation and try to avoid pushing other panel members towards their own delights, frustrations, or solutions. They must also try to make other panel members feel at ease when giving feedback. Equally, we found that it was very useful to take the various criticisms into account during the research period, although it was not always easy to note down every insight in a structured way. Therefore, we propose to conduct further research leading to a more detailed framework about methods to use during the "practical part" of the action research process. Moreover, the long-term effects of involving panel members in improving the operations of a living lab on their motivation to participate should also be investigated.

About the Authors

Sara Logghe is a Living Lab Researcher at imec – MICT – Ghent University in Belgium. She holds master's degrees in History and Communication Sciences from Ghent University, and her research interests include the potential of social media for cultural institutions, the changing library landscape, and living labs.

Dimitri Schuurman is the Team Lead in User Research at imec.livinglabs and a Senior Researcher at imec – MICT – Ghent University in Belgium. He holds a PhD and a Master's degree in Communication Sciences from Ghent University. Together with his imec colleagues, Dimitri developed a specific living lab offering targeted at entrepreneurs in which he has managed over 100 innovation projects. Dimitri is responsible for the methodology and academic valorization of these living lab projects and coordinates a dynamic team of living lab researchers. His main interests and research topics are situated in the domains of open innovation, user innovation, and innovation management. His PhD thesis was entitled *Bridging the Gap between Open and User Innovation? Exploring the Value of Living Labs as a Means to Structure User Contribution and Manage Distributed Innovation.*

References

Baccarne, B., Logghe, S., Veeckman, C., & Schuurman, D. 2013. *Why Collaborate in Long-Term Innovation Research? An Exploration of User Motivations in Living Labs.* Paper presented at the 4th ENoLL Living Lab Summer School, August 27–30, Manchester, UK.

Baskerville, R. L., & Wood-Harper, A. T. 1996. A Critical Perspective on Action Research as a Method for Information Systems Research. *Journal of Information Technology,* 11(3): 235–246. http://dx.doi.org/10.1007/978-3-319-29269-4_7

Baskerville, R., & Wood-Harper, A. T. 1998. Diversity in Information Systems Action Research Methods. *European Journal of Information Systems,* 7(2): 90–107. http://dx.doi.org/10.1057/palgrave.ejis.3000298

Baskerville, R. L. 1999. Investigating Information Systems with Action Research. *Communications of the AIS,* 2(3): Article 4.

Baskerville, R., & Pries-Heje, J. 1999. Grounded Action Research: A Method for Understanding IT in Practice. *Accounting, Management and Information Technologies,* 9(1): 1–23. http://dx.doi.org/10.1016/S0959-8022(98)00017-4

Bergvall-Kåreborn, B., & Ståhlbröst, A. 2009. Living Lab: An Open and Citizen-Centric Approach for Innovation. *International Journal of Innovation and Regional Development,* 1(4): 356–370. http://dx.doi.org/10.1504/IJIRD.2009.022727

Bogers, M., Afuah, A., & Bastian, B. 2010. Users as Innovators: A Review, Critique, and Future Research Directions. *Journal of Management,* 36(4): 857–875. http://doi.org/10.1177/0149206309353944

Checkland, P., & Holwell, S. 1998. Action Research: Its Nature and Validity. *Systemic Practice and Action Research,* 11(1): 9–21. http://dx.doi.org/10.1023/A:1022908820784

Chiasson, M., Germonprez, M., & Mathiassen, L. 2009. Pluralist Action Research: A Review of the Information Systems Literature. *Information Systems Journal,* 19(1): 31–54. http://dx.doi.org/10.1111/j.1365-2575.2008.00297.x

Donnelly, R., & O'Keeffe, M. 2013. Exploration of ePortfolios for Adding Value and Deepening Student Learning in Contemporary Higher Education. *International Journal of ePortfolio,* 3(1).

Eriksson, M., Niitamo, V. P., & Kulkki, S. 2005. *State-Of-The-Art in Utilizing Living Labs Approach to User-Centric ICT Innovation-A European Approach.* Luleå, Sweden: Centre for Distance-Spanning Technology.

Franke, N., & Piller, F. 2004. Value Creation by Toolkits for User Innovation and Design: The Case of the Watch Market. *Journal of Product Innovation Management,* 21(6): 401–415. http://dx.doi.org/10.1111/j.0737-6782.2004.00094.x

Franke, N., & Hippel, E. V. 2003. Satisfying Heterogeneous User Needs via Innovation Toolkits: The Case of Apache Security Software. *Research Policy,* 32(7): 1199–1215. http://dx.doi.org/10.1016/S0048-7333(03)00049-0

Hand, E. 2010. Citizen Science: People Power. *Nature,* 466 (7307): 685–687. http://dx.doi.org/10.1038/466685a

Kusiak, A., & Tang, C. Y. 2006. Innovation in a Requirement Life-Cycle Framework. In *Proceedings of the 5th International Symposium on Intelligent Manufacturing Systems (IMS):* 61–67, Sakarya, Turkey.

Lewin, K. 1946. Action Research and Minority Problems. *Journal of Social Issues,* 2(4): 34–46. http://dx.doi.org/10.1111/j.1540-4560.1946.tb02295.x

Logghe, S., Baccarne, B., & Schuurman, D. 2014. An Exploration of User Motivations for Participation in Living Labs. In *Proceedings of the XXV ISPIM Conference 2014,* June 8–11, Dublin, Ireland.

Logghe, S., Baccarne, B., Veeckman, C., Lievens, B., & Schuurman, D. 2014. Uit passie of voor de poen? Een exploratie van gebruikersmotivaties voor deelname aan innovatie onderzoek in Living Labs. Etmaal 14.

Logghe, S. 2015. *Getting Personal: Exploring the Usage of Persona in Order to Optimize the Involvement of a Living Lab Panel.* Paper presented at the 6th ENoLL Living Lab Summer School 2015.

McNiff, J., Lomax, P., & Whitehead, J. 2003. *You and Your Research Project.* London: Routledge.

Norton, L. S. 2009. *Action Research in Teaching and Learning: A Practical Guide to Conducting Pedagogical Research in Universities.* London: Routledge.

Reason, P., & Bradbury, H. (Eds.). 2001. *Handbook of Action Research: Participative Inquiry and Practice.* London: Sage Publications, Inc.

Reason, P. & Bradbury, H. 2008. (Eds.) *The SAGE Handbook of Action Research: Participative Inquiry and Practice.* London: Sage Publications, Inc.

Rönnerman, K. 2004. Vad är aktionsforskning? i K. Rönnerman (Red.). Aktionsforskning i praktiken–erfarenheter och reflektioner, 13–30.

Schuurman, D., Lievens, B., De Marez, L., & Ballon, P. 2012. Towards Optimal User Involvement in Innovation Processes: A Panel-Centered Living Lab-Approach. In *Proceedings of the Technology Management for Emerging Technologies PICMET '12 Conference:* 2046–2054, July 29 – August 2 Vancouver, BC.

Schuurman, D. 2015. *Bridging the Gap between Open and User Innovation? Exploring the Value of Living Labs as a Means to Structure User Contribution and Manage Distributed Innovation Doctoral dissertation.* Ghent University and Vrije Universiteit Brussel VUB.

Schuurman, D., De Marez, L., & Ballon, P. 2016. The Impact of Living Lab Methodology on Open Innovation Contributions and Outcomes. *Technology Innovation Management Review,* 1(6): 7–16. http://timreview.ca/article/956

Sein, M. K., Henfridsson, O., Purao, S., Rossi, M., & Lindgren, R. 2011. Action Design Research. *MIS Quarterly,* 35(1): 37–56.

Ståhlbröst, A. 2006. *Human-Centric Evaluation of Innovation.* Luleå, Sweden: Luleå University of Technology.

Ståhlbröst, A. 2008. *Forming Future IT the Living Lab Way of User Involvement.* Doctoral thesis. Luleå, Sweden: Luleå University of Technology.

Ståhlbröst, A., & Bergvall-Kåreborn, B. 2011. Exploring Users Motivation in Innovation Communities. *International Journal of Entrepreneurship and Innovation Management,* 14(4): 298–314. http://dx.doi.org/10.1504/IJEIM.2011.043051

Van Merriënboer, J. J. 2015. What People Say vs. What People Do. *Perspectives on Medical Education,* 4(1): 47–48. http://dx.doi.org/10.1007%2Fs40037-015-0163-2

Veeckman, C., Schuurman, D., Leminen, S., Lievens, B., & Westerlund, M. 2013. Characteristics and Their Outcomes in Living Labs: A Flemish-Finnish Case Study. In *Proceedings of the XXIV ISPIM Conference: Innovating in Global Markets: Challenges for Sustainable Growth,* June 16–19, Helsinki, Finland.

Yin, R. K. 2013. *Case Study Research: Design and Methods.* Thousand Oaks, CA: Sage Publications.

Keywords: action research, participatory action research, panel management, user research, living lab

7

Co-Creating User Stories: A Tool for Making Sense of Business Opportunities

Anna-Greta Nyström, Miia Mustonen, and Seppo Yrjölä

> " *The real goal of using stories is shared understanding.* "
>
> Jeff Patton
> Product manager and designer
> In *User Story Mapping*

This article discusses co-creation as a means of sensemaking among business-to-business actors, and presents a case study from the information and communications sector, in which the aim is to understand current and future media consumption and behaviour. We propose to expand the notion of co-creation in such a way that it also recognizes interaction and sensemaking between different stakeholders within an industry or sector, compared to the current view, in which the focus is on the consumers' role in co-creation activities. Stakeholder co-creation entails jointly creating meaning of a chosen complex phenomenon by using practical tools, such as narratives in the form of user stories. Sensemaking is a narrative process and can thus be used in combination with practical facilitation tools in order to co-create user stories or other stories, and in other contexts.

Introduction

Predicting and making sense of the future is a complex endeavour – impossible, yet at the core of every organization's developmental work. The mobile communications sector is preparing the transition from 4G to 5G, aware of possible shifts in trends and media consumption as end users will be able to access greater speed and capacity. Over the decade, we have witnessed the exponential growth of wireless communications with a vast range of diverse devices, applications, and services requiring connectivity. In particular, the number of mobile broadband subscribers and the amount of data used per user is set to grow significantly over the coming years. Recent research points at trends such as increased mobile streaming on demand, increased mobile media consumption on the go, and a visible rise in artificial intelligence and the Internet of Things (IoT) (cf. Ericsson Consumerlab, 2015). Thus, actors are trying to understand the business context within which they operate in order to make decisions on future investments.

The search for clues and signs of future trends relates to a process of sensemaking (cf. Weick, 1995; Weick et al., 2005). Sensemaking is about the process that culminates in interpretation (Craig-Lees, 2001): a social process where meanings become shared and turned into a so-called collective mind. Sensemaking thus requires interaction and communication between individuals and actors. Individuals co-create stories and narratives of the reality and surrounding context in which they exist. Co-creation of stories and alternative realities becomes a tool for making sense of a complex world. However, little is known of how firms jointly engage in a process of co-creation, and how activities along the process aid firms in identifying business opportunities, among other things. The aim of this article is, therefore, to provide insight into co-creation as a means of sensemaking, and to provide practical tools for co-creating a shared view of business opportunities.

The remainder of this article is organized as follows. We first discuss sensemaking and how it can be facilitated and enhanced through co-creation activities. We then present the data used for our analysis and the results from an empirical study of co-creating user stories within a project consortium aiming at understanding ongoing trends in media consumption, and investigating both technological and political enablers needed to facilitate those trends. We conclude by discussing our findings and their implications for theory and practice.

Sensemaking through Co-Creation

Weick (1995) argues that sensemaking exists and can be examined; sensemaking is a process in which an individual cognitively applies cultural constructs, makes sense of an inter-organizational interaction event, and ascribes meaning to it (Ivanova & Torkkeli, 2013). Sensemaking allows us to cope with enormous complexities and gives us the opportunity to turn circumstances into a situation, which we can understand explicitly in words (Bean & Hamilton, 2006; Weick et al., 2005). Sensemaking focuses on the interplay between interpretation and action (Mills, 2003; Weick et al., 2005). Without action, there is no sensemaking (Weick, 1995). The notion of action and its importance in the sensemaking process is elaborated further by Czarniawska-Joerges (1992), who states that sensemaking depends more on shared actions and less on, for example, shared goals.

Similarly, interpretation and meaning-making are at the core of "creation", or the culture of co-creation (Ind & Coates, 2012). Meaning is always co-created. Ind and Coates (2012) note that co-creation has emerged due to the coincidence of several developments, such as the mainstream adoption of Internet technologies, the orientation towards services and experiences, a more open approach to innovation, and the growth of social, collaboration, and customization technologies. Researchers often refer to co-creation as *value* co-creation (cf. Saarijärvi et al., 2013). Alves and colleagues (2016) identify research streams on co-creation, such as service theory (cf. Vargo & Lusch, 2006), innovation studies (cf. Saarijärvi et al., 2013), many-to-many marketing (cf. Bogers et al., 2010), postmodern marketing defining consumers as prosumers (cf. Bendapudi & Leone, 2003), and consumer culture theory. However, common to these research strands is a focus on the consumer as one part in the co-creation process. Even though Ind and Coates (2012) note that co-creation has become a widely used term to describe a shift in thinking from the organization as a definer of value to a more participative process where people and organizations together generate and develop meaning, the focus of researchers is seldom co-creation between organizations in an attempt to structure and make sense of the surrounding business context, changing consumption habits among end users, potential business opportunities, changing business models, etc.

Research on co-creation in a business-to-business setting is scarce. An exception can be found in industrial network theory, where value co-creation is studied in buyer–seller relationships or business network constellations (cf. Ford, 2011; Hyötyläinen et al., 2011). Co-creation has so far focused on exploiting consumers and other stakeholders who offer their time and intellect for the benefit of organizations. In a business-to-business setting, stakeholders should be engaged in a reciprocally useful way; answers or insights are not existing somewhere and waiting to be discovered, as Ind and Coates (2012) put it. Rather, they have to be discovered with others. It is the process of co-creation and the co-discovery through interaction that generates new ways of seeing the world. Thus, sensemaking materializes in co-creation of meaning; sensemaking can be considered a process of co-creation and vice versa, co-creation can be considered a process of sensemaking. In order to study the phenomenon further, we propose the notion of *stakeholder co-creation* in order to shift focus from co-creating with consumers to co-creating among business-to-business actors, in other words, representatives of industry actors co-create a shared view of the world, current business settings, phenomenon, etc.

Case Study: Co-Creating User Stories

We use a qualitative research approach in order to study stakeholder co-creation. We chose a project consortium in Finland as the case study for researching sensemaking trough co-creation, namely "The Future of UHF" (fuhf.turkuamk.fi), which is financed by the Finnish Funding Agency for Innovation, Tekes, from 2015 to 2016. The project partners represent research institutions, mobile network manufacturers, mobile network operators, broadcast network operators and broadcasters, as well as technology developers. The consortium is unique given that the participants cover different parts of the mobile communications ecosystem. The scope of the project is to study supplemental downlink (SDL) technology, which is based on 4G mobile networks, and to offer a solution to a problem with the networks' frequencies and their limited capacities. The technology enables mobile operators to transfer video content to consumers on the same spectrum band without disturbing the normal TV operations and enables flexible use of ultra-high frequency (UHF) for media and mobile broadband. This approach allows for more efficient utilization of spectrum and for the broadcasters to better reach their increasingly mobile audience. The project thus explores how to fulfill 5G needs and requirements.

One of the activities in the project is a bi-monthly facilitated workshop with a specific theme. During 2015 and the first half of 2016, nine whole-day workshops were held. Among these workshops, we focus on three in which user stories were jointly created for internal use in the project and as a basis for upcoming topics within the project: future scenarios and business models. During the workshops, notes were taken continuously. The materials produced (post-it notes, hand-written stories, drawn pictures, etc.) were stored and categorized according to: i) which one of four user stories it related to and ii) which actor the individual who produced the material represented. As the main source of data, we use the outcome of the workshops: the co-created user stories.

User stories as narratives

User stories as a concept can be found mainly within information systems and computer sciences, where they relate to different stages of system development and are used in determining system requirements. A user story consists of a few sentences in the everyday language of the end user and has the goal of capturing the essence of part of the work a user does or needs to do (with the system) (cf. Dimitrijevic et al., 2015; Jeffries, 2001). The INVEST model depicts user stories as: i) independent, ii) negotiable, iii) valuable to users and customers, iv) estimable, v) small, and vi) testable (Cohn, 2004). User stories are business oriented (Trkman et al., 2016), and typically they follow a template designed for contributing to programming, system, and software development projects. However, Lucassen and colleagues (2016) point out that user stories are limited in terms of improving quality and there is a lack of empirical studies on their use and effectiveness.

In our case study, user stories were, together with future scenarios, one of the main topics during a number of workshops in 2015. A researcher facilitated the workshops and prepared tasks and tools for the workshop participants. Co-creating user stories was one of the tasks given to the participants. Four user stories, representing "Paul", "Rita", "Minnie", and "John" as fictional users or consumers, were developed jointly and summarized in text by the facilitator (see Boxes 1 to 4). The facilitator asked each workshop participant to individually write stories on the four fictional users or consumers, answering the questions: i) who? (person), ii) what? (action), iii) where? (location, situation), iv) when? (time), and v) why? (motive). After this, the participants were paired together in order to develop the stories collaboratively. These stories were written and drawn on large sheets of paper to make the stories visible when subsequently presented to the whole group. The whole group then further developed the user stories. From the facilitator's perspective, the structure follows the method of *me-we-us,* in other words, ideation individually, in pairs, and, finally, in groups.

Findings and Discussion

The case study features the information and communications sector and a consortium working together in developing technology that aids the transition from 4G to 5G. The participants co-created four user stories: more than 50 individual narratives were presented, discussed, compared, and transformed into shared understandings of the current user of mobile devices. The participants had to interpret the ideas of competing actors and collaboratively discover the user stories, their contexts, and their features. Given that sensemaking comprises both past experiences and expectations of the future, the co-created user stories also act as representations of the future by expressing unmet customer needs and possible shifts in behaviour. For instance,

Box 1. Paul's story

Paul is 68 years old and lives with his wife. He has been retired for three years. He has an active social life, exercises several times a week, and travels with his wife.

Media behaviour in general

Paul appreciates routines and his media usage has remained the same throughout the years. Last year, Paul's children bought him a tablet, and Paul also replaced his old mobile phone with a smartphone. The new devices have changed Paul's everyday life and the way he consumes media. Paul also has a laptop, which he uses frequently. However, Paul does not want to give up TV and radio.

Devices used and for what purpose

- Laptop: for practical tasks such as paying bills, buying tickets, etc.
- Tablet: for reading newspapers and novels, and watching TV shows
- Smartphone: for calling, following real-time news and map services (navigation)
- TV: for regular TV watching, mostly sports
- Radio: the radio is always on at home, even when nobody is actively listening

the increased use of smartphones and apps with movie features pose capacity challenges for mobile broadband networks. The discussions during the workshops quickly turned towards the enabling of mobile broadband networks and how they should support diverse personalized uses, services, and applications besides offering voice and data. Drawing up these fictional users and how they use mobile devices for various purposes allowed the participants to relate end-user personas and scenarios to regulatory and technology issues. A key point is to allow the participants to discuss, share, and debate ideas and thoughts, guided by a facilitator who keeps the time schedule, helps summarize the discussion, and assures results.

The user stories were later on used as starting point for creating future scenarios of UHF. It was therefore important that all stakeholders accept and understand the collaboratively created user stories, as the sensemaking processes continue in the project with new topics. Stakeholder co-creation provides an opportunity to expand and form perceptions of ongoing trends. Stakeholder co-creation in the case study refers to *meaning-making between industrial partners* without the involvement of end users in the process. Stakeholder co-creation may, however, be a step towards engaging with end users and including them as stakeholders in the co-creation of, for instance, more versatile user stories or specific services featured in the user stories.

Box 2. Rita's story

Rita is a 45-year old single mother with two boys, 8 and 10. She works as a secretary in a multinational corporation. Most of Rita's time is dedicated to the children and running the household; she does not have much time for herself and own hobbies.

Media behaviour in general
Rita's everyday life is structured around her two boys and their hobbies, running the household, and her work. Mobile devices and services facilitate her life. Her laptop is usually at the office. Rita bought a tablet a few years ago, which she uses frequently. She also uses her smartphone frequently. She wants her boys to be able to reach her at all times; they also have smartphones. It feels like her boys learn new technology much faster than she does, but she tries to keep up with the latest apps in order to understand what the boys are talking about.

Devices used and for what purpose
- Laptop: for work, paying bills, watching movies, and shopping
- Tablet: used mainly at home in the evenings, while waiting or commuting, for managing practical tasks, watching movies and TV shows, reading news, shopping, and following social media sites
- Smartphone: for calling, reading news, receiving offers, and navigating
- TV: in the evenings, Rita watches TV with her boys.

Box 3. Minnie's story

Minnie is 32 years old and is single. She is a career woman who works as a social media strategist at a large company. She has an active social life, travels, and plays sports.

Media behaviour in general
As a social media strategist, Minnie keeps herself up to date with work at home and while travelling. She multitasks by using several devices simultaneously and has developed quite regular routines or "checking habits". Minnie feels she must be reachable at all time. Her most important tool is the laptop, which she always carries with her. The smartphone is also actively in use, as it is a simple way of staying up to date. Minnie's work requires a functioning Internet connection, for updating blogs and social media sites, and for uploading and downloading video clips and pictures. In her free time, Minnie uses different media services, especially to manage her life.

Devices used and for what purpose
- Laptop: for work; reading news, blogs, and vlogs; creating networks (LinkedIn); booking trips and hotels; listening to music (via Spotify); and watching movies
- Tablet: she carries the tablet along, but uses it seldom. She uses apps such as Netflix, Spotify, and blogs.
- Smartphone: Minnie is an active user of WhatsApp, Facebook, Twitter, and Instagram. The smartphone functions as a physical calendar, booklet for making notes, etc.

Box 4. John's story

John is 14 years old and lives with his parents and two siblings. He goes to school, plays ice hockey two times a week, and enjoys playing computer games and spending time with friends.

Media behaviour in general
John is what they refer to as "digi-native", and he uses his smartphone and content services daily (at school, with friends, while waiting on the bus, etc.). John uses the family tablet and laptop at home, mostly for schoolwork. In his free time, John plays computer games and actively participates in discussion forums related to these games. John also uses his smartphone for playing games. John would like to have the latest smartphone model, just like all of his friends. They demonstrate and recommend to each other the latest and coolest apps. John uses the smartphone for sending messages (also during lessons at school). John appreciates the fact that his teachers use online material and games in their teaching. John never listens to the radio; he wants to decide himself what kind of music he listens to, where and when.

Devices used and for what purpose
- Laptop: John uses the family laptop for doing schoolwork
- Tablet: John uses the family tablet sometimes when playing games late in the night
- Smartphone: actively used for playing games, watching videos on YouTube, and communicating with his friends using WhatsApp. Calls are made mostly using Skype.

Conclusions

From a managerial point of view, sensemaking can be viewed as an approach where actors, or stakeholders, representing different positions in the industry value network, are invited to an event or workshop, in which they co-create representations of their business context. Sensemaking can thus be used in combination with practical facilitation tools in order to co-create user stories or other stories, and in other contexts. The facilitation tools depend on the aim of the workshop or event, and should activate the participants (e.g., service design tools such as customer journey mapping, scenario writing, collective mind maps, open spaces, and world cafés).

Based on lessons learned developing the case study described in this article, we offer managers three recommendations for co-creating user stories among stakeholders:

1. Use a facilitator to aid the co-creation process and goal-setting.

2. Facilitate interaction and communication by using activation tools.

3. Create the user stories in multiple steps, allowing all participants to individually create stories, which are then collectively developed further.

From a theoretical point of view, and based on the case study, we propose the concept of *stakeholder co-creation,* by which we refer to *co-creation carried out in a business-to-business setting among industry partners.* Through stakeholder co-creation, industry actors aim at making sense not only of business opportunities and industry-related events, but also of the end user per se. Currie and Brown (2003) suggest that we can collectively make sense of our social world through jointly negotiating narratives, or stories. Stakeholder co-creation addresses this issue by shifting focus from the exploitation of consumers in developmental and sensemaking processes to the interaction and sensemaking process among business-to-business actors.

Research on creative processes among business-to-business actors is scarce. In terms of the current case study, it is limited as it studies merely one consortium and stakeholder co-creation based on user stories as the overall topic. We call for more detailed descriptions and analyses of stakeholder co-creation processes and suggest using individual stories as data for content analysis. The dialog between individuals and group conversations become data for qualitative analysis, when taped and transcribed. Also, co-creation in different settings offers a research theme to explore further; combined with new perspectives, co-creation as a concept should be broadened and developed further into subcategories, such as stakeholder co-creation.

About the Authors

Anna-Greta Nyström D.Sc. (Econ.) is a Senior Lecturer in the School of Business and Economics at Åbo Akademi University in Turku, Finland. Anna-Greta holds a doctoral degree in International Marketing from the School of Business and Economics at Åbo Akademi University. Her doctoral research focused on industry change in the Finnish telecommunications sector, with a special focus on technological convergence. Anna-Greta's current research interests include consumer behaviour in high-tech industries and changing media-consumption patterns.

Miia Mustonen is a Research Scientist at VTT Technical Research Centre of Finland. She received her MSc degree in Electrical Engineering from University of Oulu, Finland, in 2005. She has been actively involved in regulatory activities in both European and international regulatory bodies. She holds an MSc degree in Electrical Engineering from the University of Oulu, Finland, where she is currently finalizing her PhD degree on spectrum-sharing models in policy making.

Seppo Yrjölä is a Principal Innovator at Nokia Innovation Steering. Prior to joining Innovations in 2007, he headed wireless technology research at Nokia. He holds an MSc degree in Electrical Engineering from the University of Oulu, Finland, and did his postgraduate studies in the area of telecommunication and radio technology. His current focus is innovating in the area of future radios with cognitive radio business opportunities and potential disruptions.

References

Alves, H., Fernandes, C., & Raposo, M. 2016. Value Co-Creation: Concept and Contexts of Application and Study. *Journal of Business Research*, 69(5): 1626–1633. http://dx.doi.org/10.1016/j.jbusres.2015.10.029

Bean, C. J., & Hamilton, F. E. 2006. Leader Framing and Follower Sensemaking: Response to Downsizing in the Brave New Workplace. *Human Relations*, 59(3): 321–349. http://dx.doi.org/10.1177/0018726706064177

Bendapudi, N., & Leone, R. P. 2003. Psychological Implications of Customer Participation in Co-Production. *Journal of Marketing*, 67(1): 14–28. http://dx.doi.org/10.1509/jmkg.67.1.14.18592

Bogers, M., Afuah, A., & Bastian, B. 2010. Users as Innovators: A Review, Critique, and Future Research Directions. *Journal of Management*, 36(4): 857–875. http://dx.doi.org/10.1177/0149206309353944

Brown, A. D, Stacey, P., & Nandhakumar, J. 2008. Making Sense of Sensemaking Narratives. *Human Relations*, 61(8): 1035–1062. http://dx.doi.org/10.1177/0018726708094858

Cohn, M. 2004. *User Stories Applied for Agile Software Development.* Boston, MA: Addison-Wesley.

Craig-Lees, M. 2001. Sense Making: Trojan Horse? Pandora's Box? *Psychology & Marketing*, 18(5): 513–526. http://dx.doi.org/10.1002/mar.1019

Currie, G., & Brown, A. D. 2003. A Narratological Approach to Understanding Processes of Organizing in a UK Hospital. *Human Relations*, 56(5): 563–586. http://dx.doi.org/10.1177/0018726703056005003

Czarniawska-Joerges, B. 1992. *Exploring Complex Organizations: A Cultural Perspective.* Newsbury Park, CA: Sage Publications.

Dimitrijevic, S., Iovanovic, J., & Devedzic, V. 2015. A Comparative Study of Software Tools for User Story Management. *Information and Software Technology*, 57: 352–368. http://dx.doi.org/10.1016/j.infsof.2014.05.012

Ericsson Consumerlab. 2015. *10 Hot Consumer Trends 2016: An Ericsson Consumer Insight Summary Report.* Stockholm: Ericsson. https://www.ericsson.com/res/docs/2015/consumerlab/ericsson-consumerlab-10-hot-consumer-trends-2016-report.pdf

Ford D. 2011. IMP and Service-Dominant Logic: Divergence, Convergence and Development. *Industrial Marketing Management*, 40(2): 231–239. http://dx.doi.org/10.1016/j.indmarman.2010.06.035

Hyötyläinen, R., Valkokari, K., & Kalliokoski, P. 2011. Renewal Through Co-Creation in Business Networks. *Open Source Business Resource*, March 2011: 29–37. http://timreview.ca/article/426

Ind, N. & Coates, N. 2012. The Meanings of Co-Creation. *European Business Review*, 25(1): 86–95. http://dx.doi.org/10.1108/09555341311287754

Ivanova, M. & Torkkeli, L. 2013. Managerial Sensemaking of Interaction within Business Relationships: A Cultural Perspective. *European Management Journal*, 31(6): 717–727. http://dx.doi.org/10.1016/j.emj.2013.07.007

Jeffries, R. 2001. Essential XP: Card, Conversation, and Confirmation. *XP Magazine*, August 20, 2001. Accessed August 1, 2016: http://ronjeffries.com/xprog/articles/expcardconversationconfirmation/

Lucassen, G., Dalpiaz, F., van der Werf, J. M., & Brinkkemper, S. 2016. The Use and Effectiveness of User Stories in Practice. In M. Daneva & O. Pastor (Eds.), *Requirements Engineering: Foundation for Software Quality:* 205–222. Cham, Switzerland: Springer International Publishing. http://dx.doi.org/10.1007/978-3-319-30282-9_14

Mills, J. H. 2003. *Making Sense of Organizational Change.* London, UK: Routledge.

Patton, J. 2014. *User Story Mapping. Discover the Whole Story, Build the Right Product.* Sebastopol, CA: O'Reilley Media.

Saarijärvi, H., Kannan, P. K., & Kuusela, H. 2013. Value Co-Creation: Theoretical Approaches and Practical Implications. *European Business Review,* 25(1): 6–19. http://dx.doi.org/10.1177/0149206309353944

Trkman, M., Mendling, J., & Krisper, M. 2016. Using Business Process Models to Better Understand the Dependencies among User Stories. *Information and Software Technology,* 71: 58–76. http://dx.doi.org/10.1016/j.infsof.2015.10.006

Vargo, S. L., & Lusch, R. F. 2006. Service-Dominant Logic: What It Is, What It Is Not, What It Might Be. In R. F. Lusch, & S. L. Vargo (Eds.), *The Service-Dominant Logic of Marketing: Dialog, Debate, and Directions:* 43–56. Armonk, NY: ME Sharpe.

Weick, K. E. 1995. *Sensemaking in Organizations.* Thousand Oaks, CA: Sage.

Weick, K. E., Sutcliffe, K. M., & Obstfeld, D. 2005. Organizing and the Process of Sensemaking. *Organization Science,* 16(4): 409–421. http://dx.doi.org/10.1287/orsc.1050.0133

Keywords: user stories, co-creation, sensemaking, business-to-business

8

The Design-Driven Living Lab: A New Approach to Exploring Solutions to Complex Societal Challenges

Rens Brankaert and Elke den Ouden

> *If you ask customers what they want, they will tell you: 'Better, faster, and cheaper'– that is, better sameness, not revolutionary change.*
>
> Guy Kawasaki
> Marketing specialist, author, and venture capitalist

In this study, we aimed to explore the potential of a design-driven living lab as an innovative approach to addressing societal challenges. This living lab incorporates design qualities such as exploration, open-ended results, and disruption. This approach was applied in three case studies within the context of dementia, each of which explored the impact of Qwiek.up – a media system that creates an ambient experience in a room through projection and sound. A cluster analysis of the results in the three case studies showed that the system has considerable potential for people with dementia, and possibly also for other groups. In addition, the design-driven approach led to new applications in care, improved functionality, and a broader design space. Our findings show that design-driven living labs can widen the scope of innovation and improve the value proposition of an innovative solution.

Introduction

With an ageing global population, there has been a substantial increase in the number of people living with dementia around the world (Wimo et al., 2003). Dementia severely hampers an affected individual's ability to live independently, and therefore they often have to rely on both formal and informal care (Prince et al., 2013). Thus, dementia is putting considerable pressure on healthcare costs and quality (Knapp et al., 2013). Furthermore, in Europe, healthcare budgets are being cut, including those for dementia care (Prince et al., 2013). The consequence is more accidents and a lower quality of life for people with dementia. However, the search for a solution is not straightforward, because dementia is an example of a "wicked problem" (Martin, 2009), meaning it requires a multi-perspective approach. The various stakeholders each have their own perspective and often contradictory needs (Brankaert et al., 2015).

Previous studies have revealed that living labs are suitable for tackling complex societal challenges (Liedtke et al., 2012). The living lab allows for different innovation methods, such as user evaluation, to be applied in collaboration with various stakeholders to find and evaluate new solutions (Brankaert, 2016). In a living lab, the validity of results is high because the methods are applied in a real-life context. Additionally, living labs involve various stakeholders, such as end users and both public and private parties. Indeed, living labs should involve end users in constructing meaningful innovation with and for them through co-creation (Almirall et al., 2012). Moreover, involving stakeholders with a market interest in living lab activities fosters successful innovation and increases market impact (Schuurman et al., 2016). Thus, this approach supports innovation in all phases of the lifecycle and enables a rapid route to market for innovative products (Leminen et al., 2012).

Although studies on living labs focus on evaluating solutions (Veeckman et al., 2013), finding innovative solutions requires going beyond the evaluation of existing offerings. Another aim of living labs is to look at problems creatively, and this aim can be achieved by adopting a design perspective. Design helps by identifying needs, conceptualizing, prototyping, implementing, and taking different perspectives (Krogstie, 2012). In addition, designers can explore, envision, and create more disruptive scenarios (Hummels & Frens, 2011).

In the literature, it is suggested that living labs need outlines and structured, predefined goals (Korman et al., 2015). However, to deal with uncertainties in complex societal challenges, we need to take an open attitude to innovation in living labs. Moreover, different stakeholders can have different views and needs. There might be, for example, conflicting perspectives among business and care stakeholders concerning value versus revenue. To address such challenges, innovators need to apply integrative thinking and be able to hold two opposing ideas in their minds, thus creating a synthesis that contains elements of both while improving each. Designers have skills in this area, they seek factors that are not immediately obvious, and they tend to see a problem as a whole (Martin, 2007). Moreover, designers are able to generate creative resolutions in the form of new ideas.

With this article, we aim to present a new perspective on living labs by introducing the design-driven living lab and to investigate its potential by asking the following research question: *How can we implement a design-driven living lab to explore innovation for dementia care challenges?* We explored this question inductively by applying a design-driven living lab approach in three technology-based case studies.

Methodology

For this research, we conducted an explorative in-context study to investigate the effectiveness of a design-driven living lab approach. The living lab was positioned in a real-life environment, as opposed to a true lab environment, and we involved users as co-creators rather than subjects of study (Almirall et al., 2012). In addition, our design-driven focus allowed for open-ended insights into the context through an evaluation that was driven by the use of a "technology probe" (Gaver et al., 1999; Hutchinson et al., 2003), meaning that the participants were allowed to use the focal technology as they saw fit – without restrictions or prescriptions. Moreover, they were motivated to contribute to the value proposition: the promise of value to be delivered by a product. They were encouraged to experiment with the technology and to reflect on their needs and the overall context through the technology. As such, the design-driven living lab approach goes beyond the mere evaluation of technology.

Through this design-driven approach, open-ended results can be generated in collaboration with the relevant stakeholders (Sanders & Stappers, 2008). In our case study, people with dementia, care professionals, and a company were involved. In this way, a design-driven living lab can be used to navigate the "fuzzy" front-end of innovation, which is the initial phase of innovation where the context and target market segment are still unclear. The design-driven living lab involves design activities, such as identifying needs, conceptualizing, and taking different perspectives (Krogstie, 2012) within the living lab (Koskinen et al., 2011). We aim to use these activities in this study to address challenges in dementia care.

The three case studies in this article focused on the introduction of an innovative technical solution – the Qwiek.up system. By putting this device in the real-life contexts, we wished to gain an understanding of its contribution to dementia care, our understanding of dementia care itself, and potential new directions to take with the design. The results of the three case studies were compared and clustered to formulate insights into the design-driven living lab approach when applied in the context of dementia.

Technology probe: Qwiek.up

The technology probe used in the case studies is the Qwiek.up media system (Figure 1), which was developed by the company Qwiek (qwiek.eu). Research shows the importance of meaningful activities for good mental health and general wellbeing (Gold, 2013). The system addresses the problem of having insufficient suitable activities for people with dementia and therefore supports caregivers with their task. Through visual and audio output, this system creates a calming ambient experience for dementia patients in institutional care homes. The system comes with easy-to-use "experience modules", which simulate experiences such as a walk through the woods, looking up at a starry sky, visiting a farm, or viewing a custom slideshow of family photos with music. To use the system, the caregiver inserts an experience modules into the system, which then automatically initiates the corresponding experience. The system can transform a room into an experience by projecting images and video onto a wall or a ceiling (Figure 1). This device is a useful tool for professional caregivers in dementia

Figure 1. The Qwiek.up media system (right) depicting a walk through a forest on a regular wall (left)

care homes, where it can be used as a meaningful activity or as a non-pharmaceutical remedy for stress and agitation, which potentially reduces medicine use in care environments.

Several interventions that use immersive elements and sound have previously been investigated, such as for example "Snoezel(en) rooms" (tinyurl.com/go3v7as), which are controlled multisensory environments designed to create a soothing experience by stimulating the senses (Riley-Doucet & Dunn, 2013). These earlier interventions, however, did not use familiar content for patients or were limited to a single room (Jakob & Collier, 2015). The Qwiek.up system, in contrast, adds value because it is easy to use and is integrated into a wheeled stand that can easily be moved from room to room. Additionally, using a system that is beyond the prototyping phase allows the users to focus on how the technology functions within the context and the various applications it can include.

The main focus of the study was not to obtain feedback to further improve the system, but rather to explore how caregivers use the system in everyday practice. In this way, we aimed to explore the care context and reveal latent user needs. Additionally, this allowed us to gather new perspectives on the value proposition of the system. Nevertheless, insights relating to usability and the concept itself can be expected as a by-product. Therefore, our aim was consistent with our conceptual proposal of the design-driven living lab: to discover opportunities by introducing a new piece of technology in a real-life context.

In-context evaluation of the Qwiek.up system

Research on the Qwiek.up system was carried out in three care environments for people with dementia. Two studies were performed in a care home and the third one was performed within a dementia day programme where people with moderate dementia attended a facility for daily activities, yet still lived at home. The company behind the Qwiek.up system was involved in setting up this study and reflected with us on the outcomes. The characteristics of the three case studies are described in Table 1.

To carry out the study in the three locations, the following sequence of activities was adopted:

1. The system and research method were introduced to the staff.

2. Staff members were invited to use and experiment with the system during the study period. During this period, they recorded their experience with the system on an evaluation form.

3. After the study period, additional insights were collected during a focus group with the care professionals.

The evaluation form used to capture the experience requested the following details: a description of the experience, the length of engagement, its usefulness in the care context, and usability. To aggregate the findings in each of the case studies, we performed a cluster analysis with the retrieved data from both the form and the focus group (Koskinen, 2003). In this regard, the study had a dual role. We asked staff members to use the system and describe their experience, and then we instructed them to go beyond evaluation and experiment with using the device. This second role inspired new design directions and widened the value proposition. During the focus group, we encouraged them to reflect on their experience. Subsequently, the findings were aggregated into three topics to identify the role of the design-driven living lab in the innovation process. The insights gathered concerning both the system and the living lab approach were then shared with the company that created the system and interpreted with them.

Results

The three case studies delivered improvements to the system and demonstrated its value in dementia care. These results, from both the evaluation form and the focus group, are shown in Table 2. Especially in case studies 1 and 3, the system was found to be a strong addition to formal care. The system was used as an activity for individuals and groups with dementia, and it increased the efficiency of care provision by giving the nurses more time to engage in their care practices. During the focus group with users from case study 3, they even mentioned that the system could reduce the need for medication in some cases. In case study 2, the system was less well received; however, here the professional caregivers showed less willingness to experiment with the system during their general care practices.

The cluster analysis of the data from both the evaluation forms and the focus group identified three clusters of topics: new insights concerning applications in care, the functionality of the system, and possible design extensions. The business perspective on each of these topics was added to these results. The results for each of these topics are described in the subsections that follow.

Topic 1: Applications in care
By using an open-ended evaluation, we found that, although the original target context (care homes) was suitable, a different context (day programmes) also shows great potential. The care professionals even indicated it that they felt it could be used for other conditions such as autism, because it provides a comfortable and recognizable ambience. In the case study 3 in particular, multiple purposes for the system were suggested. The system had benefits as a soothing individual activity as intended by the design. However, it was also used as an interactive group activity, as an activity for quieting down a group after lunch, and for one-on-one engagement and storytelling.

The attitude of the caregivers at the day programme centre might well have played an important role in this study. During the focus group, for example, we found that users in case study 3 were more engaged compared to those in the other case studies. They were very enthusiastic and they actively contributed more feedback than expected. They even wrote a thirty-page report in addition to the evaluation form to describe their experiences with the system in detail.

In summary, the users experimented with the device in all kinds of ways and discovered new purposes while doing so. This broadened the potential market for the system and also allowed us to discover latent needs in this care context. From a business perspective, the company reflected on the study that these insights added value to the system and opened potential future markets.

Topic 2: Functionality
We also discovered the need to make improvements to the Qwiek.up system in terms of functionality, technology, and usability. For instance, the projector sometimes malfunctioned, and the company should prioritize fixing such technical issues. Additionally, some of the care professionals in case study 1 found the system hard to use. They suggested making the system's physical controls easier to use or adding a remote control. Furthermore, in some of the experience modules, the music that was preselected by the company to fit with certain videos did not match, resulting in a conflicting experience.

The company is already improving some elements of the product based on this feedback on its functionality and is investigating the potential of adding a remote control or other ways of interacting with the system by both the care professionals and the patients.

Topic 3: Design extensions
Opportunities arose in all three case studies for improving the conceptual design of the Qwiek.up. In case study 1, for example, the potential of adding interactivity was discussed to allow those with dementia to have

Table 1. Context description of the three case studies

	Case Study 1	Case Study 2	Case Study 3
Country	Netherlands	Germany	Germany
Care context	Care home	Care home	Day programme
Dementia profile	Advanced	Advanced	Moderate
Care professionals	6	4	3
Residents	14	11	28
Study period	29 days	33 days	35 days
Evaluation records	6	7	15

Table 2. Evaluative results concerning the Qwiek.up system in the three case studies

	Case Study 1	Case Study 2	Case Study 3
General response to the system	Positive	Moderate to negative	Very positive
Experience modules	Natural modules were preferred (e.g., Aquarium and Forest)	Module of starry sky was used to help individuals to sleep better; others were less appreciated.	Residents responded well to the natural modules; in some cases, there was a mismatch between image and music in the modules.
Technical difficulties	The projector had some issues with finding focus	Physical controls on the system were considered difficult to use.	None
Experience	Various use cases were found such as group sessions (max. 30 minutes) and individual use.	Positive use for one-on- one session and sleep sessions; other uses were not explored.	Many unexpected uses such as calming down a group after lunch and association games with a group.
Focus group results	System usefulness and application varied from person to person, but some users were engaged for 3–4 hours at a time. Interactivity would contribute to the experience.	Care professionals were not very fond of the system. However, the system was used as a conversation starter once, which was deemed positive.	Care professionals were enthusiastic about the system; it could easily become a standard tool in day programmes. The system was used for many purposes in various ways.

an even more engaging experience. In addition, this means that the system could cater to a wider spectrum of users, and that those, for example, in a less advanced state of dementia might be more engaged with the system with the addition of interactive aspects. This extension could also increase the engagement of people with dementia by stimulating play (Anderiesen et al., 2015). In addition, in case studies 2 and 3, we found that the selection of experience modules should be increased. The selection is currently limited and more options would be desirable.

Currently, additional experience modules are being developed and released by the company that created the system. In this way, they are able to quickly broaden their offering and improve the overall experience of the system. For more radical conceptual improvements, for example adding interactive games, development will take longer. However, based on the findings in this study, the company is already developing new conceptual extensions to their system.

Discussion

Over the course of three case studies, we demonstrated the potential of a design-driven approach to living labs. Earlier living lab studies have included active user involvement, for example through co-creation (Sanders & Stappers, 2008). However, by adding design characteristics such as exploration and dealing with uncertainty, living labs can be refocused to better deal with complex problems such as dementia care. In this study, we contributed to answering our research question: *How can we implement a design-driven living lab to explore innovation for dementia care challenges?* The design-driven approach allowed us by actively involving care professionals to broaden the value proposition of the Qwiek.up system, explore the context of dementia care, repurposing the design, and find new opportunities for innovation.

The design-driven living lab approach uses a fully operational system that is presented as a technology probe. This is a high-quality product – rather than a functional prototype – that can be used to explore latent needs, new uses, and opportunities for innovation beyond the evaluation of the current product. This different perspective on putting innovative systems in a real-life context is based on design approaches such as needsfinding, conceptualization, and opportunity seeking (Krogstie, 2012). Today's technology enables the creation of high-quality products in a shorter amount of time, making it possible for in-context studies to influence the value proposition and market release specifications – resulting in shorter innovation cycles. This approach enables fast learning, for example, as was apparent in the redesign of the experience modules or the interface of the Qwiek.up system. This work can be done in a short amount of time, leading to a market release that improves the product offering. Nevertheless, for larger design improvements, such as adding interactivity or answering new use environments such as for children with autism, follow-up design processes are needed, which will take longer to conduct.

Involvement of users in the design-driven living lab is inspired by co-creation (Bergvall-Kåreborn & Ståhlbröst, 2009). However, instead of focusing co-creation efforts in a controlled session, we apply a co-creation perspective during the entire in context study. This allowed participants greater freedom in using the concept, and it allowed us to be more open in interpreting the results. As a result, the participants maintain an open attitude throughout the study, allowing additional insights to emerge such as, for example, new applications or opportunities for meaningful activities in a dementia care context. This was notably demonstrated during the third case study, in which the system was clearly used in an unrestricted and explorative way by the users (Valk et al., 2012). In this case, the users felt they had the freedom to use the system as they saw fit and experimented with it in their specific care settings. This approach also allowed the company to find new uses for the system and orient towards new opportunities in care innovation.

Navigating the early stages of innovation often seems challenging, and this is also the case for living lab practices. However, the inclusion of design skills, such as dealing with opposing perspectives, offers concrete tools to do this (Koskinen et al., 2011). In this study, we have shown that an explorative approach with technology probes can enable the identification of needs beyond the original value proposition and inspire new innovative solutions. The suggestion to facilitate play within the system, for example, allows the company to widen the market impact of the system.

Further research

This article presented an initial application of the design-driven living lab. However, the exact characteristics of this phenomenon are unclear and need to be investigated in greater depth. For example, the system used in this study was arguably already purposed for a care context and did not allow for an open-ended exploration in the same context. The effect of different development levels of prototypes/products needs to be further evaluated.

In addition to our insights, we found some limitations in our application of the design-driven living lab approach. We see that, based on our three case studies, there is scope to improve the evaluation method and the design-driven approach. There are, as yet, no best practices for design-driven living labs, and the evaluation methods we used had some limitations. Some users even stopped filling it in regularly after a week into the study because they felt it was too boring. The participants might need more motivation to explore the new technology and more in-depth evaluation forms with fewer intervals might maintain interest. Additionally, alternative ways to capture an open-ended experience, such as diary studies, might be applied. The report provided by the users in case study 3 provided much richer insights into their experience. Also, the more open approach of the focus group methodology made it possible to do more than just capturing experiences and, in addition, allowed for discussion of other topics, such as design opportunities. An important aspect we discovered is the attitude of the users, as we saw in case study 3, where it was reflected in the participants' willingness to experiment. To clarify this aspect, further research is needed on methodologies that can be used to engage users in open and explorative studies as part of a design-driven living lab.

In the future, we aim to create more diverse forms of the design-driven living lab in order to explore its potential for innovation. In this regard, we aim to target the complexity of other societally relevant challenges and emphasize the inclusion of more diverse stakeholders.

Conclusions

Over three case studies, we demonstrated the potential of a design-driven living lab approach. Participants were invited to openly and freely use a new system in their professional context of caring for people with dementia. The results show that the type of feedback obtained can indeed go well beyond an evaluation of a particular design, however, this still needs to be explored further. New insights can be obtained for repurposing the design and defining new value propositions. A design-driven way of performing a living lab makes it possible to extend the boundaries of the approach from evaluation to include exploring innovative solutions. It shows that there is potential for more disruptive, open-ended development of innovation.

In this way, we succeeded in implementing a first iteration of a design-driven living lab. It allowed us to explore both the method and the contexts in which the intervention was applied. The evaluation results are more open-ended, with more room for input from the various stakeholders. The intervention was not evaluated on its effectiveness but rather as an opportunity and, as a result, it strengthened the value proposition and the business case. This was evident in the response by the company, which has included our findings in their system and development processes. Overall, our approach shows that there is potential for navigating the early stages of future innovation processes with developed products designed to address complex societal challenges.

About the Authors

Rens Brankaert is an Assistant Professor working on design for "Active & Healthy Ageing" within the Business Process Design group in the Department of Industrial Design at Eindhoven University of Technology (TU/e) in the Netherlands. He received his PhD early in 2016 for work related to design for people with dementia involving the application of living labs. In this work, he aimed to build a bridge between a design-based approach and current healthcare practices.

Elke den Ouden is based at the Innovation, Technology and Entrepreneurship Management Group in the Department of Industrial Engineering and Innovation Sciences at Eindhoven University of Technology (TU/e) in the Netherlands. As a TU/e Fellow and Strategic Director of TU/e LightHouse, she forges links between research and industry. Her long history at Philips, including in the role of group leader, has provided her with the network and expertise needed to do perform her current role. Elke operates as the TU/e living lab expert and regularly publishes on this topic.

References

Almirall, E., Lee, M., & Wareham, J. 2012. Mapping Living Labs in the Landscape of Innovation Methodologies. *Technology Innovation Management Review*, 2(9): 12–18.
http://timreview.ca/article/603

Anderiesen, H., Scherder, E., Gossens, R., Visch, V., & Eggermont, L. 2015. Play Experiences for People with Alzheimer's Disease. *Interactive Journal of Design*, 9(2): 156–165.

Bergvall-Kåreborn, B., & Ståhlbröst, A. 2009. Living Lab: An Open and Citizen-Centric Approach for Innovation. *International Journal of Innovation and Regional Development*, 1(4): 356–366.
http://dx.doi.org/10.1504/IJIRD.2009.022727

Brankaert, R. 2016. *Design for Dementia: A Design-Driven Living Lab to Involve People Living with Dementia and their Context.* Eindhoven, The Netherlands: University of Technology Eindhoven.

Brankaert, R., den Ouden, E., & Brombacher, A. 2015. Innovate Dementia: The Development of a Living Lab Protocol to Evaluate Interventions in Context. Info, 17(4): 40–52.
http://dx.doi.org/10.1108/info-01-2015-0010

Gaver, B., Dunne, T., & Pacenti, E. 1999. Design: Cultural Probes. *Interactions*, 6(1): 21–29.
https://doi.org/10.1145/291224.291235

Gold, K. 2013. But Does It Do Any Good? Measuring the Impact of Music Therapy on People with Advanced Dementia. *Dementia*, 13(2): 258–264.
https://doi.org/10.1177/1471301213494512

Hummels, C., & Frens, J. 2011. Designing Disruptive Innovative Systems, Products and Services: RTD Process. In D. A. Coelho (Ed.), *Industrial Design-New Frontiers:* 147–172. Rijeka, Croatia: InTech.
http://dx.doi.org/10.5772/22580

Hutchinson, H., Mackay, W., Westerlund, B., Bederson, B. B., Druin, A., Plaisant, C., Lafon-Beaudouin, M., Conversy, S., Evans, H., Hansen, H,. Roussel, N,. & Eiderbäck, B. 2003. Technology Probes: Inspiring Design For and With Families. In *Proceedings of the SIGCHI Conference on Human Factors in Computing Systems:* 17–24, April 5–10, Ft. Lauderdale, FL.
http://doi.org/10.1145/642611.642616

Jakob, A., & Collier, L. 2015. *How to Make a Sensory Room for People Living with Dementia – Developing Design Guidance for Health Care Practitioners.* Paper presented at the Third European Conference on Design4Health 2015, Sheffield, UK.

Knapp, M., Iemmi, V., & Romeo, R. 2013. Dementia Care Costs and Outcomes: A Systematic Review. *International Journal of Geriatric Psychiatry*, 28(6): 551–561.
http://dx.doi.org/10.1002/gps.3864

Korman, M., Weiss, P. L., & Kizony, R. 2015. Living Labs: Overview of Ecological Approaches for Health Promotion and Rehabilitation. *Disability and Rehabilitation*, 38(7): 613–619.
http://dx.doi.org/10.3109/09638288.2015.1059494

Koskinen, I. 2003. User-Generated Content in Mobile Multimedia: Empirical Evidence from User Studies. In *Proceedings of the 2003 International Conference on Multimedia and Expo (ICME '03)*, 2(3): 645–648.

Koskinen, I., Zimmerman, J., Binder, T., Redström, J., & Wensveen, S. 2011. *Design Research Through Practice From the Lab, Field, and Showroom.* Waltham, MA: Morgan Kaufmann.

Krogstie, J. 2012. Bridging Research and Innovation by Applying Living Labs for Design Science Research. In C. Keller, M. Wiberg, P. J. Ågerfalk, & J. S. Z. Eriksson Lundström (Eds.), *Nordic Contributions in IS Research:* 161–176. Berlin: Springer-Verlag Berlin Heidelberg.
http://dx.doi.org/10.1007/978-3-642-32270-9_10

Leminen, S., Westerlund, M., & Nyström, A.-G. 2012. Living Labs as Open-Innovation Networks. *Technology Innovation Management Review*, 2(9): 6–11.
https://timreview.ca/article/602

Liedtke, C., Welfens, M. J., Rohn, H., & Nordmann, J. 2012. LIVING LAB: User-Driven Innovation for Sustainability. *International Journal of Sustainability in Higher Education*, 13(2): 106–118.
http://dx.doi.org/10.1108/14676371211211809

Martin, R. 2007. *Opposable Mind: How Successful Leaders Win Through Integrative Thinking.* Boston, MA: Harvard Business School Press Books.

Prince, M., Prina, M., & Guerchet, M. 2013. *World Alzheimer Report 2013 – Journey of Caring: An Analysis of Long-Term Care for Dementia.* London: Alzheimer's Disease International.

Riley-Doucet, C. K., & Dunn, K. S. 2013. Using Multisensory Technology to Create a Therapeutic Environment for People with Dementia in an Adult Day Center: A Pilot Study. *Research in Gerontological Nursing*, 6(4): 225–233.
http://dx.doi.org/10.3928/19404921-20130801-01

Sanders, E., & Stappers, P. J. 2008. Co-Creation and the New Landscapes of Design. *CoDesign*, 4(1): 5–18.
http://dx.doi.org/10.1080/15710880701875068

Schuurman, D., Marez, L. De, & Ballon, P. 2016. The Impact of Living Lab Methodology on Open Innovation Contributions and Outcomes. *Technology Innovation Management Review*, 1(6): 7–16.
http://timreview.ca/article/956

de Valk, L., Rijnbout, P., Bekker, T., Eggen, B., & Schouten, B. 2012. Designing for Playful Experiences in Open-Ended Intelligent Play Environments. In *Proceedings of the IADIS International Conference Game and Entertainment Technologies 2012 (GET '12):* 3–10. IADIS Press.

Veeckman, C., Schuurman, D., Leminen, S., & Westerlund, M. 2013. Linking Living Lab Characteristics and Their Outcomes: Towards a Conceptual Framework. *Technology Innovation Management Review*, 3(12): 6–15.
http://timreview.ca/article/748

Wimo, A., Winblad, B., Aguero-Torres, H., & von Strauss, E. 2003. The Magnitude of Dementia Occurrence in the World. *Alzheimer Disease and Associated Disorders*, 17(2): 63–67.

Keywords: design, living lab, societal challenges, technology probe

9

Empowering Local Tourism Providers to Innovate through a Living Lab Process: Does Scale Matter?

David Guimont and Dominic Lapointe

" *Why don't we scale up those things that do work?* "

Tavis Smiley
Broadcaster, author, and entrepreneur

A destination management organization looking to integrate technology into its tourism offering tasked a living lab with engaging tourists and tourism providers in the process. At the end of the two-year initial funding period for an action research project, the process is a success and stakeholders are engaged in the innovation ecosystem. But what is next? By observing participants and gathering feedback from stakeholders through a Policy Delphi panel, the outcomes of the project and the intentions and actions of the tourism providers and other parties were identified. Innovation capacity has increased: spin-offs were created and stakeholders have embraced open and collaborative innovation. Now, stakeholders are determined to make the process sustainable by finding other funding sources. But what should be the level of cooperation and intervention? What level can best foster innovation and knowledge retention? A case study combining a grid of characteristics and levels of analysis for living labs was used to identify one key question: for a living lab in tourism, does scale matter? This article will explore that question and will contribute to the understanding of the living lab as a model of socio-territorial action.

Introduction

To grow and develop, businesses must innovate (Schumpeter, 1934). The tourism industry is no different: it needs to innovate so that businesses and destinations can evolve and stand out. Because of its fragmented and multi-stakeholder nature, the tourism industry is particularly well suited to open and collaborative innovation (Egger et al., 2016). Some believe that the competitiveness of destinations is determined by their capacity to reinvent themselves by offering new products and services that recombine the partnerships between the various elements of the territory's tourism offering (Pikkemaat & Weiermair, 2011). Although destinations struggle to innovate (Najda-Janoszka & Kopera, 2014), opening up the innovation process by engaging stakeholders in a living lab project could stimulate innovation (Lapointe & Guimont, 2015).

Open innovation is about increasing and improving collaboration with an ecosystem's stakeholders and engaging users in the co-creation process. The living lab approach aims to promote and structure stakeholder collaboration, and its key feature is that users become fully fledged stakeholders in the co-creation process and are involved from the very beginning, in a real-life environment. Through living labs, "tourist service providers will obtain insight into what tourists actually want and will have an opportunity to improve and develop new services targeted to different customer segments" (Pucihar et al., 2014). Such insight (from the living lab) could not only enable identification of new markets, but also spur innovation, development, and product improvement (Buhalis & Amaranggana, 2014) through more frequent interactions among stakeholders in a partnership. With interactions between users (tourists) and providers of technology and tourist services being a key catalyst for innovation (Hjalager, 2002), living lab could create increased collaboration opportunities through a common platform where stakeholders would share, discuss, assess, and design various solutions

(Pucihar et al., 2014). In addition, living labs have the potential to become innovation facilitators (Lapointe & Guimont, 2015; Schuurman et al., 2013) and thus create what destination management organizations are struggling to build: an environment that promotes co-operation among tourism industry stakeholders to enable innovation (Najda-Janoszka, 2013).

The living lab under consideration in this article deals with a geography-based activity: tourism. A living lab is not only an innovation process (Bergvall-Kåreborn & Ståhlbröst, 2009); it is also a physical space, a place where innovation happens, a territorial development tool (Lafontaine & Gallant, 2013), and a form of socio-territorial action (Doyon et al., 2015).

Living labs often use a territory as a focus for development or innovation. However, few researchers have examined the matter of selecting a scale for intervention, although some authors have defined scales of concertation and intervention (Bergvall-Kåreborn et al., 2015; Doyon et al., 2015; Franz et al., 2015,). When living labs deal with strongly territorial industries, such as tourism (Lapointe et al., 2015) and agriculture, the need to reflect on scales and levels of action emerges.

To advance reflection on these matters, we used the three-levels grid of analysis proposed by Schuurman and colleagues (2015) to describe the case of a tourism industry living lab (Lapointe et al., 2015). We are attempting to highlight how the framework applies to the reality of the pilot project in order to identify gaps, particularities, and similarities that could provide insight into the scaling and sustainability of the implemented innovation ecosystem.

The article is structured as follows. First, we will review the literature on living labs and levels of innovation. Then, we will apply the three-level framework to an action research project in living lab mode with the tourism industry. Finally, we will discuss the relevance of combining levels of innovation process with socio-territorial geographic scales.

Literature Review

With the living lab approach, users must be at the centre of research or innovation efforts. Instead of attempting to understand users through studies, some organizations now prefer to directly involve users in their actual innovation process (Westerlund & Leminen, 2011). This co-creation approach stimulates innovation and delivers a number of benefits: better grasp of consumers' latent needs, lower risk of failure in product and service design, shorter lead times for new products and services, and higher profits (Westerlund & Leminen, 2011).

Living labs usually include the following features and principles (Bergvall-Kåreborn et al., 2009):

- a technology infrastructure
- an ecosystem of stakeholders who can interact to develop and assess products, services, processes, or systems
- an innovation process that is as open as possible (e.g., regarding intellectual property and types of partnership)
- users playing a key role as co-creators of the products, services, and technologies being developed
- a human-centric design approach that involves ethnographic observation, empathy, and rapid prototyping
- sustained and meaningful interaction and community involvement
- consideration of users' natural environments

Living labs also provide "physical regions or virtual realities in which stakeholders form public–private–people partnerships (PPPP) of firms, public agencies, universities, institutes, and users all collaborating for creation, prototyping, validating, and testing of new technologies, services, products, and systems in real-life contexts" (Westerlund & Leminen, 2011).

According to Schuurman and colleagues (2015), the living lab approach combines practices from both the open innovation and user innovation paradigms: "Open Innovation can be used to study the knowledge transfers on the constellation level, whereas User Innovation can provide insights into user contribution and user involvement methods." This observation stems from the three-level model proposed by the same authors: open innovation happens at the macro level (the living lab constellation) and user innovation happens at the micro level (user contribution and engagement methods), while the living lab project (innovation project) is between the two, at the meso level.

With their three-level model (Table 1), they propose the following definition (Schuurman et al., 2015):

> *"Living Labs are an approach to innovation consisting of three separate, but interrelated levels of analysis. On the macro level, Living Labs are a Public-Private-People partnership organized to exchange knowledge and conduct innovation projects. We regard these Living Lab innovation projects, that are characterized by active user involvement, co-creation, multi-method and multi-stakeholder, as the meso level. These projects consist of different research steps that are aimed at generating user input and contribution to the innovation process, which we consider to be the micro level."*

This article will apply this three-level model to analyze a living lab project in the tourism sector, which is not technology intensive. We will combine this model with the territorial scales (macro, meso, and micro) to describe and understand how innovation level and territorial scales can be combined in a socio-territorial model of action.

We will use this grid to characterize the ongoing living lab project and to explore possible options for sustainable engagement. However, these levels refer to the processes implemented to stimulate innovation, not to geographic scales. Nevertheless, when the innovation process is integrated into an economic sector that is deeply rooted in a given territory, reflections on the correspondence between innovation levels and geographic scales can be fruitful.

Living labs and territoriality

In a paper demonstrating the transferability of the living lab approach as a model of socio-territorial action, Franz and colleagues (2015) argue that "living labs have the potential to be an instrument for the active inclusion of citizens in urban research projects investigating socio-spatial questions." They use the term "space of encounter" to describe the space where citizens converge naturally, and where the living lab is moved to take this natural tendency into account. They discuss spaces of encounter (meso level) as well as the time and nature of encounters (micro level).

According to Leminen, DeFillippi, and Westerlund (2015), who identified paradoxical tensions in the living lab approach, larger scales are better because they maximize stakeholder diversity and knowledge retention, and because larger user pools are easier to segment.

Table 1. Levels of analysis for living labs (Schuurman, 2015; Schuurman et al., 2015)

Levels	Description
Macro	Living lab constellation (ecosystem, public–private–people partnership)
Meso	Living lab innovation project
Micro	Living lab methodological research steps

However, local needs and contexts should be taken into account. These authors believe that combining several local contexts into a broader ecosystem is beneficial.

To integrate living lab characteristics into a territorial conception of innovation, we can say that a living lab is both an approach and a milieu/environment (Bergvall-Kåreborn et al., 2009), where the milieu is a continuum between the macro level (constellation) and the micro level (methodological research steps). This continuum of physical, social, and virtual spaces enables governance, concertation, and knowledge retention (at the macro level) as well as interventions, user involvement, co-creation, experimentation, and evaluation (at the meso and micro levels). In cases with a territorial dimension, such as the tourism industry, living labs would feature relatively broad concertation and cooperation territories roughly following official administrative space. It also encompasses, at a lower geographic scale, spaces of encounter and intervention (e.g., space of encounter, virtual community, real-life context) where people are in close physical or virtual proximity. These spaces following social and lived spaces at meso and micro geographic scales.

Methodology

We conducted action research to document the iterative co-creation process behind the design of a technology-enhanced tourist experience, which was conceptualized according to the needs and actions of tourists discovering a new destination. The living lab is used as "an approach to support and implement processes of open innovation in the context of academy-society collaboration projects" (Levén & Holmström, 2008). The role of lead researchers is to oversee the living lab process, prepare co-creation workshops, and facilitate co-creation both in situ and online. They lead the "experimentation" component. They describe how innovation capability is growing, and the drivers and barriers at play. The collaborating researchers support the co-creation process, deliver specific workshops on technology and the tourist experience, and help to document and characterize the growth in innovation capability. It is done using case study methodology (Yin, 2011).

In this case study, we captured user feedback iteratively and sometimes led to modifications of the processes. There were multiple opportunities for feedback and evaluation by users: before, during, and after activities; during workshops; and during field testing. Our action research in a living lab context relied on an adaptation of the FormIT approach (Bergvall-Kåreborn et al., 2009). The role of the lead researcher was to oversee the process, prepare co-creation workshops, and facilitate co-creation both in situ and online.

Two sources of data were used to describe the case and analyze the innovation processes. The first source was all the documents and notes produced by the partners involved in the process, which includes text data from a tourist panel and from an in situ observation of tourists using the innovations. The second source of data was a Policy Delphi (Linstone & Turoff, 2002) exercise run through 10 stakeholders involved in the process.

Case Study – When Technology Meets Territory: Co-Creation of a Technology-Enhanced Tourist Experience

Through this article, we continue our analysis of an innovation project in a living lab setting that included a higher-education institution, a destination management organization, a technology developer, tourism providers, and tourists. The destination management organization for Rivière-du-Loup in Quebec, Canada, wanted to improve tourists' sightseeing experience using technology. Its intention was to produce podcasts. The local higher-education and research institution, the Cégep de Rivière-du-Loup (cegeprdl.ca), offered support for living lab projects, and the destination management organization tasked the institution with bringing together all stakeholders listed above and launching an action research project. The intention of the project was then transformed by the stakeholders into getting creative distance by introducing the users' inputs into the project of co-creating a technology-enhanced tourist experience in a living lab setting (Guimont & Lapointe, 2015) for the discovery of the rural areas of the county.

The two-year project involved a community of tourism providers and tourists and used a combination of virtual and in situ collaborative processes to conceptualize, prototype, and test the tourist experience. The project provided an opportunity to achieve co-creation in a living lab setting. A mobile application (ICI Rivière-du-Loup) was conceptualized and prototyped during the first year of the project, and was tested and refined during the second year.

In terms of living lab characteristics, the project is a small, but typical living lab. The following subsections describe the levels of analysis.

Macro level: Living lab constellation

The macro level is where stakeholders in the innovation ecosystem come together and discuss the planning and governance of the area or industry where user-driven innovation is needed. The macro level is the group of stakeholders that choose the specific living lab projects to be conducted. It is where knowledge and expertise accumulate. In the ongoing project under consideration, the macro and meso levels overlap to form a single level, as is often the case in less mature living lab structures that work on a single project. Indeed, in this action research project, the governance, and the planning are done by the same key stakeholders who are leading the co-creation process at the meso level. These stakeholders include a technology provider, a destination management organization, and the local higher-education and research institution, which is also a living lab called LLio: *Living lab en innovation ouverte* (Living Lab in Open Innovation; llio.quebec).

For the moment, the research question of the living lab constellation (macro) is the research question of the living lab project (meso): *How can we create a technology-enhanced tourist experience – and how can we integrate this approach at the destination level?*

The macro level, based on the model by Schuurman and colleagues (2015), could improve transfer and retention of the knowledge accumulated through living lab projects (meso level). In the case under consideration, training could be offered to stakeholders outside the living lab project. Tools and documents could be circulated, and documents could be shared in a monitoring and curation context. In the three-level model, the macro level (the living lab constellation) is viewed as a structure that can accommodate various projects. Accordingly, the "co-creating a technology-enhanced tourist experience" project would be the first project, and others could follow, either in the same community or in a different territory or subindustry within the living lab constellation.

Even though the macro and meso levels overlap in the current living lab structure by having the same key stakeholders, some impacts at the macro level can be identified:

- increased strategic thinking/tourist intelligence (appropriation of the Destination 2.0 concept, which refers to the technological participation of the tourist destination in creating, marketing, and commenting on the destination and the tourist experience principles)
- greater capacity for innovation among tourism providers (Lapointe et al., 2015)
- knowledge retention (collaborative methods, tourist experience and co-creation concepts, and technology concepts and prototypes)
- transfer of knowledge and know-how to members of the destination management organization through technology demonstrations, document circulation, and training on technology, tourism, co-creation of tourist experiences and Destination 2.0
- identification of opportunities (e.g., technology-enhanced tourist experience concepts that the destination's stakeholders can experiment with)
- identification of research funding opportunities where the living lab engagement space becomes a forum for a national-scope project (e.g., climate change)

In the two rounds of the Policy Delphi exercise, all 10 participants rated the integration of all elements of the Destination 2.0 model (Neuhofer et al., 2014) into the destination strategy as very important or important. They identified the living lab as an enabler, but they also pinpointed the importance of extending their new capacities to the whole destination.

Meso level: Living lab innovation project
The meso level guides the innovation project conducted using a living lab approach. In the case under consideration, the meso level is the steering committee made up of representatives from three stakeholders (the teaching and research institution/living lab, the technology provider, and the destination management organization). The committee plans the key project stages and approves inputs and outputs at each stage. The lead researcher plans individual workshops in more detail. So far, the living lab project (meso level) has enabled the development of two mobile apps, has improved innovation capacity, and has provided a physical and virtual space for engagement and co-creation.

Micro level: Living lab methodologies
The micro level focuses on the timing and content of co-creation meetings and user interactions. At this level, through various in situ and online workshops and through observation, tourists' needs and expectations are identified and concepts are co-created, co-developed, and eventually tested by users in the field. These various types of interactions take place in spaces of encounter and interaction. In the living lab project under consideration, the following spaces were used:

- a collaborative space at the local college (used by the LLio living lab) for co-creation workshops and training sessions
- a tourist information centre and museum institution for field observations
- a secret Facebook group, SurveyMonkey, Google Drive, and a website for data collection and interactions

Discussion

In the project under consideration, the meso and macro levels overlap. In a more mature living lab structure, the meso level would likely be associated with a specific innovation project like the ongoing one, while the macro level – within a broader territory or industry – could supervise multiple innovation projects and optimize knowledge retention and transfer. The meso level would provide a reusable but adaptable template that could be applied to various individual cases and subterritories across the living lab constellation.

The starting point for the innovation project is local (a county), and so is the project's scale, with the local destination management organization, the regional higher-education institution, and a local technology provider. This is a meso scale of intervention. All tourism providers were also local. The tourist community was made up of people (users) from within and outside the region. Assuming that the research question and the living lab structure were applied at the macro level in order for the living lab to mature and to consolidate its operations, what would be the appropriate scale? Local, regional, supra-regional, national, or industry-wide ?

Although Schuurman and colleagues (2015) argue that knowledge and expertise retention happens at the macro level, and while Leminen and colleagues (2015) suggest that a very broad user pool increases opportunities for innovation, applying the macro level to a strongly territorial industry such as tourism raises the following question: What is the appropriate scale of intervention? Perhaps the overlapping in our case is due not only to the living lab's low maturity level, but also to the context, given that tourism practices are deeply rooted in a physical and social space. Nevertheless, the macro level could become more relevant by breaking away from the meso level and focusing on the micro level.

Accordingly, if the living lab structure were to be applied at a broader macro level, conditions for success would need to be reproduced at the micro level: spaces of encounter for co-creation and training, real-life user observation and engagement contexts, and a technology infostructure (information technology structure) to engage users remotely. Most expertise developed through the living lab project was at the micro level. However, the omnipresence of living lab coordination at all three levels of the current project raises the thorny question of transfer between levels. Such transfer needs to involve a wide range of stakeholders at the micro and meso levels while maintaining a measure of consistency at the macro level. In the current project, transfer happened organically, with stakeholders at the micro level appropriating techniques and processes in order to apply them to other projects they are involved in at the meso level. However, the ecosystem's low maturity level and the territorial specificities of the tourism industry prevent us from determining whether the macro level encompasses all projects that benefit from the spread of living lab processes.

We therefore believe that the question of territories and scales is extremely relevant when considering living lab interventions in a regional development context, especially in strongly territorial industries such as tourism. Although there is overlap in our case, such overlap does not hinder reflection. However, it is necessary to combine innovation process levels, geographic scales, and living lab characteristics to arrive at a descriptive grid that can guide reflection on both innovation and scales for a sustainable living lab process, eventually leading to the institutionalization of the living lab process at the appropriate territorial scale.

An institutionalization territory (Lévesque & Vaillancourt, 1998) is a territory where a body of social relationships can be maintained over time; it remains sustainable while continuing to evolve. The diversity of interventions and types of living labs calls for formal institutionalization levels that can promote sustainability of the relationships created by living labs. Table 2 summarizes the links between innovation process levels, territory levels, and living lab characteristics.

Table 2. Links between levels and characteristics of living labs

	Innovation Process Levels	**Territory Levels**	**Living Lab Characteristics**
Macro	Living lab constellation	Concertation / governance	• Ecosystems approach • Real-world context • Technical infrastructure • Lifespan • Scale • Level of openness
Meso	Living lab project	Intervention / steering	• Technical infrastructure • Community • Level of openness
Micro	Living lab methodologies	Socialization / encounter	• Real-world context • User role • Level of openness • Evaluation, context research, and co-creation

Conclusion

By applying the three-level living lab analysis model to our tourism living lab project, we observed that the current project's micro level matches the micro level described in the literature. However, the current project merges the macro and meso levels into a single level, with no hierarchical or territorial distinctions. Returning to Schuurman and colleagues (2015), we can state that the two levels of innovation, macro and meso, are led by the same stakeholders in different processes but sometimes at the same time, with the territorial scale being the same: the scale of the local county. We also found that a broader scale at the macro level might be preferable, although the low maturity of the living lab under consideration makes it hard for us to determine whether the ideal scale would be regional, supra-regional, or national. These results raised questions about the appropriate territorial scale for ensuring sustainability and efficiency in a living lab. The overlapping of the macro and meso level raised issues of diversity in terms of the different points of view "at the table" but also in terms of the diversity of territorial governance at play in the project.

Living labs are not working in a neutral deterritorialized context; they work on a territorial context at a scale or another. Our analysis of the action research shows that, as a territorial model of action, the living lab acts at different territorial scales. We therefore proposed a descriptive grid that merges the levels of the innovation process with the scales of the territory associated with the living lab's areas of intervention. Future research will be required to test this grid and to probe deeper into the relationships between innovation, living labs, territories, and territorial industries involved in living labs.

Acknowledgments

This research is funded by the Ministère de l'Éducation et de l'Enseignement supérieur du Québec (Programme d'aide à la recherche et au transfert – Volet innovation sociale).

An earlier version of this article was presented at the 2016 International Conference on Tourism (ICOT) Innovation Conference, which was held in Naples, Italy from June 29 to July 6, 2016, in collaboration with the International Association for Tourism Policy (IATOUR; iatour.net), a non-profit organization dedicated to promoting scientific tourism research.

About the Authors

David Guimont is a Teacher-Researcher at the Cégep de Rivière-du-Loup in Quebec, Canada, where he is associated with the Recreation and Leadership Training Department and the Living Lab in Open Innovation (LLio). He holds a master's degree in Tourism Management and Development from the Université du Québec à Montréal (UQAM).

Dominic Lapointe is a Professor in the Department of Urban Studies and Tourism at the Université du Québec à Montréal (UQAM) in Canada, where he is the Director of the Tourism and Hospitality Management Undergraduate program. His research addresses development and environmental issues with the use of critical theory, especially in the fields of tourism, conservation, and the environment. He holds a doctoral degree in Regional Development from the Université du Québec à Rimouski (UQAR).

References

Bergvall-Kåreborn, B., & Ståhlbröst, A. 2009. Living Lab: An Open and Citizen-Centric Approach for Innovation. *International Journal of Innovation and Regional Development*, 1(4): 356–370. http://dx.doi.org/10.1504/IJIRD.2009.022727

Buhalis, D., & Amaranggana, A. 2014. Smart Tourism Destinations. In Z. Xiang & L. Tussyadiah (Eds.), *Proceedings of the International Conference on Information and Communication Technologies in Tourism:* 553–564. Switzerland: Springer International Publishing. http://dx.doi.org/10.1007/978-3-319-03973-2_40

Doyon, M., Rochman, J., Fontan, J. M., Klein, J. L., Ducruc, S., Xiao, J., Yorn, C., Fortin, J., & Dugré, S. 2015. L'approche Living Lab et l'aménagement des espaces ouverts agricoles. Un exemple en région métropolitaine de Montréal. *Articulo – Journal of Urban Research*, Special Issue 6. http://dx.doi.org/10.4000/articulo.2662

Egger, R., Gula, I., & Walcher, D. 2016. Towards a Holistic Framework of Open Tourism. In R. Egger, I. Gula, & D. Walcher (Eds.), *Open Tourism: Open Innovation, Crowdsourcing and Co-Creation Challenging the Tourism Industry:* 3–16. Berlin: Springer. http://dx.doi.org/10.1007/978-3-642-54089-9_1

Franz, Y., Tausz, K., Thiel, S., & Yunus, M. 2015. Contextuality and Co-Creation Matter: A Qualitative Case Study Comparison of Living Lab Concepts in Urban Research. *Technology Innovation Management Review*, 5(12): 48–55. http://timreview.ca/article/952

Guimont, D., & Lapointe, D. 2015. Co-Creation of a Tourist Experience Enhanced by Technology in the Context of a Living Lab. *e-Review of Tourism Research.* ENTER 2015 (6).

Hjalager, A.-M. 2002. Repairing Innovation Defectiveness in Tourism. *Tourism Management,* 23(5): 465–474. http://dx.doi.org/10.1016/S0261-5177(02)00013-4

Lafontaine, D., & Gallant, N. 2013. *Des laboratoires vivants pour des territoires innovants.* Centre facilitant la recherche et l'innovation dans les organisations (CEFRIO). http://www.cefrio.qc.ca/media/uploader/Laboratoires_vivants_rapport_synthese.pdf

Lapointe, D., & Guimont, D. 2015. Open Innovation Practices Adopted by Private Stakeholders: Perspectives for Living Labs. *info,* 17(4): 67–80. http://dx.doi.org/10.1108/info-01-2015-0003

Lapointe, D., Guimont, D., & Sévigny, A. 2015. The Living Lab Approach to Raise Innovation Capability Among Tourism Practitioners. *Tourism Dimension,* 2(2): 18–27.

Leminen, S., DeFillippi, R., & Westerlund, M. 2015. Paradoxical Tensions in Living Labs. In *Proceedings from the XXVI ISPIM Conference,* June 14–17, 2015, Budapest, Hungary.

Levén, P., & Holmström, J. 2008. Consumer Co-Creation and the Ecology of Innovation: A Living Lab Approach. *Proceedings of IRIS31,* August, 10–13. http://www.diva-portal.org/smash/record.jsf?pid=diva2%3A311009&dswid=-102

Lévesque, B., & Vaillancourt, Y. 1998. *Les services de proximité au Québec: de l'expérimentation à l'institutionnalisation.* Montréal: Université du Québec à Montréal.

Linstone, H. A., & Turoff, M. (Eds.). 2002. *The Delphi Method: Techniques and Applications. Upper Saddle River, NJ: Addison-Wesley.*

Marez, L. D., Schuurman, D., & Ballon, P. 2015. *Bridging the Gap between Open and User Innovation? Exploring the Value of Living Labs as a Means to Structure User Contribution and Manage Distributed Innovation.* Ghent: Ghent University. https://biblio.ugent.be/record/5931264

Najda-Janoszka, M., & Kopera, S. 2013. Exploring Barriers to Innovation in Tourism Industry – The Case of Southern of Poland. *Procedia - Social and Behavioral Sciences,* 110: 190–201. http://dx.doi.org/10.1016/j.sbspro.2013.12.862

Najda-Janoszka, M. 2013. Innovative Activity of Small Tourist Enterprises – Cooperation with Local Institutional Partners. *Journal of Entrepreneurship, Management and Innovation,* 9(1): 17–32. http://mpra.ub.uni-muenchen.de/id/eprint/52991

Neuhofer, B., Buhalis, D., & Ladkin, A. 2014. A Typology of Technology-Enhanced Tourism Experiences. *International Journal of Tourism Research,* 16(4): 340–350. http://dx.doi.org/10.1002/jtr.1958

Pikkemaat, B., & Weiermair, K. 2007. Innovation through Cooperation in Destinations: First Results of an Empirical Study in Austria. *Anatolia,* 18(1), 67–83. http://dx.doi.org/10.1080/13032917.2007.9687036

Pucihar, A., Malešič, A., Lenart, G., & Borštnar, M. K. 2014. User-Centered Design of a Web-Based Platform for the Sustainable Development of Tourism Services in a Living Lab Context. In L. Caporarello, B. Di Martino, & M. Martinez (Eds.), *Lecture Notes in Information Systems and Organisation,* 7: 251–266. Switzerland: Springer International Publishing. http://dx.doi.org/10.1007/978-3-319-07040-7_24

Santoro, R., & Conte, M. 2009. Living Labs in Open Innovation Functional Regions. In *Proceedings from the 15th International Conference on Concurrent Enterprising,* June 22–24, 2009. http://dx.doi.org/10.1109/ITMC.2009.7461431

Schaffers, H., Guzmán, J. G., & Merz, C. 2008. An Action Research Approach to Rural Living Labs Innovation. In P. Cunningham and M. Cunningham (Eds), *Proceedings of the Collaboration and the Knowledge Economy: Issues, Applications, Case Studies:* 617–624. Amsterdam: IOS Press.

Schuurman, D. 2015. *Bridging the Gap between Open and User Innovation? Exploring the Value of Living Labs as a Means to Structure User Contribution and Manage Distributed Innovation.* Doctoral dissertation. Ghent University and Vrije Universiteit Brussel VUB.

Schuurman, D., De Marez, L., & Ballon, P. 2013. Open Innovation Processes in Living Lab Innovation Systems: Insights from the LeYLab. *Technology Innovation Management Review,* 3(11): 28–36. http://timreview.ca/article/743

Schuurman, D., De Marez, L., & Ballon, P. 2015. *Living Labs: A Structured Approach for Implementing Open and User Innovation.* Paper presented at the 13th Annual Open and User Innovation Conference, July 13–15, 2015, Lisbon, Portugal. http://hdl.handle.net/1854/LU-6888241

Schumpeter, J. 1934. *The Theory of Economic Development.* Boston: Harvard University Press.

Westerlund, M., & Leminen, S. 2011. Managing the Challenges of Becoming an Open Innovation Company: Experiences from Living Labs. *Technology Innovation Management Review,* 1(1): 19–25. http://timreview.ca/article/489

Yin, R. K. 2011. Applications of Case Study Research. London: SAGE Publications.

Keywords: living labs, tourism, destination management, innovation, scale, level, territoriality, socio-territorial action, action research

10

Methods for Supporting Older Users in Communicating Their Emotions at Different Phases of a Living Lab Project

Sonja Pedell, Alen Keirnan, Gareth Priday, Tim Miller, Antonette Mendoza, Antonio Lopez-Lorca, and Leon Sterling

> *"The great secret that all old people share is that you really haven't changed in seventy or eighty years. Your body changes, but you don't change at all. And that, of course, causes great confusion."*
>
> Doris Lessing (1919–2013)
> Writer and Nobel Laureate in Literature (2007)

In this article, we focus on living lab methods that support the elicitation of emotions – a key success factor in whether a design solution will be accepted and taken up over the long term. We demonstrate the use of emotional goal models to help understand what is relevant for a target user group in the early phases of design. We promote animations and storyboards to envision the context of use and to gain an understanding of how design ideas can integrate into people's lives. For the evaluation of ideas and to further understand user needs, we show how technology probes facilitate natural interactions with a suggested solution concept. All methods have in common that they enable older adults without design or development experience to participate in the design process and work towards a meaningful solution by helping to communicate feelings and goals that are often hard to define. Lastly, we present a process model that demonstrates our emotion-led design toolkit at various phases of a living lab process.

Introduction

Each design and research process consists of varied methods that are fundamental to realizing user goals. In this article, we demonstrate how our methods can be used within three generic living lab design phases: exploration, experimentation, and evaluation (Schuurman et al., 2016). We apply our methods in a living lab project to give older adults a strong voice to share and describe their experiences and emotions and to explore how these insights can be captured for design purposes. The project objective was to develop an innovative personal emergency alarm that evokes positive emotions in older adults and reduces the feeling of "being monitored". We ask the question:

> *"What methods cater for the goals and emotions of older adults in a co-design process to develop innovative solutions?"*

We propose an emotion-led design toolkit with several artefacts: motivational goal models, animations, and technology probes. We argue that using these artefacts at different phases of a living lab project cycle facilitates effective communication between participant stakeholders and contributes to both innovation and service design methodologies. Service design and user-driven design methods are increasingly important aspects of living labs, recognized as two means of increasing user acceptance of innovations (e.g., the FormIT methodology [Ståhlbröst &Holst, 2013] and citizen-driven innovation [Eskelinen et al., 2015; Gray et al., 2014]. This view is in alignment with Muller (2007) who argued a decade ago that user engagement is too often one-directional, creating applications of technology rather than solutions to user problems.

The following sections report on emotions in designing for health, followed by a review of existing living lab processes. We show how our emotion-led design toolkit can support the phases of a typical living lab process, demonstrated through a case study of personal alarm systems.

Emotions in Designing for Health

In the discipline of design, emotions influence decisions about the look and feel of products and services. For a personal topic such as health, people's emotions play a major role in the success of a technology, and they afford an opportunity to increase compliance (Lo Bianco et al., 2015).

Yet, design of systems in the domain of health services is still functionality-driven rather than emotion-driven, particularly when institutions, as main stakeholders, fulfill government policies and focus on compliance and liability towards patients rather than patients' feelings. Here, we complement functional-driven design with users' desired emotions in order to develop innovative products with a high uptake (see Figure 1).

Personal alarm systems are an example of technology that has high impact potential but neglects the emotional needs of older people (Miller et al., 2015). Personal alarm systems typically have two features: i) a wearable personal device – the user can raise an alarm if they require emergency attention, for example by pushing a button on a wristband or a pendant worn around the user's neck; and ii) a wellbeing check – the user informs the service provider that they are fine, usually on a daily basis, for example, by pushing a button on a base station connected to a telephone line. In the second case, if no indication of wellbeing is received during the specified period, the service provider initiates checks on the user (Pedell et al., 2014). In this article, the term "personal alarm system" will be used to describe both features: the wellbeing check and the personal device.

The Living Lab Phases: Exploration, Experimentation, and Evaluation

Many studies have noted the importance of real-life contexts and the involvement of end users in living lab innovation processes (Almirall et al., 2012; Leminen, 2015; Veeckman et al., 2013). The end users of potential innovations are seen as "co-creators" (Veeckman et al., 2013) in the innovation process rather than subjects of study. Dell'Era and Landoni (2014) highlight the importance of the research-led aspect of living labs while also emphasizing the importance of users as active co-creators and the real-life context as a factor that modifies the users' needs.

Although individual living labs have different overall approaches (Almirall et al., 2012), there are many similarities. The FormIT methodology (Ståhlbröst & Holst, 2013) illustrates a typical approach, which emphasizes user involvement in the innovation lifecycle from ideation through to eventual commercialization. Schuurman and colleagues (2016) have identified three generic phases that are common to many living labs: exploration (idea/concept), experimentation (prototyping), and evaluation (pre-launch, launch, and post-launch). Their study also shows that the more closely a living lab approach follows this "ideal" approach, including multi-method user involvement, the greater the positive impact on the final outcome. Leminen and colleagues (2015) note four different user roles within a living lab – informant, tester, contributor, and co-creator – although they indicate that each user can perform multiple roles. The informant contributes an understanding about the users' life, problems, and needs. The tester

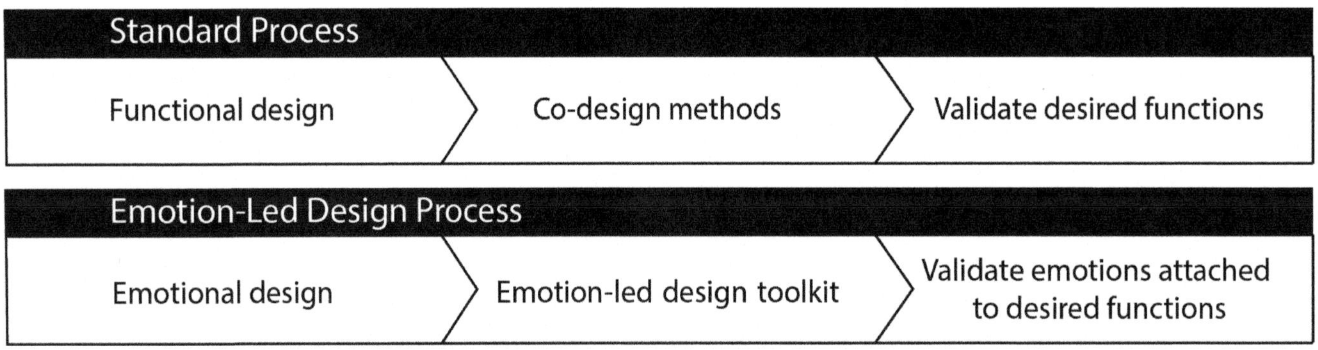

Figure 1. Comparison model showing the difference between a standard process and an emotion-led design process with emphasis on understanding users' emotions

evaluates innovations in the environment. The contributor collaborates with the other stakeholders in the development of a service. Finally, the co-creator creates and develops actual solutions with the other stakeholders. Leminen and colleagues (2015) show that the first three roles are most common in living lab projects. Co-creative methods have been established in research as "people who are not professional technology designers may not be able to define what they want from a design process, without knowing what is possible. A process of mutual learning for both designers and users can inform all participants' capacities to envisage future technologies and the practices in which they can be embedded" (Robertson & Simonsen, 2012). However, recent literature has recognized the challenge of actively engaging older adults in design processes and have come up with methods to do this (Edlin-White et al., 2012; Lindsay et al., 2012; Vines et al., 2012; Waycott et al., 2012; Waycott et al., 2013). What we contribute here is a means to specifically integrate emotions in a co-creative living lab process. Table 1 summarizes our emotion-led methods, aligned with the three typical phases of living lab methodologies identified by Schuurman and colleagues (2016).

Phase 1: Exploration

Our case study focuses on the use of methods to enable the exploration and evaluation of emotions with older adults around personal alarm systems and their wider context of use as a means of innovating both the function and the service offering. In the conceptual phase, we report on the development of an early goal model, with emotions captured from initial exploration and in-depth interviews where the users are operating in the role of an informant.

Participants, data collection, and analysis

Twelve in-depth interviews were conducted, categorized into three groups: i) older people who lived alone (with one exception) and who either currently have or previously have had a personal alarm system installed in their home; ii) family members of older adults who either currently have or previously have had a personal alarm system installed into their home; and iii) older people who never have had a personal alarm system installed in their home. The interviewees in the first group were older than the interviewees in the last group: those who had experience with personal alarm systems ranged from 85 to 91 years of age, whereas those who did not have experience with personal alarm systems ranged from 66 to 79 years of age.

The interviews explored three key questions: What should an alarm technology do (functions)? How should it be (qualities)? and How should it feel (emotional response)? We transcribed the data using content analysis according to Patton (2002) and derived common themes from the data.

Results: Emotions around personal alarm system use

Our interviews revealed that some older people perceived that their feelings were not being taken into consideration. They viewed the wearable pendants as "cowbells" forced onto them:

> *"She always would joke about her cowbell, and complain about it. 'Look at what my kids are making me do,' kind of comment, a slight resentfulness about it. And it was kind of against her independence."* [Participating relative]

The pendants were perceived by the wearers as having a "stigma" attached to them – a perception that others be-

Table 1. A process table outlining the phases in which emotion-led methods are employed during a standard living lab process, in context of a personal alarm living lab project

Phase 1: Exploration	Phase 2: Experimentation	Phase 3: Evaluation
Initial investigation	Scenario design	Tech probe design
In depth interviews	Co-evaluation of scenario	Define user needs
Early goal model	Goal model refinement	Concept refinement

lieve the wearers are no longer are able to care for themselves, resulting in the pendants not being worn. The considerations and social environment around the decision of whether or not to wear the pendant were complex and loaded with emotions. The complexity is expressed well by the nephew of one of the users:

"So we did have the discussion and she sort of admitted that she didn't want to wear it and she didn't think that she should and she understood the risks and she was prepared to take the risks and that she didn't want to upset me and she didn't want me to feel like she wasn't cooperating with me. And so she said [mimicked aunts voice] 'so at least I wore it some of the time'. You know these times when she was wearing it was when someone was there and she didn't really need it. But for her, that was her compromise." [Participating relative]

It became clear that personal alarm systems need to consider more than just safety aspects. The pendant limits mobility in that the alarm only works in the owner's house, so that the wearer might be hesitant to leave their home. Interestingly, most pendants have an effective range of 300 metres, but despite the strong feeling of being confined to a small space, older users are told that the maximum range is about only 70 metres in order to better pinpoint a wearer's location in an emergency. The limitations this information poses to the older person in their everyday life apparently are not considered by service providers.

The wellbeing check (the second component of the system) requires the user to remember to push a button each day. Otherwise, the service provider calls to check upon the client, which leads many older people to feel they are a burden, despite paying for the service. Others feel that they are perceived by their families as suffering from memory loss:

"And no matter what system I try [claps with hand on his knee in frustration and enforcement several times] I still manage out of 10 days that I miss out 2 or 3 times by completely forgetting and that is what ANNOYS [emphasis] me." [Older user]

Pressing the button on the wellbeing check base station does not convey any meaning to the older person and is therefore forgotten. Additionally, the wellbeing check provides no feedback indicating whether the button has been pressed on a particular day. Pressing the button on the wellbeing check a second time on the same day initiates an inquiry to the service providers and is perceived as a signal of an emergency. Hence some older people do not feel confident using the system. Further, the wellbeing check is not easily configurable, for example, users cannot adjust the time of day when the wellbeing button should be pushed, which leaves users feeling that they are not in control of the system.

The older people we interviewed indicated a desire for the personal alarm system, in particular the wellbeing check, to evoke feelings of independence, safety, being in touch with other people, control, and integration. Most importantly, they wanted to feel cared about. Table 2 provides an overview of the captured emotions.

Based on the emotions we captured, we integrated the emotional goals into a motivational goal model according to the notation of Sterling and Taveter (2009), extended by Marshall (2014), as shown in Figure 2. In this model, emotions (hearts) are attached together with desired qualities (cloud shapes) to functional goals (parallelograms). The emotions were used as high-level specifications in the following phases of the design process to develop a prototype and the final design of the personal alarm system.

Table 2. Overview of current and preferred emotions surrounding personal alarm systems

Current Emotions about the Personal Alarm System		Preferred Emotions about the Personal Alarm System	
Resentfulness	A burden on other people	In touch	Integrated
Not independent	Not confident	Independent	Cared about
Stigmatizing	Not in control	Safe	In control

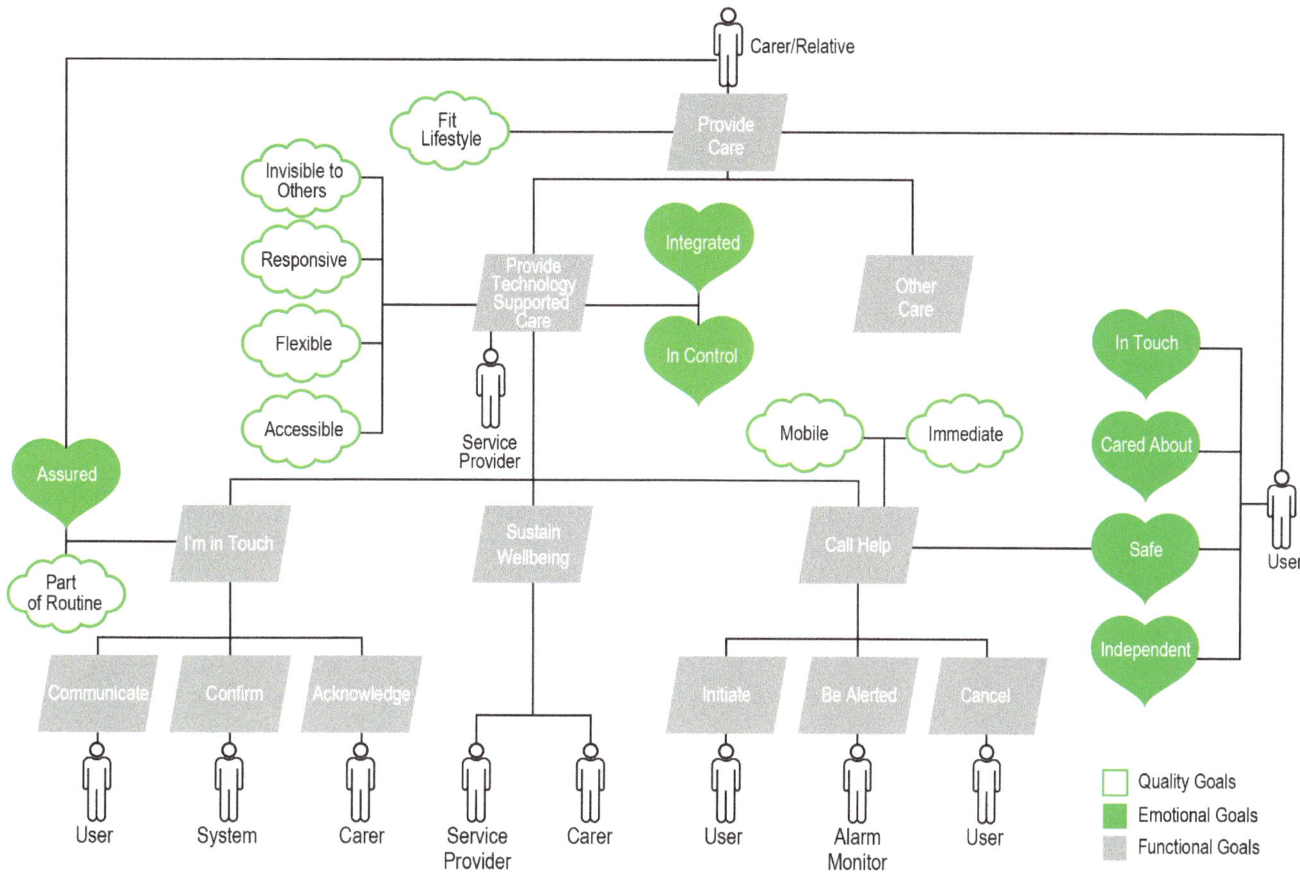

Figure 2. Mapping functional, quality, and emotional goals with stakeholder roles and system interactions

In order to ensure a smooth transition from the exploration to the experimentation stage, we suggest that the outcomes of the exploration phase, namely the emotional goal model, should be validated in its context of use with the prospective end users. The goal models are intended as high-level specifications for designers and developers to create a prototype.

Phase 2: Experimentation

In technology development, we face the challenge of anticipating how technology will be adopted and integrated into people's lives. Technology itself changes our lives — how we perceive and handle situations (described by Carroll and Rosson (1992) as the task-artefact-cycle). Although user-centered design and experience design (Buxton, 2007) help us envision future use by better understanding users' lives, designers still face a problem of validating future use scenarios before creating a solution. Future users themselves often have no clear understanding of the implications new technologies will have on their lives and are thus limited in giving input into design decisions. When changing existing situations into preferred ones (Simon, 1982), a designer with input from end users should strive to understand contexts, issues, relationships, environments, and emotions where a design problem is situated. Scenarios one of many useful tools in a designer's toolkit (Loke et al., 2005) and are used to understand the complexities of the design context (Iacucci et al., 2002). Here, we use scenarios, or imagined stories of events, during the experimentation phase as means to explore design options, anticipate future problems, and describe contexts of user experiences with products (Lim & Sato, 2005; Mathews & Heinemann, 2012). We therefore develop a new approach to animate current and future use scenarios with emphasis placed on the early emotional goal model described in the exploration phase. In doing so, we aim to envision and visualize future technology use. We suggest that animated scenarios can be used as a tool to co-evaluate earlier insights of user research with participants, particularly surrounding sensitive issues such as feelings and personal life goals. We also expect that, by co-evaluating insights from the conceptual phase using animated scenarios, participants can express their own emotions using personas and those emotions can be better expressed in animations rather than written or sketched scenarios.

Scenario design: Creating the animations
We created three animations (Figure 3) based on the existing interview data, including emotions around personal alarm system use, from the exploration phase. These animations were used to co-evaluate the goal model and validate the barriers and reasons older adults do not use current personal alarm systems.

Our three aims for the use of the animations for co-evaluation as a means to explore future scenarios were:

1. To determine whether the problems presented in the scenarios were identified and interpreted in similar ways by the participants reflecting on their own situation.

2. To determine if the animated scenarios reflected a realistic story (context) of a user involved with personal alarm systems, with particular focus on both operating the pendant and conducting daily well-being checks.

3. To encourage participant feedback to help redesign a new personal alarm system once a shared understanding about the relevance of the scenarios was established.

In a co-design workshop with four older people, we discussed these animations. According to Massimi and colleagues (2007), a participant number of four was considered to be suitable due to the personal nature of the topic, creating a familiar environment of "having tea together". People with and without alarm pendants discussed the scenario in pairs, including feelings about the personas, intervention points in the scenario, and design ideas to improve feelings and living situation of the people depicted in the scenarios. In this case, users act as both informants, who correct understanding of the situation, and contributors, who collaborate to develop the service design.

Results using animated scenarios and storyboards
When shown the animations, participants were engaged with the plot of the story. This engagement became particularly clear after participants commented on the animations after viewing them, because they related feelings of the animated personas to their own life situations. We confirmed that the three scenarios and the depicted emotions were perceived as realistic and something the participants could relate to in their own lives, but we were also able to create an atmosphere of openness that provided a foundation for engaging the participants in co-creative design activities. Using printed storyboards (Figure 4), the participants identified and commented on aspects of the animations in which personas needed to be better understood by their relatives and service providers with implications for design.

Design ideas were directly put into the context of the scenario. The ideas were adjusted until the scenario reflected a true reality for a personal alarm user, as shown by the annotations in Figure 4. For example, the size of the pendant was not problematic for participants, but merely its appearance. Figure 5 shows an example of one of the ideas generated in the workshop: the pendants could be redesigned to be worn as a piece of jewellery.

Figure 3. Screen captures of animated scenarios. Left: "I forgot", middle: "Cow Bell", right: "Dress Code"

Figure 4. An example of an animation storyboard with annotations from workshop participants

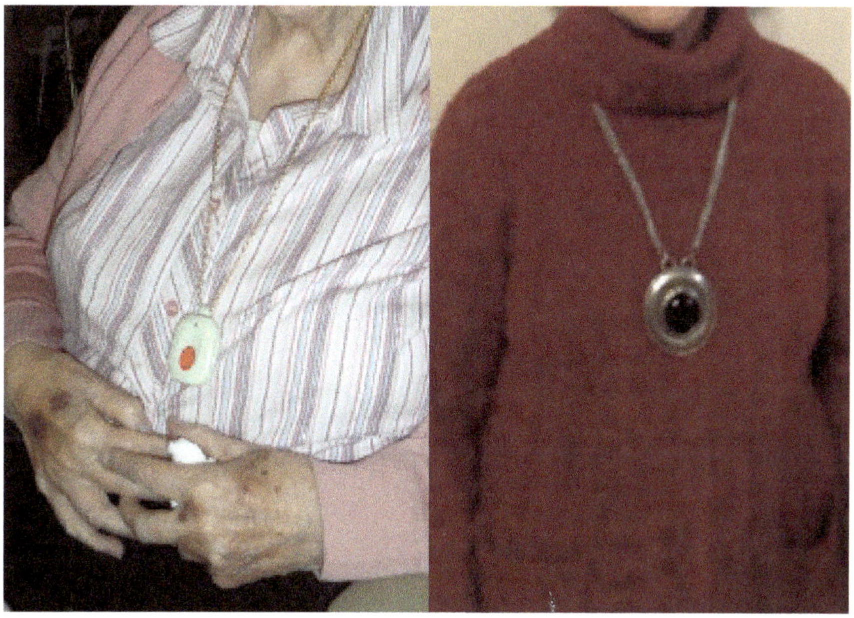

Figure 5. A user wearing an alarm pendant (left) and a workshop participant showing an item of her own jewellery as a pendant redesign idea (right)

We found that the animated scenarios helped us to better include older people in the decision and design processes and to validate some of the imagined everyday scenarios of use. The older adults' emotions and their motivations were key in this process. Animations are well suited to expressing such emotions. The storyboards were a good way to capture comments and ideas generated by these animations.

Phase 3: Evaluation

Probes are particularly suited to investigating people's everyday life in situations difficult to reach with traditional social science methods such as questionnaires, interviews, focus groups, or participant observations. Rather than relying on the presence and intervention of the researcher, probes are designed to encourage and empower subjects to collect data themselves (Arnold, 2004). The participants use probes to provide some insight, at their discretion, about their daily lives. Often, challenges and opportunities are only discovered when the technologies are used and evaluated with users in real-world settings (Doyle et al., 2010; Waycott et al., 2012). Personal information and story generation are two important benefits that we see here in the use of probes as artefacts contributing to users' point of view. Due to the logging functionality, technology probes ensure that participation of a user is visible and re-countable (Graham & Rouncefield, 2007).

The technology probe was seen as instance of the goal model and had logging capabilities (as is typical of technology probes) to monitor and record the use of the application. At the beginning of the field study, none of the researchers, designers, or older adults had a clear idea about how the final personal alarm system technology would look. It was particularly important to first engage the participants in simple technology use so they could confidently handle the interaction with their family members. Future design is thus grounded in a thorough understanding of users' experiences, requirements, and preferences (Lindsay et al.,2012).

Generally, technology probes can collect data about use to inform a better understanding, not so much about how to improve the technology but rather about actual needs in supporting specific activities (in our case activities evolving around building and maintaining interactions of older adults and their relatives to communicate wellbeing) (also see Hutchinson et al., 2003). Hence, technology probes are conducted prior to actual prototyping of the future system. In these latter phases, the users are acting as testers of the technology probe as well as collaborators influencing the evolving design as the functional and emotional goal models are refined through the supporting interview processes.

Prototype development: Technology probe for the wellbeing check

The technology probe for the wellbeing check was motivated by the goal model (see Figure 1) and facilitated the involvement with the user. The technology probe focused on the daily wellbeing check, rather than the pendant. We collaborated with a software company that followed an agile development approach. The emotional goals were communicated to the company and they defined their development goals in alignment with the emotional goals and mirrored their daily progress on each of them. The technology probe development and communication about alternatives was driven by the emotional goals. The technology shown in Figure 6 was developed and implemented in nine households.

The prototype in Figure 6 enabled relatives to send photos with captions. The user had could then scroll through photos and send messages back to their relatives. Only when no interaction takes place over a defined period of time does the app ask the user to indicate their wellbeing (e.g., "You haven't been in touch. Are you ok?"), and the user responds by pressing a button in the app. Thus, the wellbeing check in this prototype is the existence or absence of this "ping" as monitored by the backend systems of a service provider.

Figure 6. IPad with picture app used as base station for wellbeing check

Evaluation: Results from implementing the technology probe for the wellbeing check

Older people and relatives (nine older participants and five relatives) were interviewed after a four-week trial with the implemented technology probe for the wellbeing check. Overall, the participants liked the social and personalized aspect of the application and the feedback was positive. One relative commented:

"It's really fantastic. Because it's not masquerading or trying to pretend it's something that it's not. It's harnessing that activity, or harnessing that interaction to mine it for really useful data, so it's not [...] the teddy bear with the hidden camera in it monitoring what's going on in the room, it's not presenting in that way. It's very upfront." [Relative]

However, although one participant was happy with the social aspect, they were not comfortable with the monitoring aspect of the alarm and still felt that control was taken away from her.

"I wouldn't want to have any automatic checking on me. I want to be in control of whether someone is coming. I want to make a conscious decision. Last year, I had really high blood pressure and I went to bed and thought 'either I will wake up or not and that is fine'." [Older person]

The technology probe for the wellbeing check, coupled with interviews, enabled us to view the goal qualities in the light of the user activities.

However, it was very difficult for us to meet the expectations of some older people to truly feel "cared about". The following quote summarizes the different expectations:

"I think that that's the conflict because, for me as a relative and a carer, the assuredness was related to the functional aspects of the device, whereas for the user, their assurance isn't related to that at all. Their assurance is much more around the emotional ideas and that idea around the connectivity. And I think that was the clash, in that what I emotionally needed was very different to what my aunt emotionally needed.[...] and the reason that we implemented the system was... I mean to put it really bluntly was farming out a task." [Relative]

In our solution, we expect a certain commitment of relatives and carers to spend some time in communicating with the older person. It was difficult to find people in this trial that would send a photo to the older person every day. Although we try to meet the emotional goals of older people, we are aware that we rely on other people whose emotions or time allowance might not be in alignment with those of the older person.

The Emotion-Led Design ToolKit

Here, we summarize the use of the different emotion-led methods used in the three phases of the living lab process. The first phase of the process (exploration) begins with the designer conducting user research around a design problem or theme. Our design problem focused on personal alarm systems for older adults. Collections of insights are formed during this phase and then are translated into a goal model with a focus on emotions (Figure 7).

Emotions gathered during the exploration phase are represented in the goal model and are crucial in moving towards a designed outcome. It is important for researchers to evaluate these emotional goals to ensure that they reflect the true nature of the design problem. During the exploration phase, the researcher begins to create individual scenarios that show the emotional goals of the user and the functional goals of the system in context. It is presented to end users in the format of

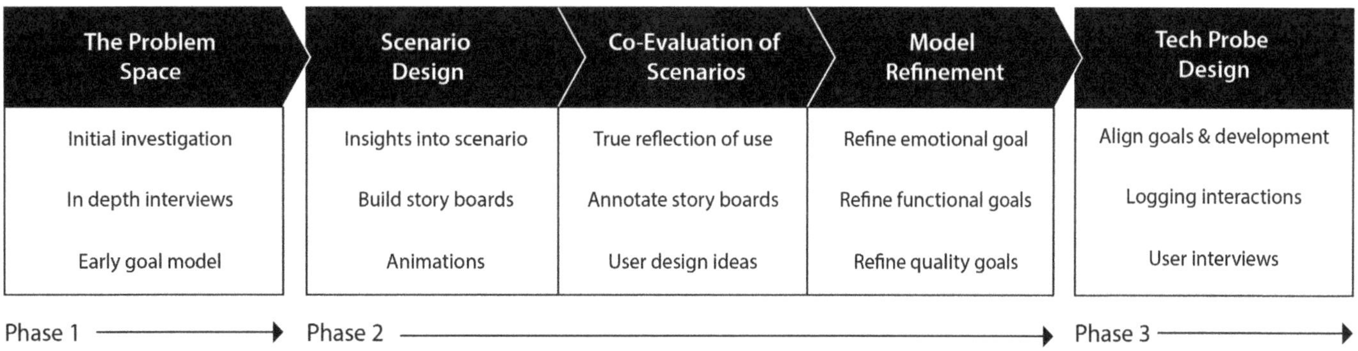

Figure 7. A process flow model illustrating the emotion-led design toolkit to supports the living lab process from exploration (Phase 1), to experimentation (Phase 2), and then to evaluation (Phase 3)

animations to co-evaluate initial user research from the conceptual phase to ensure the goals reflect the key concerns and emotions of the users that need to be addressed in the design solution. In our case, we focused on the older adults as the main users of the system. We suggest that it would be useful to repeat the co-evaluation with other main stakeholders, such as relatives and carers. Evaluation is the last phase of the working model. It is here, after the scenarios have been refined and themes and insights have been developed and evaluated with the user, that a designer can work towards a designed outcome.

Conclusion

In this case study, our approach conforms to an overall model that is typical of living labs. We take a similar iterative developmental approach from exploration, to experimentation, and then to the early stages of evaluation. Our users took on the typical living lab roles of informants, testers, contributors, and co-creators and were involved in multi-method approaches throughout the lifecycle. Our study proposes an emotion-led design toolkit that can be added to similar approaches in other living labs. This approach captures the emotional and quality goals of older adults (in this case) and translates them into actionable requirements that sit alongside functional goals. We show that motivational goal models are a suitable way to express field data derived from interviews – in particular emotions of users. These models are part of a development methodology and can be combined with scenarios to express user's emotions, motivations, and roles (Marshall, 2014; Sterling & Taveter, 2009), each of them describing and providing context of the domain. The goal models provide a place where abstract design concepts can be collected and represented (Pedell et al., 2009), but it will also be possible to evaluate final solutions against these goals. They are a lens through which use activities can be analyzed and recorded and then discussed among researchers and older adults. In reflection of our initial aim to give older adults a strong voice in the design process, the technology probes facilitated natural interactions between family members and yielded useful insights into how they used the newly designed system. Data gathered using technology probes are fragmentary and unstructured, which makes the process of translation from field data to abstract generalization for development difficult. A process of combining technology probe data collection and motivational goal models allowed us to talk about intangible outcomes with users that can be surprising, complex, and subtle.

To conclude, we emphasize the contribution of this research, which is a demonstration of how goal models, animations, and technology probes can be used to refine, link, and strengthen the transitions from different phases of a living lab process.

Acknowledgements

We would like to thank both the Smart Services CRC and the Australian Research Council (DP160104083) for partial funding of the research. We also extend a big thank you to all our participants for sharing their experiences and providing input into the design process.

About the Authors

Sonja Pedell is an Associate Professor in the Department of Communication Design and Digital Media Design and Director of the Future Self and Design Living Lab at Swinburne University of Technology in Melbourne, Australia, where she contributes extensive knowledge of human–computer interaction (HCI) to the co-creation of innovative technologies. Her research interests are user-centred design methods, scenario-based and mobile design, domestic technology development for health and wellbeing, and the design of engaging novel technologies for various user groups, in particular for the ageing population. Sonja holds a Master of Psychology degree from the Technical University of Berlin, Germany and for several years was employed in industry as an interaction designer, usability consultant, and product manager.

Alen Keirnan is a Post-Doctoral Research Fellow in the Centre for Design Innovation at Swinburne University of Technology in Melbourne, Australia, where he received his PhD in Industrial and Product Design. Working on a variety of health and ageing related projects in the Future Self and Design Living Lab, he has a strong interest in co-creation and journey-mapping techniques. He embeds his interests of co-creation and journey mapping into collaborative research projects between academia and industry, affording rich user insights appropriate for human-centered design outcomes. His current projects include developing services for retirement park managers to better communicate with their clients, envisioning the waiting room of the future and, evaluating technologies with older adults.

Gareth Priday is a foresight practitioner, researcher, and entrepreneur. He is Co-Director of the Australian Living Labs Innovation Network and recently supported the development of Swinburne University of Technology's Future Self and Design Living Lab in Melbourne. In 2014, Gareth led a Financial Resilience Living Lab pilot project and presented at the ENoLL Summer School. He held a futures research position with the Queensland University of Technology (Smart Services CRC). He has published in the *Journal of Futures Studies* and has presented at a number of Futures and Innovation conferences. Gareth holds a Master of Strategic Foresight degree from Swinburne University of Technology. His first career was in the financial services sector working for large international banks in the UK and Australia (UBS Warburg, Macquarie, ABN Amro, Royal Bank of Scotland) where he delivered on large-scale global projects.

Tim Miller is a Senior Lecturer in the Department of Computing and Information Systems at the University of Melbourne, Australia. He holds a PhD in Software Engineering from the University of Queensland and spent four years at the University of Liverpool, UK, as a Postdoctoral Researcher Associate in the Agent ART group. Tim's primary interests are in artificial intelligence and human–AI interaction.

Antonette Mendoza is a Lecturer in the Department of Computing and Information Systems at the University of Melbourne, Australia. Her research expertise includes how users interact and adopt technology; how systems can be better designed to support that interaction; and once deployed, what we can do to facilitate adoption and sustainability of technologies. She has extensive experience in software engineering, IT project management, and qualitative methods of research. She is currently collaborating with researchers on ARC and NHMRC projects in the health care and homelessness environments. She is also involved in local and international collaborations with researchers on value realization of e-learning platforms and tools. Her achievements include Teaching Excellence Awards in the Melbourne School of Engineering and in the Department of Computing and Information Systems.

Antonio Lopez-Lorca is a Lecturer at the Computing and Information Systems Department at the University of Melbourne, Australia. Antonio holds a PhD in Information Technology from the University of Wollongong, Australia. He first studied ways of applying semantic web techniques to agent-oriented models to validate them prior to software development. His research then shifted to looking at ways of introducing design processes into software engineering by considering the emotional needs of users throughout the complete lifecycle of the system. Much of his teaching effort focuses on supervising industry projects and developing the soft skills of software engineering students, particularly around communication with clients.

Leon Sterling received a BSc(Hons) from the University of Melbourne and a PhD in Pure Mathematics from the Australian National University. He has worked at universities in the UK, Israel, the US, and Australia. His teaching and research specialties are artificial intelligence, software engineering, and logic programming. Leon had a range of roles for 15 years at the University of Melbourne, including Professor of Software Innovation and Engineering and Director of e-Research. He served at Swinburne University of Technology as Dean of Information and Communication Technologies from 2010–2013 and as Vice Chancellor (Digital Frontiers) from 2014–2015. He also has served as President of the Australian Council of Deans of ICT.

References

Almirall, E., Lee, M., & Wareham, J. 2012. Mapping Living Labs in the Landscape of Innovation Methodologies. *Technology Innovation Management Review*, 2(9): 12–18. http://timreview.ca/article/603

Arnold, M. 2004. The Connected Home: Probing the Effects and Affects of Domesticated ICTs. In A. Bond, A. Clement, F. de Cindio, D. Schuler, & P. van den Besselar (Eds.), *Proceedings of Participatory Design Conference (PDC 2004):* 183–186. Pala Alto, Ca: CPSR.

Buxton, B. 2007. *Sketching User Experiences: Getting the Design Right and the Right Design.* San Francisco, CA: Morgan Kaufmann.

Carroll, J. M. & Rosson, M. B. 1992. Getting around the Task-Artifact Cycle: How to Make Claims and Design by Scenario. *ACM Transactions on Information Systems (TOIS)*, 10(2): 181–212. http://doi.org/10.1145/146802.146834

Dell´Era, C., & Landoni, P. 2014. Living Lab: A Methodology between User-Centred Design and Participatory Design. *Creativity and Innovation Management,* 23(2): 137–154. http://dx.doi.org/10.1111/caim.12061

Doyle, J., Skrba, Z., McDonnell, R., & Arent, B. 2010. Designing a Touch Screen Communication Device to Support Social Interaction amongst Older Adults. In *Proceedings of the 24th BCS Interaction Specialist Group Conference:* 177–185. Swindon, UK: BCS Learning & Development Ltd.

Edlin-White, R., Cobb, S., Floyde, A., Lewthwaite, S., Wang, J., & Riedel, J. 2012. From Guinea Pigs to Design Partners: Working with Older People in ICT Design. In P. Langdon, J. Clarkson, P. Robinson, J. Lazar., & A. Heylighen (Eds.), *Designing Inclusive Systems:* 155–164. London: Springer-Verlag. http://dx.doi.org/10.1007/978-1-4471-2867-0_16

Eskelinen, J., Robles, A., Lindy, I., Marsh, J., & Muente-Kunigami, A. 2015. *Citizen-Driven Innovation: A Guidebook for City Mayors and Public Administrators.* World Bank and European Network of Living Labs.

Graham, C., Rouncefield, M., Gibbs, M., Vetere, F., & Cheverst, K. 2007. How Probes Work. In *Proceedings of the 19th Australasian Conference on Computer-Human Interaction: Entertaining User Interfaces:* 27–37. New York: ACM. http://doi.org/10.1145/1324892.1324899

Gray, M., Mangyoku, M., Serra, A., Sánchez, L., & Aragall, F. 2014. Integrating Design for All in Living Labs. *Technology Innovation Management Review,* 4(5): 50–59. http://timreview.ca/article/793

Hutchinson, H., Mackay, W., Westerlund, B., Bederson, B. B., Druin, A., Plaisant, C., Beaudouin-Lafon, M., Conversy, S., Evans, H., Hansen, H., Roussel, N., & Eiderbäck, B. 2003. Technology Probes: Inspiring Design for and with Families. In *Proceedings of the SIGCHI conference on Human factors in Computing Systems:* 17–24. New York: ACM. https://doi.org/10.1145/642611.642616

Iacucci, G., Iacucci, C., & Kuutti, K. 2002. Imagining and Experiencing in Design: The Role of Performances. In *Proceedings of the 2nd Nordic Conference on Human-Computer Interaction:* 167–176. New York: ACM.

Leminen, S. 2015. Q&A. What Are Living Labs? *Technology Innovation Management Review,* 5(9): 29–35. http://timreview.ca/article/928

Leminen, S., Nyström, A.-G., & Westerlund, M. 2015. A Typology of Creative Consumers in Living Labs. *Journal of Engineering and Technology Management,* 37: 6–20. http://dx.doi.org/10.1016/j.jengtecman.2015.08.008

Lim, Y.-K., & Sato, K. 2006. Describing Multiple Aspects of Use Situation: Applications of Design Information Framework (DIF) to Scenario Development. *Design Studies,* 27(1): 57-76. http://dx.doi.org/10.1016/j.destud.2005.04.004

Lindsay, S., Jackson, D., Schofield, G., & Olivier, P. 2012. Engaging Older People Using Participatory Design. In *Proceedings of the SIGCHI Conference on Human Factors in Computing Systems:* 1199–1208. New York: ACM. http://doi.org/10.1145/2207676.2208570

Lo Bianco, M., Pedell, S., Renda, G., & Kapoor, A. 2015. HCI Methods for Empowering Discussion on Person-Centered Fall Prevention with Older Adults. In *Proceedings of the Annual Meeting of the Australian Special Interest Group for Computer Human Interaction:* 255–263. New York: ACM. http://doi.org/10.1145/2838739.2838767

Loke, L., Robertson, T., & Mansfield, T. 2005. Moving Bodies, Social Selves: Movement-Oriented Personas and Scenarios. In *Proceedings of the 17th Australia Conference on Computer-Human Interaction: Citizens Online: Considerations for Today and the Future.* Canberra, Australia: Computer-Human Interaction Special Interest Group (CHISIG) of Australia.

Marshall, J. 2014. Agent-Based Modelling of Emotional Goals in Digital Media Design Projects. *International Journal of People-Oriented Programming (IJPOP),* 3(1): 44–59. http://dx.doi.org/10.4018/ijpop.2014010103

Massimi, M., Baecker, R. M., & Wu, M. 2007. Using Participatory Activities with Seniors to Critique, Build, and Evaluate Mobile Phones. In *Proceedings of the 9th International ACM SIGACCESS Conference on Computers and Accessibility:* 155–162. New York: ACM. http://doi.org/10.1145/1296843.1296871

Matthews, B., & Heinemann, T. 2012. Analysing Conversation: Studying Design as Social Action. *Design Studies,* 33(6): 649-672. http://dx.doi.org/10.1016/j.destud.2012.06.008

Miller, T., Pedell, S., Lopez-Lorca, A. A., Mendoza, A., Sterling, L., & Keirnan, A. 2015. Emotion-Led Modelling for People-Oriented Requirements Engineering: The Case Study of Emergency Systems. *Journal of Systems and Software,* 105(C): 54–71. https://doi.org/10.1016/j.jss.2015.03.044

Muller, M. J. 2008. *Participatory Design: The Third Space in Human-Computer Interaction: Development Process.* Mahway, NJ: Erlbaum.

Patton, M. Q. 2002. *Qualitative Evaluation and Research Methods.* Thousand Oaks, CA: Sage Publications, Inc.

Pedell, S., Miller, T., Vetere, F., Sterling, L., Howard, S., & Paay, J. 2009. Having Fun at Home: Interleaving Fieldwork and Goal Models. In *Proceedings of the 21st Annual Conference of the Australian Computer-Human Interaction Special Interest Group: Design: Open 24/7:* 309–312. New York: ACM. http://doi.org/10.1145/1738826.1738882

Pedell, S., Lopez-Lorca, A. A., Miller, T., & Sterling, L. 2015. Don't Leave Me Untouched: Considering Emotions in Personal Alarm Use and Development. In M. Tavana, A. Ghapanchi, & A. Talaei-Khoei (Eds.), *Healthcare Informatics and Analytics: Emerging Issues and Trends:* 96–127. Hershey, PA: IGI Global. http://dx.doi.org/10.4018/978-1-4666-6316-9

Robertson, T., & Simonsen, J. 2012. Participatory Design: An Introduction. In J. Simonsen & T. Robertson (Eds.), *Routledge International Handbook of Participatory Design:* 1-18. London: Routledge.

Schuurman, D., De Marez, L., & Ballon, P. 2016. The Impact of Living Lab Methodology on Open Innovation Contributions and Outcomes. *Technology Innovation Management Review,* 1(6): 7–16. http://timreview.ca/article/956

Simon, H. A. 1982. *The Sciences of the Artificial* (2nd edition). Cambridge, MA: MIT press.

Ståhlbröst, A., & Holst, M. 2013. *The Living Lab Methodology Handbook.* Luleå, Sweden: Luleå University of Technology and Centre for Distance-Spanning Technology (CDT).

Sterling, L., & Taveter, K. 2009. *The Art of Agent-Oriented Modeling.* Cambridge, MA: MIT Press.

Veeckman, C., Schuurman, D., Leminen, S., & Westerlund, M. 2013. Linking Living Lab Characteristics and Their Outcomes: Towards a Conceptual Framework. *Technology Innovation Management Review,* 3(12): 6-15. http://timreview.ca/article/748

Vines, J., Blythe, M., Lindsay, S., Dunphy, P., Monk, A., & Olivier, P. 2012. Questionable Concepts: Critique as Resource for Designing with Eighty Somethings. In *Proceedings of the SIGCHI Conference on Human Factors in Computing Systems:* 1169–1178. New York: ACM. http://doi.org/10.1145/2207676.2208567

Waycott, J., Pedell, S., Vetere, F., Ozanne, E., Kulik, L., Gruner, A., & Downs, J. 2012. Actively Engaging Older Adults in the Development and Evaluation of Tablet Technology. In *Proceedings of the 24th Australian Computer-Human Interaction Conference:* 643–652. New York: ACM. http://doi.org/10.1145/2414536.2414633

Waycott, J., Vetere, F., Pedell, S., Kulik, L., Ozanne, E., Gruner, A., & Downs, J. 2013. Older Adults as Digital Content Producers. In *Proceedings of the SIGCHI Conference on Human Factors in Computing Systems:* 38–48. New York: ACM. http://doi.org/10.1145/2470654.2470662

Keywords: co-design methods, emotion-led design, expressing emotions, living lab, personal alarm systems, aging well

11

Needsfinding in Living Labs: A Structured Research Approach

Louise Savelkoul and Murk Peutz

> *"Cycling is possibly the greatest and most pleasurable form of transport ever invented. It's like walking only with one-tenth of the effort. Ride through a city and you can understand its geography in a way that no motorist, contained by one-way signs and traffic jams, will ever be able to. You can whiz from one side to the other in minutes. You can overtake $250,000 sports cars that are going nowhere fast. You can park pretty much anywhere. It truly is one of the greatest feelings of freedom one can have in a metropolitan environment. It's amazing you can feel this free in a modern city."*
>
> Daniel Pemberton
> In *The Book of Idle Pleasures*

Living labs enable innovations to be facilitated and implemented quickly and efficiently. A key element of the living lab approach is the active involvement of users. In this article, we examine a structured needsfinding phase of a living lab infrastructure project within the context of bicycle commuting. Given that effectuation costs are high, it is essential for the lab to focus on tackling the right user needs. Thus, the living lab's needsfinding phase aims to identify user needs and wants, as measured by bicycle commuting intention. We examined intention in a structured way by following the theory of planned behaviour. The results show that bicycle commuting intention can be explained by the variables of our model (R square=0.808). The specific insights arising from the needsfinding phase are an important focus for the activities and experiments in the later phases of the living lab. The generalized insights are also relevant to innovation experts outside the area of cycling.

Introduction

Promoting cycling is increasingly considered as a solution for complex mobility problems. However, there is relatively scarce research available about the motivations of bicycle commuters and their needs. Cycling differs from other means of transport and therefore existing large bodies of scientific literature can only be used indirectly (Harms et al., 2014; Wardman et al., 1997). For example, weather and physical effort have more impact on cyclists than on other means of transport. However, a large part of the decision to cycle can be explained by personal factors, allowing the use of more commonly used psychological models (Titze et al., 2008).

This article shares insights from a living lab in the Netherlands that focused on the personal factors influencing bicycle commuting intention. Partners from academia, government, and business wished to identify and understand the needs and requirements of cyclists and potential cyclists. A particular focus was "fast cycling routes", which are routes built to solve the problem of traffic congestion and as an opportunity for organizations to encourage more employees to commute by bike. For this reason, the living lab was called the Living Lab Fast Cycling Routes. A related topic of interest was the availability of electric bicycles (e-bikes), which contribute to this development by expanding the practical commuting range, enabling even more people to consider commuting to work by bicycle.

This article elaborates on the first phase of the living lab: the needsfinding phase. The needsfinding process is developed to "frame" the needs, goals, and values of (prospective) bicycle commuters, their employers, and other stakeholders. This article focuses on the needsfinding phase because it may yield interesting outcomes for the other phases of the living lab and for other living labs in general.

The living lab approach has its origins in the experimentation with technology together with users (Niitamo et al., 2006). Whereas it is quite easy to change information quickly and cheaply, this is not the case for infrastructure projects. Furthermore, transport infrastructure projects often do not perform as promised; cost escalations appear to be the rule rather than the exception (Flyvbjerg et al., 2003). The structured focus in the beginning of the living lab is needed because of the high investment costs for this infrastructure project. Outcomes from a freeform living lab method could easily become too expensive to implement in terms of the required infrastructure. By using a structured method, costs are kept to the minimum as the users' wants and needs are measured in a structured way before the roads are being built. Hence, the users' needs and wants are implemented right away and roads do not have to be adjusted at high cost after construction.

The way to integrate this structured approach into the user involvement of the living lab is by a combination of a questionnaire, interviews, and focus group activities. Structuring the needsfinding process in this way makes it easier to respond to the needs of the users in later phases and contributes to a successful living lab. This approach is in line with the lean startup methodology, which suggests that, by validating hypotheses of customer's problems, startups find a solution that indicates there is business potential in solving the problem (Hokkanen et al., 2016). In this living lab, customer's needs and wants are also incorporated before the project/product has its final form.

The needsfinding phase has several steps that follow a logical sequence:

1. Identify technology, incentives, methods, and facilities through survey questionnaires, exploratory interviews, literature reviews, and research at the bicycle highway.

2. Triangulate the data from the questionnaires with other studies.

3. Prioritize the user problems on the basis of effects and costs in later phases.

4. Strengthen selected priorities with interviews and expert sessions.

In this article, we describe how this needsfinding phase was grounded in the theory of planned behaviour to predict bicycle commuting intention in the living lab. In the next section, we outline the theory and show how it was applied to the living lab case. Then, we present our specific methodology and results. Finally, we conclude with a discussion of these results and explore avenues for future research.

The Theory of Planned Behaviour

The theory of planned behaviour has been widely applied for predicting behavioural intention. In applying this theory, the personal motivational factors of individuals are used as determinants for the likelihood of expressing a specific behaviour (Glanz et al., 2008). Thus, the theory is useful in situations where it is not possible or practical to measure actual behaviour. It has strong predictive power, meaning that behavioural intention predicts actual behaviour (Ajzen, 1991; Armitage & Conner, 2001; Conner & Armitage, 1998; Sheppard et al., 1988). Although the theory is often used to predict health-related behaviour and exercise behaviour (Smith et al., 2007), it can be used for other types of behaviour as well.

Behavioural intention can be seen as the willingness of a person to perform a certain behaviour (Fishbein & Ajzen, 2011). The underlying idea is that behavioural intention encompasses the subjective probability that a person will perform a certain behaviour (Ajzen, 1991). In the current study, the theory of planned behaviour was selected to measure bicycle commuting intention. A positive bicycle commuting intention means that a commuter intends to cycle to work instead of using another means of transport.

The theory of planned behaviour comprises three independent variables that can explain the dependent variable of behavioural intention: i) attitude, ii) subjective norms, and iii) perceived behavioural control. These variables are related to each other in explaining behavioural intention (Fishbein & Ajzen, 2011), but depending on the case, only one or two variables may be important (Ajzen, 1991). Figure 1 describes the model and shows the relationships between the three variables, which are described in greater detail below, including how they apply to the context of bicycle commuting intention:

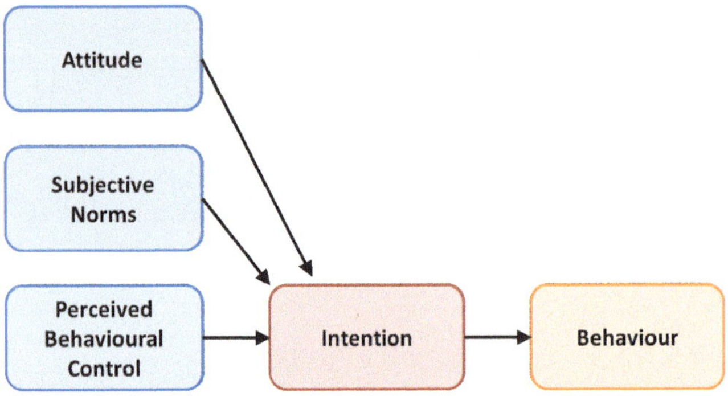

Figure 1. General model underpinning the theory of planned behaviour (Ajzen, 1991)

1. *Attitude:* this variable encompasses an individual's idea of the consequences of the behaviour and its importance to the individual (Ajzen, 1991; Fishbein & Ajzen, 2011). In the Netherlands, the attitudes regarding cycling are relatively positive compared to other nations: cycling is part of the Dutch way of living (Harms et al., 2014). A positive cycling attitude increases the likelihood of commuting by bicycle (Dill & Voros, 2008), and given that cycling levels are so high in the Netherlands, we expect that this positive attitude will lead to a high bicycle commuting intention.

2. *Subjective norms:* this variable refers to social pressure applied by others (Ajzen, 1991). It differs from attitude in that it focuses on the perception of pressure from the (social) environment instead of the personal perception of the behaviour (Glanz et al., 2008). Within the context of cycling, it is expected that bicycle commuting intention will be determined by an individual's family, friends and colleagues as they create a norm for cycling behaviour (Terry & Hogg, 1996). For this reason, we expect that high subjective norms will lead to high bicycle commuting intention. Respondents were also asked how important the pressure from family/friends and colleagues was for them.

3. *Perceived behavioural control:* this variable refers to the belief that a person has control over a specific behaviour (Francis et al., 2004). A difference can be made between internal and external control (Manstead & Eekelen, 1998), where internal control reflects the skills and capabilities of individuals (e.g., motivation and ability) and external control focuses on the available resources and difficulty of the task (Manstead & Eekelen, 1998; Terry, 1993). We expect that greater perceived behavioural control will lead to higher behavioural intention.

Added variables to account for contextual factors

Although the theory of planned behaviour is a valid and often used method for explaining behavioural intention, it only includes personal factors (Glanz et al., 2008). Within the living lab under study here, the user needs and wants are a central element. It is therefore necessary for our model to also include variables relating to the employees' organizational context. In the context of cycling, contextual factors play a large role. For example, Buehler (2012) has demonstrated that bicycle parking, cyclist showers and free car parking are determinants of cycling to work. It seems that the bicycle infrastructure of organizations and the bicycling incentives play a facilitating role in behavioural intention (Chatterjee et al., 2013). Accordingly, we added two variables to the conceptual model, as shown in Figure 2:

4. *Facilitating conditions:* Perceived facilitating conditions refer to individual beliefs about infrastructures that can remove barriers to the use of a system or can support its use (Venkatesh et al., 2008; Venkatesh et al., 2003). Therefore, it is relevant to consider employees' *perceptions* of the facilities available at their workplaces rather than simply documenting what facilities are actually available. In particular, we expect that bicycle facilitating conditions will have a positive effect on bicycle commuting intention.

5. *Incentive systems:* This variable can be described as the variety of incentives that organizations offer to their employees, such as financial incentives for buying or riding a bicycle (e.g., compensation for every kilometre cycled) (Heinen, 2011; Knoke, 1988). The growing interest in financial incentives for promoting cycling reflects findings that incentives are effective, even to the extent that they can almost double cycling levels (Dickinson et al., 2003; Handy & Xing, 2011; Wardman et al., 1997). For this reason, we expect bicycling-related incentives to have a positive influence on bicycle commuting intention.

Methodology

Bicycle commuting intention was measured by investigating three organizations in the vicinity of a "fast cycling route" in the Netherlands. The participating organizations were selected on the basis of their location but have different characteristics (public and private) and are from different sectors (healthcare, education, and industry). Employees of these organizations completed a questionnaire by which we measured their bicycle commuting intention. The study had a cross-sectional design. Before the actual questionnaire was

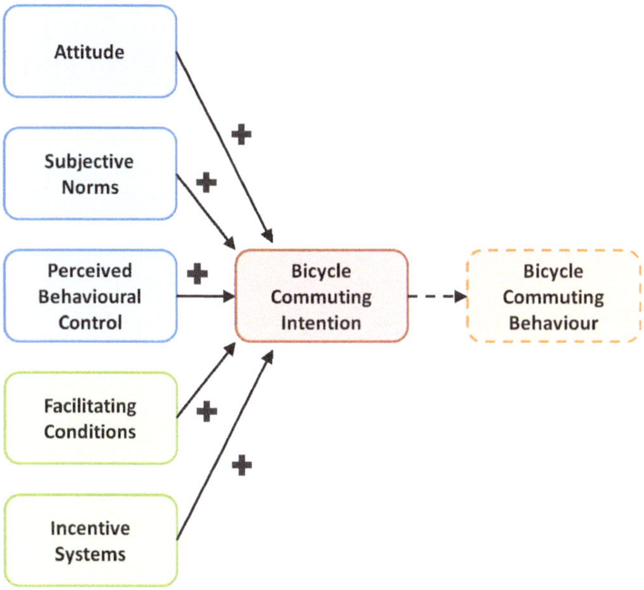

Figure 2. Conceptual model with added variables

sent to the participating organizations, it was sent to a pre-test organization and the feedback was incorporated into the final questionnaire. Only employees who lived within 30 kilometres of their workplace received the questionnaire. This distance was chosen to reflect a maximum commuting distance that would be practical by (electric) bicycle.

The gathering of data was part of the first step in the needsfinding, as described earlier. The bounded population included 3998 employees from the organizations under study. It was not necessary to do further sampling because everyone within the distance of 30 kilometres could fill in the questionnaire (convenience sampling). The questionnaire was based on literature sources measuring the variables of the theory of planned behaviour and the contextual variables and from questionnaires designed to evaluate bicycle commuting. A unipolar scale with five response categories was used because it has been shown to be valid for measuring behavioural intention (Thomas et al., 2004).

Data measurement and analysis

To measure bicycle commuting intention, multiple questions should be included (Ajzen, 2002). We included questions that followed Gibbons' (2005) splitting of behavioural intention into subdimensions to better predict actual behaviour: implementation intentions (planning behaviour), behavioural expectations (predicting behaviour), and behavioural willingness (openness to risk).

Attitude was measured according to the sub-items "direct benefit", "awareness", and "safety". To ensure validity, confirmatory principal component analysis (PCA) was performed on the nine items of attitude. The pattern matrix showed that the component "safety" was part of the component "direct benefit" and was therefore subsequently combined within it. This led to the following distribution of items: i) direct benefit including comfortability, lifestyle, pleasant/nice, time saving, and safety and ii) awareness including health benefits, cost saving, environmental benefits, and pleasantness.

Subjective norms were measured separately by normative beliefs and importance. PCA showed that the components differed from Heinen (2011). But, because the measurement of Heinen (2011) had a clear distribution (colleagues, family and friends), her distribution was used.

Perceived behavioural control was measured by two internal control items (confidence and trust) and one external control variable (constraint).

Facilitating conditions included questions about bicycle facilities and other facilities. The participating organizations' incentive systems were measured by financial incentives; other incentives were excluded due to insufficient correlation.

The following control variables were taken into account: gender, age, working hours, physical condition, car ownership, household structure, distance to work, luggage, frequency of past behaviour, and organization. Dummy variables were made for the control variables household structure (household structure 3 is the reference category), time block (multiple answers possible), and organization (company A is the reference category).

Missing values and outliers were checked and negatively formulated items were reversed. Thereafter, PCA was performed along with descriptive statistical analyses. Next, multiple hierarchical regression analyses were performed (see Appendix 1). For reasons of brevity, the control variables are presented only once without independent variables (model 1). First, the variables associated with the theory of planned behaviour were measured in total (model 2). Thereafter, the variables were separately measured (models 3, 4, and 5). Then, the contextual variables were measured (models 6 and 7). Model 8 includes all the independent variables. The additional multiple regressions for the three variables are included in models 9, 10, and 11.

Results

Multiple regression and hypothesis testing

We received 647 successfully completed questionnaires, which corresponds to a 16% response rate. Missing values were excluded pairwise. There were no striking outliers. Checks were carried out to ensure a sample size large enough for multicollinearity and singularity, distribution normality, linearity, homoscedasticity, and outliers. All the assumptions were met. Only the Mahalonobis distance was exceeded when outliers were checked. However, because the Cook distance was less than 1, there was no need to delete cases. Therefore, multiple regressions were appropriate.

The model including all the variables had an R square of 0.808 which is extremely high for testing the model of the theory of planned behaviour. Even without the contextual variables, the R square was still 0.807. Meta-analyses have shown that the explaining variance of behavioural intention by use of the theory of planned behaviour normally lies between 40% and 50% (Sutton, 1998). The high explaining variance means that this theory is a valid way to measure bicycle commuting intention.

The standardized coefficients were used to compare the contributing variables of bicycle commuting intention.

Variables

Appendix 1 shows the findings of the multiple regression analyses, which are summarized in Table 1.

Attitude had a statistically significant impact on bicycle commuting intention F Change (2.567)=49.203, p<.001. The dimensions of attitude were also tested. The dimension "awareness" did not have a significant effect on bicycle commuting intention, whereas the dimension "direct benefit" had a significant beta value (0.298, p<0.001).

The findings show that, for the variable *subjective norms,* only family had a significant contribution to bicycle commuting intention (bfamily=0.231, p<0.001). Importance of subjective norms did not have a significant contribution to bicycle commuting intention.

Perceived behavioural control had a significant impact on bicycle commuting intention. The contribution of perceived behavioural control is large in comparison to the other two variables.

For *facilitating conditions,* the R square changes from 0.713 to 0.714 (see model 6) but the change is not significant F change (2.595)=0.683, p>0.05).

Concerning *incentive systems,* only financial incentives of the organization were incorporated in the multiple regression. Financial incentives explained an additional two percent of variance in bicycle commuting intention but not a significant level F change (1.596)=3.240, p>0.05).

The control variable *past behaviour* had a significant influence on bicycle commuting intention in all models.

Table 1. Observed effects of variables and sub-items on bicycle commuting intention

Variable	Observed Effect	Significance	Conclusion
Attitude	Positive	Direct benefit significant, awareness not significant	Partially confirmed
Subjective norms	Positive	Subjective norms significant, importance not significant	Partially confirmed
Perceived behavioural control	Positive	Significant	Confirmed
Facilitating conditions	Negative	Not significant	Rejected
Incentive systems	Positive	Not Significant	Rejected
Past behaviour	Positive	Significant	Control variable

Discussion and Conclusion

The ultimate goal of the living lab "Fast Cycling Routes" is to encourage cycling among employees of organizations in close proximity of fast cycling routes. This goal is aligned with the intention of the Dutch government's to build 675 kilometres of fast cycling routes by 2025. But, enabling new physical infrastructure to be built will by itself not lead to more cycling. The intention to use it must be understood and addressed.

This research demonstrates that applying a structured approach as part of the needsfinding phase is promising. This research confirms by its high explaining variance that the model derived from the theory of planned behaviour is a valid way to research behavioural intention (R square = 0.808). The study builds upon earlier work regarding bicycle commuting intention and extends it with a combination of personal and contextual factors. Earlier work concentrated on either personal factors or contextual factors. Therefore, a more comprehensive overview of bicycle commuting intention is achieved by combining these factors into one study. One explanation for why the results for facilitating conditions and financial bicycle incentives (the contextual variables) were not significant could be because perceived conditions were measured instead of actual facilities and incentives. Future applied research could elaborate further on these results to understand if this explanation is appropriate.

Furthermore, this research shows that for the development of fast cycling routes, a personal or psychographic factor should be included (e.g., perceived behavioural control, subjective norms of the family, or the attitude dimension direct benefit). This means that the intrinsic drivers of bicycle commuting intention should receive more attention because they explain a large part of the intention to commute by bicycle. This is especially valid when the variables are split up into sub-items.

This finding led to input for the following steps in the needsfinding phase. The interviews with stakeholders were more focused on psychographic elements than would have been the case without the findings of this study. Also, the focus sessions with experts addressed the behavioural side of cycling. Without these results, the behavioural side would not have been discussed. In this way, it helped the researchers to avoid the themes for which the effect on behaviour were not significant but provided living lab actors with proven themes for brainstorming in the needsfinding phase.

Moreover, by mapping the needs of the end users, a better plan can be made for the later phases in the living lab. The researchers now know that prototypes of innovations have to include behavioural components. For example, knowing that the family plays an important role in the bicycle commuting intention, researchers can ensure that they involve families in the research. These findings give the opportunity to test this effect in a later phase (after the needsfinding phase). For example, by including families of employees in an experiment, the lab can assess how the family can stimulate a family member to commute to work by bicycle.

Another measurement in the test phase may consist of promotion campaigns that are focused on direct benefit, as this seemed to have a significant influence on bicycle commuting intention. Often in the Netherlands, promotion campaigns about cycling are focused on safety. However, this research shows that the items of direct benefit have to be emphasized in order to stimulate bicycle commuting intention. These items concern: the comfortability of the route, the pleasure derived from riding a route, and how easy it can be to fit cycling into most lifestyles. Given that these elements have a significant influence on bicycle commuting intention for potential cyclists, communication geared towards these elements would lead to a higher bicycle commuting intention. However, in the Netherlands, home and work are not fully integrated. Therefore, results from country to country may differ and promotion campaigns should be designed accordingly. For this living lab in the Netherlands, emphasis will be placed on providing more direct benefit in campaigns.

Furthermore, this research has also clarified what type of people (users) should be selected for the subsequent phase in the living lab. On the one hand, the control variable past behaviour had a direct effect on bicycle commuting intention and can therefore be seen as an additional independent variable. It is therefore important for employers to give their employees a good experience when they cycle to work for the first time, because the intention to cycle more often is directly dependent on (positive) past behaviour. On the other hand, the variable with the largest effect on bicycle commuting intention is perceived behavioural control. This variable had an interaction effect with working distance. The influence of working distance decreases if employees are provided with more control in cycling to work.

Both findings can be tested in a new phase of the living lab. One way to test this is by introducing employees to an electric bicycle and test whether their perceived behavioural control changes over time. Also, introducing the electric bicycle to (new) employees would make commuting easier (past behaviour). Alternatively, by using an electric bicycle, the time of arrival at work becomes more predictable as compared to public transport, which also contributes to perceived behavioural control.

All in all, this research provides insights in how to tackle the next phases in the living lab. The findings give guidance in an unstructured process of user requirements and needs. Accordingly, topics are being discussed in conversations with users that would not have been discussed without these findings. The findings show that changes should be made in the regular approach of cycle programs offered by governmental institutions and employers to encourage employees to cycle to work. Not only contextual factors are making a difference in the intention to cycle to work. In fact, this research shows that personal factors are particularly important. For this reason, organizations should reevaluate their strong focus on contextual interventions and put more focus on personal factors in their bicycle policy programs and living labs. Future researchers should keep an eye on the process of a living lab, and provide a structure to measure the stakeholders' findings, for example by dividing the living lab into phases.

About the Authors

Louise Savelkoul is a Consultant at Equator Research and is a PhD candidate studying user innovation at Tilburg University in the Netherlands, where she also holds Bachelor and MSc degrees in Organizational Studies. At Equator Research, her focus is on Living Labs. She is currently Lab Manager of the regional cycle lanes initiative in the Netherlands, enabling new innovations in infrastructure and the built environment. Recently she co-developed the flexible living for health care lab in the Southern Netherlands.

Murk Peutz is Director of Equator Research, a consulting firm focused on innovation management and the use of living labs as an effective tool for cocreation and collaborative innovation. He graduated from Delft University with a degree in Mechanical Engineering, and he holds a Business Law degree from Leiden University and an MBA from INSEAD. He has also worked in industry (Tate & Lyle PLC) and management consulting (The Boston Consulting Group). In 2004, he took up responsibility for Innovation Consulting to Small and Medium Enterprises as Director of the Syntens Foundation before founding Equator Research in 2013. Murk is also a non-executive director of several companies in the Netherlands.

References

Ajzen, I. 1991. The Theory of Planned Behavior. *Organizational Behavior and Human Decision Processes*, 50: 179–211. http://dx.doi.org/10.1016/0749-5978(91)90020-T

Ajzen, I. 2002. *Constructing a TpB Questionnaire: Conceptual and Methodological Considerations.* Amherst, MA: University of Massachusetts.

Armitage, C. J., & Conner, M. 2001. Efficacy of the Theory of Planned Behaviour: A Meta-Analytic Review. *British Journal of Social Psychology*, 40(4): 471–499. http://dx.doi.org/10.1348/014466601164939

Buehler, R. 2012. Determinants of Bicycle Commuting in the Washington, DC Region: The Role of Bicycle Parking, Cyclist Showers, and Free Car Parking at Work. *Transportation Research Part D: Transport and Environment*, 17(7): 525–531. http://dx.doi.org/10.1016/j.trd.2012.06.003

Chatterjee, K., Sherwin, H., & Jain, J. 2013. Triggers for Changes in Cycling: The Role of Life Events and Modifications to the External Environment. *Journal of Transport Geography*, 30: 183–193. http://dx.doi.org/10.1016/j.jtrangeo.2013.02.007

Conner, M., & Armitage, C. J. 1998. Extending the Theory of Planned Behavior: A Review and Avenues for Further Research. *Journal of Applied Social Psychology*, 28(15): 1429–1464. http://dx.doi.org/10.1111/j.1559-1816.1998.tb01685.x

Dickinson, J. E., Kingham, S., Copsey, S., & Pearlman Hougie, D. J. 2003. Employer Travel Plans, Cycling and Gender: Will Travel Plan Measures Improve the Outlook for Cycling to Work in the UK? *Transportation Research Part D: Transport and Environment*, 8(1): 53–67. http://dx.doi.org/10.1016/S1361-9209(02)00018-4

Dill, J., & Voros, K. 2008. Factors Affecting Bicycling Demand: Initial Survey Findings from the Portland, Oregon, Region. *Transportation Research Record*, 2031(1): 9–17. http://dx.doi.org/10.3141/2031-02

Fishbein, M., & Ajzen, I. 2011. *Predicting and Changing Behavior: The Reasoned Action Approach.* New York: Taylor & Francis.

Flyvbjerg, B., Skamris holm, M. K., & Buhl, S. L. 2003. How Common and How Large Are Cost Overruns in Transport Infrastructure Projects? *Transport Reviews*, 23(1): 71–88. http://dx.doi.org/10.1080/01441640309904

Francis, J., Eccles, M. P., Johnston, M., Walker, A. E., Grimshaw, J. M., Foy, R., Kaner, E. F. S., Smith, L., & Bonetti, D. 2004. *Constructing Questionnaires Based on the Theory of Planned Behaviour: A Manual for Health Services Researchers.* Newcastle upon Tyne: University of Newcastle.

Gibbons. F. X. 2005. *Behavioral Intentions, Expectations and Willingness.* Bethesda, MD: National Cancer Institute, Division of Cancer Control & Population Sciences. http://cancercontrol.cancer.gov/brp/research/constructs/intent-expect-willingness.html

Glanz, K., Rimer, B. K., & Viswanath, K. (Eds.). 2008. *Health Behavior and Health Education: Theory, Research, and Practice.* San Francisco, CA: Jossey-Bass

Handy, S. L., & Xing, Y. 2011. Factors Correlated with Bicycle Commuting: A Study in Six Small U.S. Cities. *International Journal of Sustainable Transportation,* 5(2): 91–110. http://dx.doi.org/10.1080/15568310903514789

Harms, L., Bertolini, L., & te Brömmelstroet, M. 2014. Spatial and Social Variations in Cycling Patterns in a Mature Cycling Country Exploring Differences and Trends. *Journal of Transport and Health,* 1(4): 232–242. http://dx.doi.org/10.1016/j.jth.2014.09.012

Heinen, E. 2011. *Bicycle Commuting.* Delft, Netherlands: Delft University Press.

Hokkanen, L., Kuusinen, K., & Väänänen, K. 2016. Minimum Viable User EXperience: A Framework for Supporting Product Design in Startups. In H. Sharp & T. Hall (Eds.), *Agile Processes, in Software Engineering, and Extreme Programming. XP 2016. Lecture Notes in Business Information Processing,* 251: 66–78. Cham, Switzerland: Springer. http://dx.doi.org/10.1007/978-3-319-33515-5_6

Knoke, D. 1988. Incentives in Collective Action Organizations. *American Sociological Review,* 53(3): 311–329. http://www.jstor.org/stable/2095641

Manstead, A. S. R., & Eekelen, S. A. M. 1998. Distinguishing Between Perceived Behavioral Control and Self-Efficacy in the Domain of Academic Achievement Intentions and Behaviors. *Journal of Applied Social Psychology,* 28(15): 1375–1392. http://dx.doi.org/10.1111/j.1559-1816.1998.tb01682.x

Niitamo, V.-P., Kulkki, S., Eriksson, M., & Hribernik, K. A. 2006 Stateof-the-Art and Good Practice in the Field of Living Labs. In K. S. Thoben, Pawar, M. Taisch, & S. Terzi (Eds.). *Proceedings of the 12th International Conference on Concurrent Enterprising: Innovative Products and Services through Collaborative Networks:* 349–357.

Smith, J. R., Terry, D. J., Manstead, A. S. R., Louis, W. R., Kotterman, D., & Wolfs, J. 2007. Interaction Effects in the Theory of Planned Behavior: The Interplay of Self-Identity and Past Behavior. *Journal of Applied Social Psychology,* 37(11): 2726–2750. http://dx.doi.org/10.1111/j.1559-1816.2007.00278.x

Sutton, S. 1998. Predicting and Explaining Intentions and Behavior: How Well Are We Doing? *Journal of Applied Social Psychology,* 28(15): 1317–1338. http://dx.doi.org/10.1111/j.1559-1816.1998.tb01679.x

Terry, D. J. 1993. Self-Efficacy Expectations and the Theory of Reasoned Action. *International Series in Experimental Social Psychology,* 28(2): 135–151. http://dx.doi.org/10.1207/s15326985ep2802_3

Terry, D. J., & Hogg, M. A. 1996. Group Norms and the Attitude-Behavior Relationship: A Role for Grop Identification. *Personality and Social Psychology Bulletin,* 22(8): 776–793. http://psycnet.apa.org/doi/10.1177/0146167296228002

Thomas, R. K., Terhanian, G., & Bayer, L. R. 2004. Behavioral Intention Measurement: International Findings. In *Proceedings of International Sociology Association's 6th International Conference on Social Science Methodology:* 16–20. Amsterdam, Netherlands: International Sociology Association.

Titze, S., Stronegger, W. J., Janschitz, S., & Oja, P. 2008. Association of Built-Environment, social-Environment and Personal Factors with Bicycling as a Mode of Transportation among Austrian City Dwellers. *Preventive Medicine,* 47(3): 252–259. http://dx.doi.org/10.1016/j.ypmed.2008.02.019

Venkatesh, V., Brown, S. A., Maruping, L. M., & Bala, H. 2008. Predicting Different Conceptualizations of System Use: The Competing Roles of Behavioral Intention, Facilitating Conditions, and Behavioral Expectation. *MIS Quarterly,* 32(3): 483–502. http://www.jstor.org/stable/25148853

Venkatesh, V., Morris, M. G., Davis, G. B., & Davis, F. D. 2003. User Acceptance of Information Technology: Toward a Unified View. *MIS Quarterly,* 27(3): 425–478. http://www.jstor.org/stable/30036540

Wardman, M., Hatfield, R., & Page, M. 1997. The UK National Cycling Strategy: Can Improved Facilities Meet the Targets? *Transport Policy,* 4(2): 123–133. http://dx.doi.org/10.1016/S0967-070X(97)00011-5

Appendix 1. Multiple regression analyses

A. Main analyses

Step 1: Control Variables	**1**	**2**	**3**	**4**	**5**	**6**	**7**	**8**
Gender	-0.006	-0.023	-0.024	-0.005	-0.015	-0.008	-0.004	-0.025
Age	-0.009	-0.036	-0.034	-0.002	-0.034	-0.008	-0.008	-0.037
Working distance	-0.136***	-0.014	-0.117***	-0.094***	-0.018	-0.135***	-0.135***	-0.012
Household structure 1†	-0.040	-0.016	-0.030	-0.030	-0.025	-0.040	-0.041	-0.017
Household structure 2†	0.019	0.035	0.019	0.037	0.027	0.020	0.018	0.032
Household structure 4†	-0.006	-0.001	-0.011	-0.017	0.012	-0.002	-0.006	0.001
Household structure 5†	-0.008	0.003	-0.009	-0.005	0.005	-0.008	-0.007	0.004
Household structure 6†	-0.015	-0.016	-0.017	-0.009	-0.020	-0.015	-0.014	-0.015
Household structure 7†	0.025	0.033	0.019	0.034	0.033	0.026	0.024	0.031
Working hours	-0.030	-0.017	-0.001	-0.030	-0.034	-0.030	-0.030	-0.016
Timeblock1 (06.00u-12.00u)	-0.041	-0.035	-0.035	-0.039	-0.039	-0.039	-0.044	-0.037
Timeblock2 (12.00u-18.00u)	0.064	0.038	0.046	0.032	0.065*	0.062	0.061	0.038
Timeblock3 (18.00u-24.00u)	0.005	0.029	0.021	0.015	0.018	0.007	0.006	0.030
Timeblock4 (00.00u-06.00u)	-0.047	-0.011	-0.039	-0.037	-0.015	-0.047	-0.045	-0.010
Timeblock5 (no usual time)	0.022	0.013	0.019	0.016	0.016	0.024	0.019	0.015
Luggage	-0.006	0.010	-0.002	-0.008	0.015	-0.004	-0.006	0.010
Frequency of past behaviour	0.750***	0.355***	0.566***	0.631***	0.452***	0.750***	0.743***	0.352***
Physical condition	0.053*	0.005	0.007	0.044*	0.031	0.053*	0.055*	0.006
Car ownership	0.005	0.037	0.041	0.008	0.019	0.006	0.006	0.041
Organization 2‡	0.047	0.036	0.043	0.047*	0.035	0.039	0.058*	0.035
Organization 3‡	0.037	-0.009	0.005	0.036	0.000	0.022	0.053	-0.022

Step 2: Main Effects	**1**	**2**	**3**	**4**	**5**	**6**	**7**	**8**
Attitude: direct benefit		0.179***	0.298***					0.179***
Attitude: awareness		0.005	0.015					0.004
Subjective norms: colleagues		0.052		0.056				0.051
Subjective norms: family		0.116**		0.231***				0.111**
Subjective norms: friends		-0.049		-0.054				-0.046
Subjective norm importance		0.012		0.015				0.011
Perceived behavioural control		0.355***			0.470***			0.360***
Facilities bike						-0.025		-0.033
Facilities other						0.021		-0.005
Financial incentives bike							0.044	0.021

R2	0.713	0.807	0.756	0.748	0.780	0.714	0.715	0.808
Adj. R2	0.703	0.797	0.746	0.737	0.772	0.703	0.704	0.797
F change	70.774***	38.401***	49.203***	19.917***	175.785***	0.683	3.24	27.199***
F	70.774***	83.025***	76.303***	69.177***	93.516***	64.611***	67.958***	75.108***
part correlation			0.179	0.134	0.258			

N= between 595 and 647

*p<0.05; **p<0.01 ; ***p<0.001

†Household structure 3 is the reference category

‡ Organization 1 is the reference category

Appendix 1. Multiple regression analyses (continued)

B. Additional analyses

Step 1: Control Variables	**9**	**10**	**11**
Gender	-0.029	-0.004	-0.016
Age	-0.043	-0.007	-0.035
Working distance	-0.148***	-0.095**	0.003
Household structure 1†	-0.019	-0.031	-0.026
Household structure 2†	0.016	0.032	0.025
Household structure 4†	-0.017	-0.020	0.000
Household structure 5†	-0.009	-0.005	-0.007
Household structure 6†	-0.014	-0.011	-0.025
Household structure 7†	0.017	0.035	0.036
Working hours	-0.005	-0.034	-0.038*
Timeblock1 (06.00u-12.00u)	-0.023	-0.037	-0.015
Timeblock2 (12.00u-18.00u)	0.044	0.032	0.043
Timeblock3 (18.00u-24.00u)	0.017	0.011	0.009
Timeblock4 (00.00u-06.00u)	-0.038	-0.033	-0.027
Timeblock5 (no usual time)	0.014	0.016	0.019
Luggage	0.003	-0.009	0.009
Frequency of past behaviour	0.533***	0.632***	0.386***
Physical condition	-0.008	0.044*	0.036
Car ownership	0.029	0.006	0.033
Organization 2‡	0.039	0.051*	0.031
Organization 3‡	-0.012	0.039	-0.001

Step 2: Main Effects	**9**	**10**	**11**
Attitude lifestyle	0.136***		
Attitude time saving	0.042		
Attitude comfortable	0.082*		
Attitude pleasantness	0.139***		
Subj.n. colleague: expect		0.053	
Subj.n. colleague: encourage		-0.006	
Subj.n. colleague: appreciate		0.028	
Subj.n. family: expect		0.129**	
Subj.n. family: encourage		0.061	
Subj.n. family: appreciate		0.082*	
Subj.n. friends: expect		-0.125**	
Subj.n. friends: encourage		0.058	
Subj.n. friends: appreciate		0.002	
PBC opportunity			0.124***
PBC trust			0.472***
PBC reversed hampered			0.009

R2	0.771	0.752	0.819
Adj. R2	0.761	0.739	0.812
F change	36.421***	9.967***	113.749***
F	78.491***	58.565***	110.196***

N= between 595 and 647
*p<0.05; **p<0.01 ; ***p<0.001
†Household structure 3 is the reference category
‡ Organization 1 is the reference category

Keywords: living lab, needsfinding, cycling, commuting, intention, theory of planned behaviour

12

Initiating a New Research Phase in the Field of International Entrepreneurship: An Interview with Professor Nicole Coviello

Nicole Coviello and Stoyan Tanev

> "*International entrepreneurship research seems to be at the cusp of moving from... a 'poorly understood emergence phase' of identity development to one that is better understood.*"
>
> Nicole Coviello (2015)
> Professor of Marketing, International Business, and Entrepreneurship

Introduction by Stoyan Tanev

The internationalization of businesses has become a pervasive phenomenon worldwide. In the international business literature, firms that internationalize early in their life cycle are referred to as "born globals" (Rennie, 1993), "international new ventures" (Oviatt & McDougall, 1994), "global startups" (Oviatt & McDougall, 1995), or "instant internationals" (Fillis, 2001). The term born global, however, has acquired a dominant position despite being defined in different ways by different scholars. Although many firms target global niches and develop a global presence from or within one or two years of their founding, they may not fit any of the existing definitions of "born global". One of the reasons for the variety of perspectives could be that research on born-global firms and international new ventures has necessarily brought together two different research domains – international business and entrepreneurship – and, accordingly, there has been the quick emergence of the international entrepreneurship research field.

In addition, research on born-global firms and international new ventures has become highly relevant in the context of technology entrepreneurship and has attracted the attention of scholars and practitioners in this area. Today, many new technology firms are in a better position to target competitive positions in international markets by working with global partners to innovate and pilot new disruptive technologies. The latest research suggests that the information age has enabled the adoption of a new techno-economic vision for the emerging structures and processes in international business in general. The reality of the newly emerging paradigm includes the changing nature of the competitive advantages of places, the competitive advantages and strategies of firms, and the governance structure of international business networks (Alcácer et al., 2016). Unfortunately, the terminological inconsistency in international business and international entrepreneurship research does not help practitioners to develop design principles for the creation and launching of new, globally competitive high-tech startups.

In a recent publication, Nicole Coviello (2015) emphasized the need to re-think existing research on international entrepreneurship and, more specifically, research on born-global firms. She pointed out that the main value of a critical review lies in initiating a new research phase focusing on the development of a more consistent collective research identity for international entrepreneurship through an ongoing process of terminological refinement and clarification. Here, I present my recent interview with Nicole Coviello, which was inspired by her recent publication. It focuses on issues that could be of relevance for new technology firms aiming at an international or global engagement from their very inception.

Interview with Nicole Coviello

Tanev: *Prof. Coviello, your call for a terminological refinement suggests our first question: What is your definition of a born-global firm? How does it compare to other such definitions and what is the key difference that you would like to emphasize?*

Coviello: I think the point might be more about clarification than refinement. I have long resisted using the term "born global" for two reasons. First, it was preceded in academic literature by the "international new venture typology from Tricia McDougall and Ben Oviatt: a typology that identified four different forms of international new venture. To me, that typology was a clear signal that multiple organizational forms are relevant in international research and labelling all international new ventures as "born globals" misses that point. Second, the term "born global" presumes the firm is founded on the global stage. However, although research has been talking about born globals since 1996, I believe it is only now that digital technology truly facilitates "born global-ness". I'd argue that most of our past research in international entrepreneurship has *not* been about born globals but rather, born internationals or born regionals (and even those labels imply the firms were founded with the intent to pursue international or regional expansion). The term "born global" has been over-used and mis-used because it's a catchy phrase.

My opinion (and practice) is that if one wants to study a born-global firm, then that firm should have been founded with the intent to serve global markets and then done so. Of note – this means being able to understand intent at founding: was globalization part of the founding intent or did the firm just happen to get pulled into international markets? The distinction is important because the behaviours of a proactive versus reactive firm are quite different. Furthermore, I believe it's important to clarify where the firm is relative to founding. For example, data collected on a 2–3 year-old firm is very different from that collected from a 10 year-old firm, even if you make the argument they are both "born global". The latter have survived early internationalization (that may or may not have been rapid). The former may still be in the process of internationalizing. They are not equivalent.

In a nutshell, the label "born global" was a lot more brandable than "international new venture". There are other differences too but, if I could, I would wave a magic wand so that every study in international entrepreneurship clearly described their firms and founding intent *without* using the born-global label.

Tanev: *My interest in the field of international entrepreneurship turns around new technology-based firms that engaged into a global path from their very inception. The logic behind this interest is the following: If technology firms that are born global are more successful than firms that are not born global, we should focus on defining and implementing design principles that incorporate key attributes of born-global firms and use these principles to launch and grow new companies. Our logic is based on the assumption that born-global technology startups are expected to be more successful than technology firms that did not engage into a global path from the very beginning. What does international entrepreneurship research say about that – how solid is the empirical evidence suggesting a basis for such assumption?*

Coviello: Before I answer your question, I want to make a few clarifying points. First, until recently, very few firms truly had the possibility of engaging on a global path. Global means global. International is not quite as broad in scope. Regional is closer to home. Only now, with the emergence of what I call finger-push firms, do we have a wide range of companies with the potential to serve global markets. That's because their offer is fully digitized, and all that is required for them to reach/communicate with/serve markets around the global is the push of a button on a computer. With 3D printing, even the sourcing of product becomes obsolete. Second, another point to keep in mind when conceptually discussing a born-global firm pertains to international expansion speed. Speed is another term that has been used too casually. Most speed research isn't about speed at all but rather it is about timing. My opinion is that, to study speed, you need to consider whether you are interested in: i) timing (e.g., time to first foreign market entry); ii) pace of ongoing internationalization, or iii) the rhythm of internationalization over time. Those are three very different concepts.

So where does success come in? I have two thoughts. First, no, I don't think that we have sufficient evidence in international entrepreneurship research to suggest that born globals are more successful. We assume they are. We also assume that internationalization speed is beneficial. But true born globals and speed need more attention in research. Second, we are only now able to compare born globals and non-born-globals because we can compare, for example, software-as-a-service (SaaS) firms with software developers. Both fall into the "software" sector but only one (SaaS) has the potential to be born global; the other is much more likely to be born regional or born international. Again, we don't yet have sufficient empirical evidence on firms that are truly born global versus masquerading as such (because of how authors have labelled the firms they study).

Tanev: *If you were to distill the practical implications of a decade's work on born globals and your re-thinking of the concept, what are the key points the individual entrepreneur should take away and apply in starting a new global business today?*

Coviello: Get into SaaS if you want to be "global". If you are in technology manufacturing, exploring the Internet of Things or artificial intelligence, or developing software, know there is a global market *but* your business is likely to be strongest by serving select markets and customers carefully and well. You don't need to be global to be successful.

Tanev: *Recently, you have also pointed out that, for scholars working at the interface of international business and entrepreneurship, an understanding of entrepreneurial business models is essential (Coviello, 2016). Where do you see the potential and challenges of integrating research into business models, born globals, and international new ventures?*

Coviello: It's important to remember that international entrepreneurship research and practice sit at the intersection of international business and entrepreneurship. The relevance of the business model literature comes from entrepreneurship and while we have happily adopted concepts like entrepreneurial orientation into the international entrepreneurship literature, we've been slow to recognize how international entrepreneurship parallels entrepreneurship arguments regarding business models. Your work is one of the very few exceptions (Rasmussen & Tanev, 2015).

Tanev: *In your discussion of the global entrepreneurial potential of business model research, you have also referred to the growing interest in the lean startup entrepreneurial approach and the possibility of considering lean global startups as a new type of organization/firm (Rasmussen & Tanev, 2015). By "lean startup approach", we mean the specific hypothesis-driven entrepreneurial approach that "favors experimentation over elaborate planning, customer feedback over intuition, and iterative design over traditional 'big design up front' development" (Blank, 2013), which is based on several concepts such as minimum viable product and pivoting. Do you personally see a reason for the adoption of the lean startup approach within the context of born-global firms? Does integrating lean startup and born global research allow for the identification of a new type of firm, or at least a unique growth mode, that deserves to be considered separately from others?*

Coviello: From a practical perspective, it is good news that the lean startup approach has been widely popularized. Practitioners like it and lean captures all the advantages of being nimble and with little to unlearn. Steve Blank has done a great job of popularizing concepts that have been core to entrepreneurship theory for some time (e.g., arguments from Howard Stevenson or Saras Sarasvathy). That is, "lean" isn't really new – it's just that Blank makes the ideas digestible and, I think, more tangible for practitioners. It does, however, set the stage well for internationalization. At any rate, I don't consider a lean global startup to be a new form of organization or growth mode. Lean is a way of operating. In the same way that we study organization structure, orientation, etc., we should study the influence of lean (or effectual logic) as young firms internationalize.

Tanev: *There seems to be an ongoing tendency to consider the field of innovation management in the broader context of business model innovation. We believe that there is a point in such claim, given that an innovation usually affects all the components of a business model. Your call for integrating international entrepreneurship and business model research should therefore inspire more research focusing on innovation management in a global entrepreneurial context. But why is there so little international entrepreneurship research focusing on innovation management? The few articles in this area do not seem to have a major impact. What are the challenges, theoretical or conceptual, the addressing of which would help in integrating early internationalization and innovation research with a focus on new entrepreneurial firms?*

Coviello: I think one of the challenges comes from the heritage of international entrepreneurship. Most researchers have come from an international-business-related area or from entrepreneurship whereas innovation is a distinct third discipline. Marian Jones and I tried to bring innovation into the international entrepreneurship discussion with our paper in the *Journal of International Business Studies* (Jones & Coviello, 2005), but you are right, there is more to be done.

Tanev: *Please tell us more about your understanding of the priorities of the new international entrepreneurship research phase you are calling for. In your recent paper, you refer to the need for more international entrepreneurship and international business research at the level of the individual entrepreneur and especially in the context of the next generation of young founders emerging from our entrepreneurship programs. But where do you see other similar needs, what type of future research is needed, and are there other research themes that have become urgent and need to be addressed?*

Coviello: My interest in the "people of international entrepreneurship" comes from arguments that any decision or action at the firm level stems from an individual or team. As a result, I encourage multi-level research and attention to decision makers in the firm. In terms of young founders, I think they are different in terms of how they relate to technology and the world; these differences need to be understood. They don't see the technological, cultural, or national barriers that previous generations did, and I suspect this will influence how they lead their firms. Returning to the concept of a true born global (i.e., a finger-push firm that can instantly set up a global operation), they might turn all our traditionally accepted wisdom on its head. We need to study these firms.

Tanev: *Finally, do you see any undergoing changes in the global business environment? What changes or trends in the business environment may need to be addressed in our future research? Where do you think international entrepreneurship research will be 10 years from now?*

Coviello: I think my previous answer leads into this one. Understanding, for example, ownership, locational, or internalization advantage was essential when firms expanded by geographic footprint. Things change when expansion is by "finger-push". Our new research needs to recognize that major global players are likely to be located in one region (e.g., main technology centres) and serve the world from either that location or perhaps with small groups of employees working remotely. I hope that 10 years from now, international entrepreneurship research will be a richer blend of international business, entrepreneurship, plus innovation and technology management research.

Tanev: *Thank you very much, Prof. Coviello. We greatly appreciate your time and insights for TIM Review readers worldwide.*

Summary by Stoyan Tanev

There are several points in the above interview to be highlighted in the context of the present special issue:

1. Using the term "born global" is not recommended because of the improper semantic implications generated by the term "born." Very few firms were really "born" on a global scene and, as a result, most of international entrepreneurship research so far has not been really studying born globals but rather, born internationals or born regionals. In this sense, the global startup terminology is more appropriate.

2. According to Coviello, it is only now that digital technology truly facilitates "born global-ness". An example of true born global is a finger-push firm that can instantly set up a global operation. Such firms might disrupt our traditionally accepted wisdom. International entrepreneurship research needs to focus on studying these firms.

3. If you want to be "global", get into SaaS. If you are in technology manufacturing, exploring the Internet of Things or artificial intelligence, or developing software, know there is a global market but remember that you do not need to be global in order to be successful.

4. International entrepreneurship research needs to embrace business model frameworks that could enhance the conceptualization of all business aspects in an international/global context.

5. The lean startup approach is not really new but it made entrepreneurial ideas more digestible, more tangible, and more actionable for practitioners, setting the stage well for early internationalization. In this sense, a lean global startup does not seem to be a new form of organization or special type of growth mode: it is more like a specific way of operating. For a more detailed discussion of this point, please see my article in this special issue: "Is there a Lean Future for Global Startups?" (Tanev, 2017).

6. The challenges of managing innovation in a global context come from the fact that most international entrepreneurship researchers have come from an international business related area or from entrepreneurship whereas innovation is a distinct third discipline. The international entrepreneurship research community has to work on further integrating innovation management scholarship in parallel to business model design and innovation frameworks and theories.

7. Multi-level research and attention to the actual decision makers of global startups should be strongly encouraged. Young founders are different in terms of how they relate to technology and the world. They deserve a special attention and need to be further studied.

8. In the near future, international entrepreneurship research will be a richer blend of international business, entrepreneurship, innovation, and technology management research.

Nicole Coviello is the Lazaridis Research Professor and Professor of Marketing at Wilfrid Laurier University in Waterloo, Canada. She received her PhD in Marketing and International Business from the University of Auckland, New Zealand, in 1994. She also holds an MSc (Technology Management) and BComm Hons (Marketing) from the University of Saskatchewan in Saskatoon, Canada. Dr. Coviello began her career at the University of Waikato, New Zealand, and has since held positions at the University of Calgary, Canada, and the University of Auckland. In 2010, she received an Honorary Doctorate from the Turku School of Economics in Finland. Dr. Coviello has published in the top journals across three disciplines – marketing, international business, and entrepreneurship. In 2014, she was ranked as one of the top scholars in International Marketing and among the top 20 International Business scholars. She is Field Editor for both marketing and international entrepreneurship at the *Journal of Business Venturing.*

Stoyan Tanev is an Associate Professor in the Department of Technology and Innovation, Faculty of Engineering, Southern Denmark University (SDU) in Odense. Dr. Tanev is leading the Technology Entrepreneurship stream of the Master Program of Product Development and Innovation at SDU. He is also an Adjunct Research Professor in the Sprott School of Business at Carleton University in Ottawa, Canada, where he is associated with the Technology Innovation Management Program. He has a MSc and a PhD in Physics jointly from the University Pierre and Marie Curie, Paris, France, and the University of Sofia, Bulgaria, a PhD in Theology from the University of Sofia, Bulgaria, an MEng in Technology Innovation Management from Carleton University, Canada, and an MA from the University of Sherbrooke, Canada. He has multidisciplinary research interests with a focus on the fields of global technology entrepreneurship, technology innovation management, business model design, and value co-creation. Dr. Tanev is Senior IEEE member, as well as member of the editorial boards of the *Technology Innovation Management Review*, the *International Journal of Actor-Network Theory,* and *Technological Innovation.*

References

Alcácer, J., Cantwell, J., & Piscitello, L. 2016. Internationalization in the Information Age: A New Era for Places, Firms, and International Business Networks? *Journal of International Business Studies,* 47(5): 499–512.
http://dx.doi.org/10.1057/jibs.2016.22

Bailetti, T. 2012. What Technology Startups Must Get Right to Globalize Early and Rapidly. *Technology Innovation Management Review,* 2(10): 5–16.
http://timreview.ca/article/614

Blank, S. 2013. Why the Lean Start-Up Changes Everything. *Harvard Business Review,* 91(5): 63–72.

Coviello, N. E. 2015. Re-Thinking Research on Born Globals. *Journal of International Business Studies,* 46(1): 17–26.
http://dx.doi.org/10.1057/jibs.2014.59

Coviello, N. E. 2016. Implications of Business Models for IE Research. *ie-scholars.net.* Accessed May 1, 2017:
http://ie-scholars.net/blog/2692

Fillis, I. 2001. Small Firm Internationalisation: An Investigative Survey and Future Research Directions. *Management Decision,* 39(9): 767–783.
http://dx.doi.org/10.1108/00251740110408683

Jones, M. V., & Coviello, N. E. 2005. Internationalisation: Conceptualising an Entrepreneurial Process of Behaviour in Time. *Journal of International Business Studies,* 36(3): 284–303.
http://dx.doi.org/10.1057/palgrave.jibs.8400138

Oviatt, B. M., & McDougall, P. P. 1994. Toward a Theory of International New Ventures. *Journal of International Business Studies,* 25(1): 45–64.
http://dx.doi.org/10.1057/palgrave.jibs.8490193

Oviatt, B. M., McDougall, P. P., & Loper, M. 1995. Global Start-Ups: Entrepreneurs on a Worldwide Stage. *The Academy of Management Executive,* 9(2): 30–44.
http://www.jstor.org/stable/4165256

Rasmussen, E. S., & Tanev, S. 2015. The Emergence of the Lean Global Startup as a New Type of Firm. *Technology Innovation Management Review,* 5(11): 12–19.
http://timreview.ca/article/941

Rennie, M. W. 1993. Global Competitiveness: Born Global. *McKinsey Quarterly,* 4: 45–52.

Tanev, S. 2017. Is There a Lean Future for Global Startups? *Technology Innovation Management Review,* 7(5): 6–15.
http://timreview.ca/article/1072

Keywords: international entrepreneurship, international business, born global, international new ventures, lean and global startups, internationalization, definitions

13

The Defining Characteristics of Urban Living Labs

Kris Steen and Ellen van Bueren

> *"Cities have the capability of providing something for everybody, only because, and only when, they are created by everybody."*
>
> Jane Jacobs (1916–2006)
> Journalist, author, and activist
> In *The Death and Life of Great American Cities*

The organization of supported and sustainable urban interventions is challenging, with multiple actors involved, fragmented decision-making powers, and multiple values at stake. Globally, urban living labs have become a fashionable phenomenon to tackle this challenge, fostering the development and implementation of innovation, experimentation, and knowledge in urban, real-life settings while emphasizing the important role of participation and co-creation. However, although urban living labs could in this way help cities to speed up the sustainable transition, urban living lab experts agree that, in order to truly succeed in these ambitious tasks, the way urban living labs are being shaped and steered needs further research. Yet, they also confirm the existing variation and opaqueness in the definition of the concept. This article contributes to conceptual clarity by developing an operationalized definition of urban living labs, which has been used to assess 90 sustainable urban innovation projects in the city of Amsterdam. The assessment shows that the majority of the projects that are labelled as living labs do not include one or more of the defining elements of a living lab. In particular, the defining co-creation and development activities were found to be absent in many of the projects. This article makes it possible to categorize alleged living lab projects and distill the "true" living labs from the many improperly labelled or unlabelled living labs, allowing more specific analyses and, ultimately, better targeted methodological recommendations for urban living labs.

Introduction

By 1969, before the term "sustainable development" was commonly known, the United States Congress had already emphasized the need to "create and maintain conditions that permit fulfilling the social, economic and other requirements of present and future generations" (U.S. Congress, 1969). It is apparent that the current systems of urban life and development do not succeed in achieving this sense of sustainability in cities, thereby demonstrating the need for urban innovation to contribute to the sustainability transition of cities (Joss, 2015; Wheeler & Beatley, 2015).

However, many sustainable urban innovations at present do not systemically find their way to the market, inhibiting their wider implementation and limiting their positive impact potential in the context of this urban sustainability transition (Ashuri & Durmus-Pedini, 2010; van Bueren & Broekhans, 2013). The living lab approach is believed to close this gap between production and actual market uptake by directly involving all end users and other urban stakeholders into the development of new products. This would ensure a match with the actual needs and aspirations of the users while taking into account the local and institutional contexts, cultures, and creativity potentials (ENoLL, 2006; Kresin, 2009; Lemke, 2009; Lesnikowski et al., 2016; van Bueren & De Jong, 2007). Furthermore, the new partnerships and the inclusion of new, multidisciplinary stakeholders in the development process is believed to provide new insights and offer new, more integrated solutions to the investigated problems, thereby advancing the generation of innovations (ENoLL, 2006; Vincent, 2016).

Given these alleged benefits, it is no surprise that, in recent years, the "urban living lab" has become a popular development approach on which hope is fixed to accelerate the generation and adoption of sustainable innovations in the urban system in the light of the urban sustainability transition. However, evaluative accounts of urban living labs in practice indicate that many of these experience difficulty in achieving the full benefits of this approach (Bulkeley & Castán Broto, 2013; Franz et al., 2015; Karré et al., 2015; Schuurman et al., 2016). These authors point to a lack of methodological deepening of the living lab approach as a reason for such difficulties. Yet, in order to provide targeted methodological recommendations, it is necessary to take a closer and more precise look at the aim of urban living labs. Analysis of the (urban) living lab literature and the variety of existing urban living labs, for example those presented at the ENoLL website (openlivinglabs.eu), shows that there is no consensus on this issue. The way an urban living lab is defined and used is diffuse, with many different and often abstract definitions of the concept, leaving participants involved in urban living labs in the dark as to how the "living lab dimension" makes this project different from other urban innovation projects.

This article focuses on establishing a definition of urban living labs, which will show what can be expected from urban living labs and which will provide guidance to actors engaged in urban living labs or contemplating starting one. The definition presented in this article has been based on the existing living lab literature. Additionally, a scan of 90 Amsterdam-based sustainable urban innovation projects has provided insights on the representation of "true" urban living labs, fostering cocreative innovation among the more traditional sustainable urban innovation initiatives. We conclude by identifying some key challenges to the implementation of urban living labs.

Defining "Urban Living Lab"

The term "urban living lab" is to refer to a variety of local experimental projects of a participatory nature. It is often used interchangeably with the terms "testing ground", "hatchery", "incubator", "making space", "testbed", "hub", "city laboratory", "urban lab", or "field lab". Although there are many publications on living labs and urban living labs, even in these texts, this notion of the urban living lab has not been defined clearly. In literature discussing the "theory" of (urban) living labs, they have been explain ed as a methodology (Eriksson et al., 2005), as an environment (Ballon et al., 2005; Schaffers et al., 2007), as a system (CoreLabs, 2007), and as a governance approach (Bulkeley et al., 2016).

To explore the variety of definitions and understandings of the concept, we have analyzed articles on (urban) living labs published in this journal, the *Technology Innovation Management Review* (TIM Review) and other journals. The articles in the TIM Review are particularly relevant because this journal has played a considerable role in the transformative debate on the essence, role, and shape of urban living labs, with, in addition to separate articles on this topic, special issues attributed to this topic in 2012, 2013, 2015, 2016, and 2017. Appendix A presents an overview of the definitions used in these issues.

Almost all the articles we found on (urban) living labs, including those published in other journals, referred to the variation and opaqueness in the definition of the concept. Often, these articles, which are presented in Appendix B, adopt existing definitions, such as the one used by the European Network of Living Labs (ENoLL): "Living labs are defined as user-centered, open innovation ecosystems based on systematic user co-creation approach, integrating research and innovation processes in real life communities and settings" (ENoLL, 2016). These definitions aim to grasp the essence of living labs in one sentence, mentioning the main aspects and the relations between these aspects, often using complex concepts. Besides these single-sentence definitions, there are authors who provide a rich description of the phenomenon, but do not define it. Other authors distinguish different "types" of living labs (e.g., Leminen, 2013) or imply that living labs are environments offering a "multitude of different projects" (Tuija Hirvikoski, President of ENoLL, in an open discussion on February 13, 2017). Finally, there are authors that simply refer to an empirical example (e.g., Femenías & Hagbert, 2013) or skip the usual explanatory paragraph altogether and discuss "living labs" as an already accepted notion (e.g., Coorevits & Jacobs, 2017). We see only a few examples where authors operationalize this definition by formulating key principles, defining elements, or building blocks of living labs (Schuurman et al., 2013; Ståhlbröst, 2012; Veeckman et al., 2013). Yet, gradually, in the current living lab literature, a stabilized conception of living labs with roughly similar characteristics has emerged.

Characteristics

As described in the subsections below, we have drawn on the living labs articles published in the TIM Review and in other journals and documents to capture the defining characteristics of urban living labs. In total, we identified nine characteristics in four dimensions: i) aim ii) activities, iii) participants, and iv) context. For each dimension, the defining characteristics of urban living labs are described in the subsections that follow, and the complete list of defining characteristics of urban living labs, presenting the full urban living lab definition, is summarized in Table 1.

Aim

The overall aim of living labs is to learn and experiment, by integrating processes of research and innovation (ASC, 2016; Bijsterveldt, 2016; ENoLL, 2016). The innovation aspect refers to the development of new products (i.e., an object, service, technology, application, process, or system) and to the discovery of new solutions to existing problems. Learning and experimenting (McCormick, 2016; Pallot & Pawar, 2012; Schaub, 2016; Vincent, 2016) refers to the production and exchange of knowledge among participants (ASC, 2016; Bijsterveldt, 2016; Friedrich et al., 2013; Heijden, 2016; McCormick, 2016; Pallot & Pawar, 2012). Thus, the aim is not only to learn from the experiences from the particular lab environment, but also to replicate the innovation elsewhere, in real life, or to further future innovation (ASC, 2016; Bijsterveldt, 2016; Franz et al., 2015; Juujärvi & Lund, 2016). It is this emphasis on formalized knowledge production – lessons that are formulated and that can be disseminated – that sets living labs apart from other policy experiments and niches of innovation (Evans & Karvonen, 2014).

"Urban" living labs distinguish themselves from living labs by unanimously displaying an explicit territorial focus on finding local sustainable solutions addressing wicked problems that tend to be global, such as climate change and energy transition. The use of cities or parts of cities as laboratories is well in line with the current emphasis on the city as the impactful governance level for economic development (e.g., Glaeser, 2011; Barber, 2013; Katz & Wagner, 2014) and for sustainable development (e.g., van Bueren et al., 2012; Bulkeley & Betsill, 2013). It also responds to calls for citizen empowerment (e.g., Saurugger, 2010; Fung, 2015).

Table 1. Overview of the defining characteristics of living labs

Aspects	Characteristics
Aims	Aimed at **innovation**
	Aimed at **formal learning** for replication
	For urban living labs: Aimed at increasing **urban sustainability**
Activities	**Development** (all phases of the product development process)
	Co-creation
	Iteration (feedback, evaluation, and improvement)
Participants	**Public actors, private actors, users** and **knowledge institutes** participate in the living lab activities
	All actors involved have **decision-making power**
Context	The living lab activities take place in the **real-life use context** of the innovation. In many urban living labs, this is a territory or a space-bound place.

Activities
There are a number of activities that are supposed to take place in living labs. Living labs should comprise the *development* of a product, which can be all sorts of artifacts, including process innovations, and not only, for example, testing or implementation (Budweg et al., 2011; Coenen et al., 2014; Feurstein et al., 2008; Gray et al., 2014; Leminen, 2015; Schaffers et al., 2007; Tang & Hämäläinen, 2012). A key element in this development process is *co-creation* (Bijsterveldt, 2016; ENoLL, 2016; Feurstein et al., 2008; Franz et al., 2015; Gómez-Barroso et al., 2009; Heijden, 2016; Schaub, 2016). The essence of a living lab is that the solution is sought together *with* the user, rather than just applying a fixed solution and involving the user only for testing. To qualify as co-creation, the targeted users need to be involved in the various development phases of the living lab process: not only should they be asked for their opinions, they should have decision-making power throughout the phases (Prahalad & Ramaswamy, 2004). Furthermore, the development process of living labs is iterative, which implies that, after being designed, the (prototype) product is used and evaluated by the stakeholders. The feedback and evaluation gathered from these steps are used to further develop and improve the product (Feurstein et al., 2008; Pallot & Pawar, 2012; Pierson & Lievens, 2005).

Participants
The literature further shows that the living lab is a systematic innovation approach in which all stakeholders – all actors who have a stake in the developed product and the process leading to it – participate directly in the development process. Stakeholder participation, including users, is an integral part of the development process in living labs (Feurstein et al., 2008).

The actors who need to participate in the living lab activities are, at a minimum: users (the end users of the final product that will be developed; in many cases, citizens), private actors (businesses, firms, companies), public actors (governments and public institutions), and knowledge institutes. The participation of knowledge institutes emphasizes the aim of formalized knowledge production (Krueger & Buckingham, 2009; Perry, 2006). As mentioned, these actors not only need to participate in these activities but also need to have power to influence the process (Prahalad & Ramaswamy, 2004). This power allows these actors, including the end users of the product, to be active partners in the innovation and development during the whole process, rather than just passive consumers and receivers and subjects of R&D activities (Almirall et al., 2012; Feurstein et al., 2008; Friedrich et al., 2013; Higgins & Klein, 2011; Leminen, 2015; Schaub, 2016).

Context
Authors on living labs unanimously agree that a defining characteristic of living labs is that they take place in real-life environments. The complexity and multi-contextuality that is connected to real-life environments is part of the challenge of living labs (Feurstein et al., 2008; Friedrich et al., 2013). Whether these real-life environments are limited to physical environments (Feurstein et al., 2008; Friedrich et al., 2013; Kieboom, 2013; Leminen, 2015; McCormick, 2016; Westerlund & Leminen, 2011) or also comprise virtual realities (Feurstein et al., 2008; Leminen, 2015) is contested. In this research, the criterion used is that the product central to the living lab is being developed and evaluated in its real-life *use context.* Given that this product can also be a virtual product, such as a digital data-gathering system, virtual realities can also form the context of living labs. More often, however, the living labs will take place in a physical place, such as a neighbourhood, city, or other area. The wide attention for *urban* living labs emphasizes the need or desire of involved stakeholders to capture the real-life context in all its complexity while assuming that such experiments, despite the highly uncontrolled conditions, nevertheless produce useful, transformative knowledge (Evans & Karvonen, 2013).

An overview of the defining living lab characteristics mentioned in the living lab definitions presented in the literature is provided in Figure 1. This figure shows a high level of agreement among the authors of the texts studied on the characteristics of living labs, leading to the synthesised definition presented in Table 1. In addition, Appendix B lists how authors have worded these living lab characteristics, which shows the original variety of concepts to which authors referred to convey these characteristics and demonstrating the need for operationalization of these concepts to be of guidance to stakeholders involved in urban living labs.

Living labs vs. urban living labs
The characteristics of living labs and *urban* living labs are very similar, and we assume that the general characteristics of living labs and recommendations for their design and operation are generally applicable to urban living labs as well. The difference between living labs and urban living labs is the explicit focus on finding solutions meant to increase urban sustainability. The inclusion of this explicit aim of increasing urban sustainability in the problem and goal statement of a living

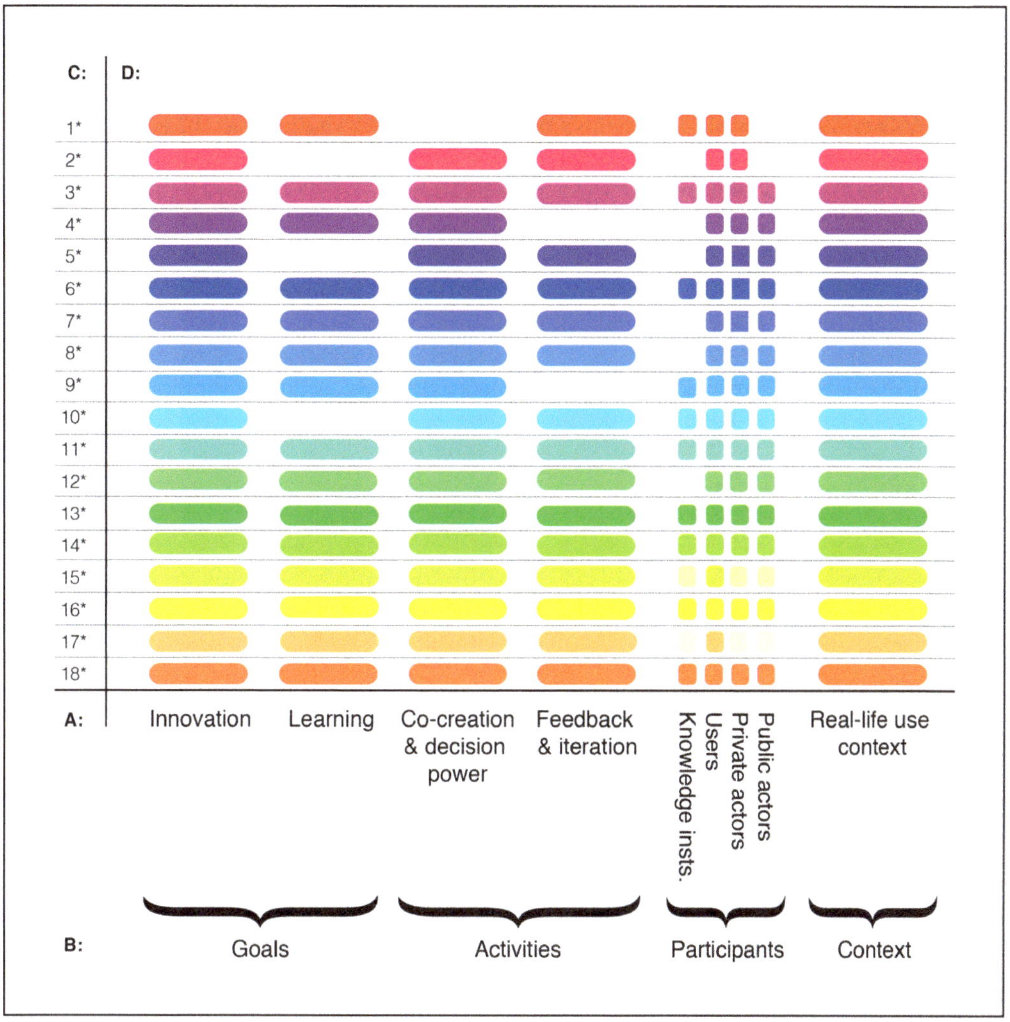

A = Proposed defining characteristics of living labs

B = Dimensions applicable to the defining characteristics

C = The definitions of 'living lab' provided in literature

D = Presence of the proposed defining characteristics in the living lab definitions provided in literature

1*: William Mitchell (MIT), ca. 2003
2*: Niitamo, Kulkki, Eriksson & Hribernik, 2006
3*: Ballon, Pierson, & Delaere, 2005
4*: European Network of Living Labs, 2006
5*: Kusiak, 2007
6*: Schaffers, Cordoba, Hongisto, Kallai, Merz & Van Rensburg, 2007
7*: Almirall & Wareham, 2008
8*: Feurstein, Hesmer, Hribernik, Thoben & Schumacher, 2008
9*: Bergvall-Kåreborn & Ståhlbröst, 2009
10*: Westerlund & Leminen, 2011
11*: Higgins & Klein, 2011
12*: Pallot and Pawar, 2012
13*: Friedrich, Karlsson & Federley, 2013
14*: Leminen, 2015
15*: Franz, Tausz & Thiel, 2015C
16*: Voytenko, McCormick, Evans, & Schliwa, 2016
17*: Hakkarainen & Hyysalo, 2016
18*: GUST, 2016

Figure 1. Overview of the presence of the proposed defining living lab characteristics in the autonomous living lab definitions provided in literature, focusing only on definitions introducing self-worded determinations of an (urban) living lab (not or besides quoting previously proposed definitions)

lab is a concrete feature of urban living labs – a feature that is often vaguely referred to by urban living lab authors as the "urban" dimension. This view is also consistent with the intuitive conception of scholars and practitioners that there *is* a difference between "normal" living labs and "urban" living labs in terms of increased complexity. The intricate number of variables and relationships influencing the process and outcome of urban living labs compared to "normal" living labs is exactly the learning environment that urban stakeholders tend to look for and appreciate. It helps them to develop real-world solutions to real-world problems while emphasizing the need for these solutions to work.

The use of living labs in cities may also raise questions about the inclusiveness and democratic accountability of the selection of areas for living labs, and about the focus of the innovation. Indeed, many innovations tend to be technological and are driven by ICT and decentralized technologies with ambitions to become "smart" and "low carbon". These are highly relevant questions, but they are beyond the focus of this article, which focuses on the living lab phenomenon itself.

Living Labs in Practice

Despite the popularity of urban living labs, stakeholders being engaged in living labs or wanting to start a living lab are struggling with what it actually is that they are supposed to do in an urban living lab. To answer this question, we have identified and quickly scanned the urban living labs in the Amsterdam metropolitan region. Given that urban living labs are not necessarily labelled as such, the identification of urban living labs in Amsterdam was not based on the projects calling themselves a living lab or an urban living lab, but on projects *potentially* being classified as an urban living lab based on their characteristics. This approach led to a sample of 90 sustainable urban innovation projects (for details, see Steen & van Bueren, 2017) with attention for the citizens or end users in their project descriptions. The projects were identified through snowball sampling using a mixture of sources: policy documents; research papers; the Internet; and suggestions from experts working at the municipality of Amsterdam, Amsterdam SmartCity (a public organization aimed to collect and disseminate knowledge on smart city initiatives in the Amsterdam region), AMS (the Amsterdam Institute for Advanced Metropolitan Solutions), and other Amsterdam-oriented researchers. Based on the project descriptions found in online and offline sources, one researcher verified which of the defining living lab characteristics (Table 1) were present in each of the 90 projects.

During this search, some notable observations were made. Of the nine characteristics of urban living labs, the characteristics of *development* taking place in the lab and of *co-creation* applying to the living lab activities were found to be the two urban living lab characteristics most frequently absent in sustainable urban innovation projects in Amsterdam. Of the 90 scanned projects, only 12 displayed these two characteristics. Of these 12 projects, all of them contained the remaining seven urban living lab characteristics (Table 1), thus qualifying as urban living labs. Surprisingly, most of these projects are not the ones calling themselves "living labs" or "labs". In fact, only three of the 14 projects with "living lab" in their name and two of the 14 projects with "lab" in their name met the identified defining characteristics of an urban living lab. The other seven projects from the 90 urban innovation projects in Amsterdam that were identified as urban living labs did not use any reference to "lab" of "living lab" in their name.

In this article, we hone in on the challenges experienced with these two urban living lab characteristics of "developing an innovation" and "co-creation", which have been identified as the two urban living lab characteristics most frequently absent in sustainable urban innovation projects.

Innovation challenges

The analysis of 90 sustainable urban innovation projects in Amsterdam shows that, besides development of an innovation, which is a criterion for urban living labs, a number of other innovation-related activities can be central in the projects.

From the literature on the innovation process – whether it concerns new product development (Cooper, 1988), agile development (Beck et al., 2001), or user-driven (open) innovation processes (Von Hippel, 2005; Mikkela, 2008) – we have identified five overarching phases in the innovation process and used them to categorize the projects:

1. *Research:* "An investigative process of revising current knowledge employed to reach understanding of a subject for the purpose of making decisions" (Cambridge Dictionary: dictionary.cambridge.org).

2. *Development:* "The process of the product being created, growing, and changing in order to become more advanced" (Business Dictionary: businessdictionary.com).

3. *Testing:* "Application of a product for the purpose of observing and assessing its functionality" (Merriam-Webster: merriam-webster.com).

4. *Implementation:* Implementation of a finished product in a real-life environment.

5. *Commercialization of the developed product:* Demonstrating a previously developed and implemented product.

The first four phases largely coincide with the natural stages of product maturation in the product development process (Cooper, 1998). The fifth phase is emphasized in the literature on user-driven, open innovation, and new product development, where the convincing of users who have not been involved in the development process needs more attention. It should be noted that this phasing is presented as a linear process. In reality, the sequence of these phases is not fixed, and processes take place in iteration.

Ideally, an innovation project would cover all these phases. In practice, we observed that each of the projects focussed on a particular phase of the innovation process (according to the pattern visualised in Figure 2). The projects aimed specifically at researching, developing, testing, implementing, or commercializing a solution. According to the living lab characteristics, only the projects (also) conducting *development* activities qualify as urban living labs.

User involvement challenges

Even though innovation in urban living labs theoretically should take place in co-creative processes in which participants can participate and influence decisions in all phases of the innovation process, this was rarely seen in the 90 investigated sustainable urban innovation projects in Amsterdam. When looking at the literature, we consider Sherry Arnstein's "ladder of citizen participation" (1969) the most usable elementary scale for measuring the degree of user involvement in the projects studied, with the three main levels of involvement being decision-making power, tokenism, and non-participation. As in the literature on co-creation (Prahalad & Ramaswamy, 2004), Arnstein associates true participation with the possession of (decision-making) power. The second level of participation identified by Arnstein is tokenism: the action of symbolically allowing people to "participate" without actually providing them with power in the decision-making process. The third main level distinguished is non-participation.

When projecting Arnstein's model onto the 90 potential living lab projects, we simply considered the first level as "participation" and the other two levels (tokenism

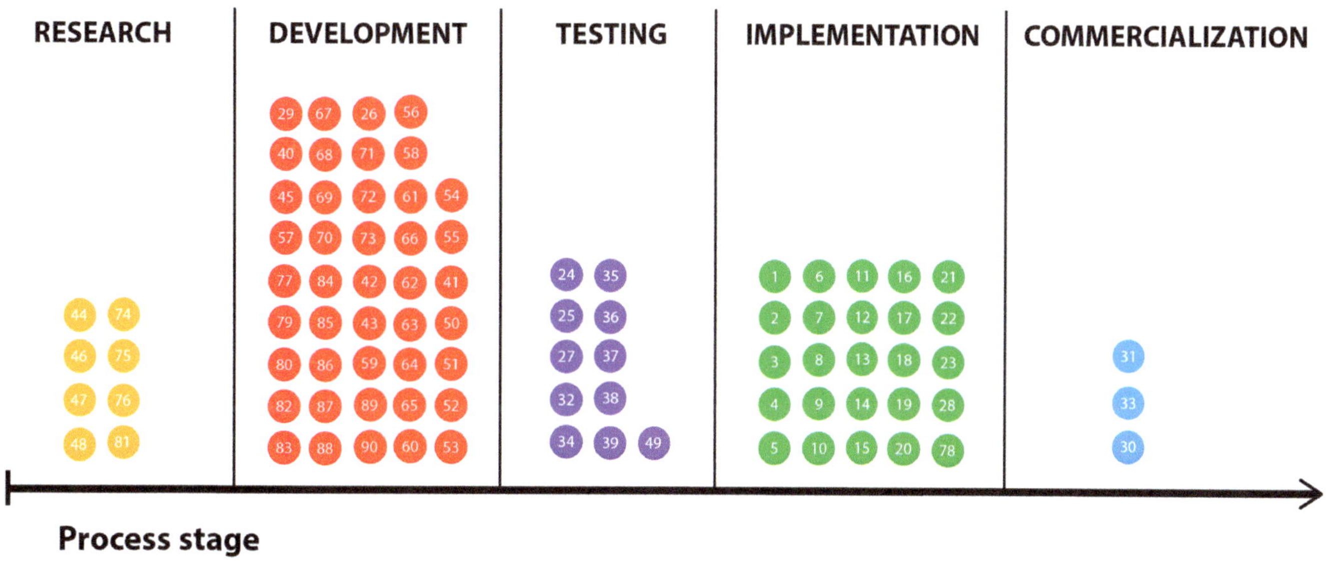

Figure 2. Classification of innovation process phase of 90 potential living lab projects in the Amsterdam region

and non-participation) as "no participation", because decision-making power by stakeholders, including users, has been shown to be conditional for living labs. Following this criterion, user participation only occurred in 51 of the 90 urban innovation projects, and, although this may sound as if co-creation takes place, user participation in the project activities is in fact quite common, if not inherent, when the project activities concerned are testing, implementation, and demonstration. However, *development* with the user (i.e., with the user participating and having decision-making power in the *development* process of the developed product) is the only form of user involvement that actually enables co-creation, which is one of the characteristics of living labs. The scan of the 90 sustainable urban innovation projects in Amsterdam found development with the user in only 12 of the projects.

When looking more closely at the sustainable urban innovation projects in Amsterdam associated with a living lab approach *not* including user participation, these projects do display a significant focus on user-related activities (Figure 3). Two-thirds of these projects either clearly adopt a user perspective and specifically aim to deliver solutions for the *end user* rather than for public or industry partners, or they perform their project activities with the help of digital user data that the user actively or passively provides, for example through iBeacons, smartphone applications, or smart meters. The first form of this user focus – in which the user does not directly participate (i.e., there is no co-creation) but the user is included in other ways – can be referred to as "user oriented", following the definition of "user-oriented design" by Veryzer and de Mozota (2005). The second form can be referred to as "user sourced", which indicates that project activities are performed using virtual user data, which is actively or passively provided by the user.

Conclusions

Although urban living labs are widely implemented in urban contexts and are popular projects to allegedly catalyze sustainable urban innovation and the sustainable transition, their definition remains unclear, both in real life as well as in the literature. This article has set up a demarcating definition of the concept of urban living labs in order to allow researchers to further investigate this topic. After a literature review on the common denominators of living lab projects, a framework has been set up outlining the defining features of living labs in terms of goals, activities, participants, and context, making it possible to categorize and diversify alleged living lab projects and distill the real living labs from the many improperly labelled or unlabelled living labs, allowing more specific analyses and recommendations.

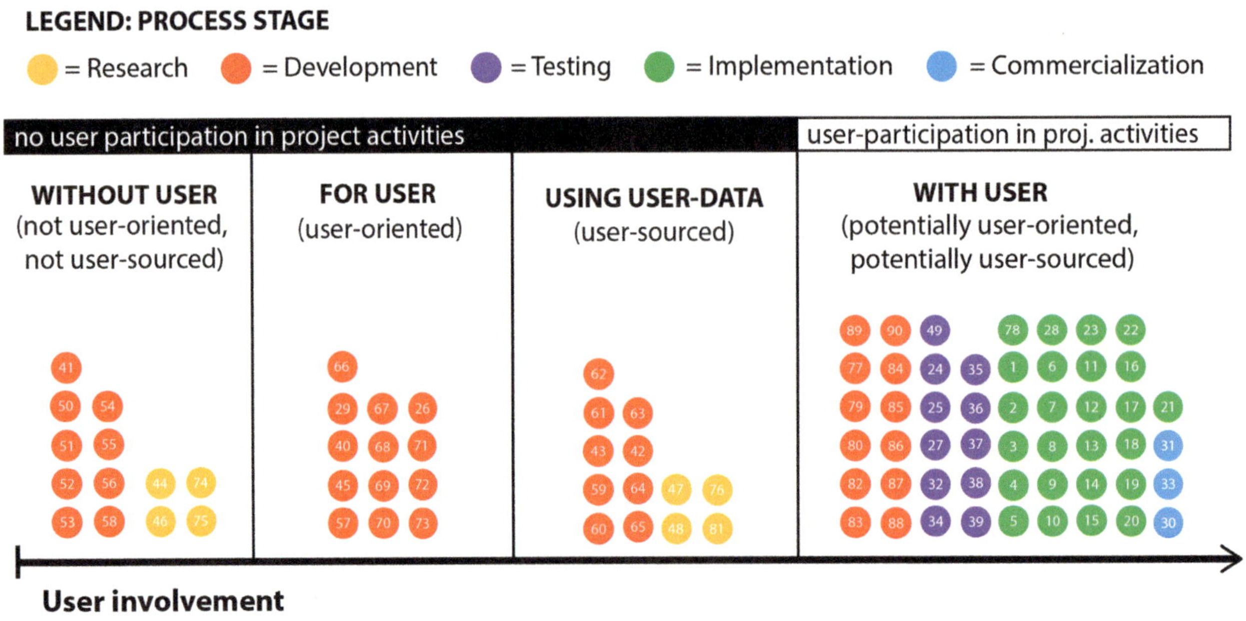

Figure 3. Degree of user involvement in the 90 potential living lab projects in Amsterdam

Based on a scan of 90 sustainable urban innovation projects and alleged living lab projects in Amsterdam, the majority of the projects that are labelled as living labs in fact do not include one or more of the defining elements of living labs, and sometimes are not even so much different from traditional top-down product development processes. However, given that many of the benefits attributed to living labs are linked to precisely to these defining characteristics, with co-creation and innovation as the most distinguishing ones, the absence of some of these basic components of living labs will also lead to disappointing performance in the field of these alleged benefits.

In order to develop an urban living lab design that does achieve the sustainable outcomes that the texts and definitions on living labs outline, the methodology of urban living labs should be more closely studied. For example, in this context, it would be valuable to extract the details of the methods recommended in the living lab literature and how they compare to the methods being used in practice, what the drivers and barriers are for the emergence and implementation of sustainable urban innovations, and which other methodological lessons can be learned from previous experiences with urban living labs. Reformulating these findings in the form of recommendations regarding the methodology, design, and governance of urban living labs would enhance the ability of urban living labs to achieve successful development, adoption, and replication of sustainable urban innovations. These aspects were beyond the scope of the current study, but could be addressed in future research.

In closing, we emphasize that today's popularity of living labs, also among subsidizing programmes preferring or demanding living labs as a research method, sometimes leads to a choice for a living lab approach without relating it to the project aims. With their specific co-creation requirement, living labs are certainly not applicable to every urban innovation project. Especially with regards to projects developing highly technological innovations, often pursued in the search for smart and sustainable cities, co-creative development, as is expected in living labs, can pose large challenges while not necessarily contributing to the quality of the solution.

Acknowledgements

This article is an outcome of a research project run under the auspices of the Amsterdam Institute for Advanced Metropolitan Solutions (AMS). We would like to thank the partners of the AMS Institute for their contribution to this research.

About the Authors

Kris Steen is a Research Fellow at the Amsterdam Institute for Advanced Metropolitan Solutions and the Chair of Urban Development Management at the Delft University of Technology, the Netherlands, where her academic work focuses on the design of concepts and strategies that produce and promote an urban environment that meets the demands of sustainability, such as "urban living labs". In 2016, she graduated Cum Laude in Architecture, Urbanism and Building Sciences at the Delft University of Technology.

Ellen van Bueren is Professor of Urban Development Management and Principal Investigator at the Amsterdam Institute for Advanced Metropolitan Solutions, and board member of the Leiden-Delft Erasmus Centre of Sustainability in the Netherlands. Her research and teaching focuses on the development of tangible concepts, tools, and principles for an integrated, area-centred approach to contemporary urban challenges. She is the author of scientific publications and a member of the editorial board of scientific journals in the field of policy management and urban development, and she teaches in various (inter)disciplinary educational settings for students and professionals.

References

Almirall, E., Lee, M., & Wareham, J. 2012. Mapping Living Labs in the Landscape of Innovation Methodologies. *Technology Innovation Management Review,* 2(9): 12–18.
http://timreview.ca/article/603

ASC. 2016. Urban Living Labs. *Amsterdam Smart City (ASC): Amsterdam Economic Board,* January 14, 2017. Accessed July 1, 2017:
https://www.amsterdameconomicboard.com/projecten/amsterdam-smart-city-2

Ashuri, B. & Durmus-Pedini, A. 2010. An Overview of the Benefits and Risk Factors of Going Green in Existing Buildings. *International Journal of Facility Management,* 1(1).

Ballon, P., Pierson, J., & Delaere, S. 2005. Test and Experimentation Platforms for Broadband Innovation: Examining European Practice. *SSRN Electronic Journal.*
http://dx.doi.org/10.2139/ssrn.1331557

Bergvall-Kåreborn, B., Ihlström Eriksson, C., & Ståhlbröst, A. 2015. Places and Spaces within Living Labs. *Technology Innovation Management Review,* 5(12): 37–47.
http://timreview.ca/article/951

Bergvall-Kåreborn, B., & Ståhlbröst, A. 2009. Living Lab: An Open and Citizen-Centric Approach for Innovation. *International Journal of Innovation and Regional Development,* 1(4): 356–370.
https://dx.doi.org/10.1504/IJIRD.2009.022727

Brankaert, R., & den Ouden, E. 2017. The Design-Driven Living Lab: A New Approach to Exploring Solutions to Complex Societal Challenges. *Technology Innovation Management Review,* 7(1): 44–51.
http://timreview.ca/article/1049

Broto, V. C., & Bulkeley, H. 2013. A Survey of Urban Climate Change Experiments in 100 Cities. *Global Environmental Change,* 23(1): 92–102.
https://dx.doi.org/10.1016/j.gloenvcha.2012.07.005

Budweg, S., Schaffers, H., Ruland, R., Kristensen, K., & Prinz, W. 2011. Enhancing Collaboration in Communities of Professionals Using a Living Lab Approach. *Production Planning & Control,* 22(5-6): 594–609.
http://dx.doi.org/10.1080/09537287.2010.536630

Buhr, K., Federley, M., & Karlsson, A. 2016. Urban Living Labs for Sustainability in Suburbs in Need of Modernization and Social Uplift. *Technology Innovation Management Review,* 6(1): 27–34.
http://timreview.ca/article/958

Bulkeley, H., & Betsill, M. M. 2013. Revisiting the Urban Politics of Climate Change. *Environmental Politics,* 22(1): 136–154.
http://dx.doi.org/10.1080/09644016.2013.755797

Bulkeley, H., & Castán Broto, V. 2013. Government by Experiment? Global Cities and the Governing of Climate Change. *Transactions of the Institute of British Geographers,* 38(3): 361–375.
http://dx.doi.org/10.1080/09537287.2010.536630

Bulkeley, H., Coenen, L., Frantzeskaki, N., Hartmann, C., Kronsell, A., Mai, L., & Palgan, Y. V. 2016. Urban Living Labs: Governing Urban Sustainability Transitions. *Current Opinion in Environmental Sustainability,* 22: 13–17.
https://dx.doi.org/10.1016/j.cosust.2017.02.003

Coenen, T., van der Graaf, S., & Walravens, N. 2014. Firing Up the City-A Smart City Living Lab Methodology. *Interdisciplinary Studies Journal,* 3(4): 118.

Coenen, T., & Robijt, S. 2017. Heading for a FALL: A Framework for Agile Living Lab Projects. *Technology Innovation Management Review,* 7(1): 37–43.
http://timreview.ca/article/1048

Cooper, R. G. 1988. The New Product Process: A Decision Guide for Management. *Journal of Marketing Management,* 3(3): 238–255.
http://dx.doi.org/10.1080/0267257X.1988.9964044

Coorevits, L., & Jacobs, A. 2017. Taking Real-Life Seriously: An Approach to Decomposing Context Beyond "Environment" in Living Labs. *Technology Innovation Management Review,* 7(1): 26–36.
http://timreview.ca/article/1047

CoreLabs. 2007. *Building Sustainable Competitiveness-Living Labs Roadmap 2007-2010.* Luleå, Sweden: Luleå University of Technology, Centre for Distance-Spanning Technology.

ENoLL. 2006. What is a Living Lab? *European Network of Living Labs (ENoLL),* Accessed August 2, 2016:
http://www.openlivinglabs.eu/FAQ

Eriksson, M., Niitamo, V.-P., & Kulkki, S. 2005. *State-of-the-Art in Utilizing Living Labs Approach to User-Centric ICT Innovation – A European Approach.* Luleå, Sweden: Luleå University of Technology, Centre for Distance-Spanning Technology.

Evans, J., & Karvonen, A. 2014. 'Give Me a Laboratory and I Will Lower Your Carbon Footprint!'—Urban Laboratories and the Governance of Low-Carbon Futures. *International Journal of Urban and Regional Research,* 38(2): 413–430.
https://dx.doi.org/10.1111/1468-2427.12077

Femenías, P., & Hagbert, P. 2013. The Habitation Lab: Using a Design Approach to Foster Innovation for Sustainable Living. *Technology Innovation Management Review,* 3(11): 15–21.
http://timreview.ca/article/741

Feurstein, K., Hesmer, A., Hribernik, K. A., Thoben, K., & Schumacher, J. 2008. Living Labs: A New Development Strategy. In J. Schumacher &V.P. Niitamo (Eds.), *European Living Labs – A New Approach for Human Centric Regional Innovation:* 1–14. Berlin: Wissenschaftlicher Verlag.

Franz, Y., Tausz, K., & Thiel, S. K. 2015. Contextuality and Co-Creation Matter: A Qualitative Case Study Comparison of Living Lab Concepts in Urban Research. *Technology Innovation Management Review,* 5(12): 48–55.
http://timreview.ca/article/952

Friedrich, P., Karlsson, A., & Federley, M. 2013. *SubUrbanLab Report D2.1: Boundary Conditions for Successful Urban Living Labs.* SubUrbanLab.
http://suburbanlab.eu/tulokset/?lang=en

Georges, A., Schuurman, D., & Vervoort, K. 2016. Factors Affecting the Attrition of Test Users During Living Lab Field Trials. *Technology Innovation Management Review,* 6(1): 35–44.
http://timreview.ca/article/959

Gómez-Barroso, J.-L., Feijóo, C., Pascu, C., & van Lieshout, M. 2009. User-Led, Citizen Innovation at the Interface of Services. *info,* 11(6): 82–96.
https://dx.doi.org/10.1108/14636690910996731

Gray, M., Mangyoku, M., Serra, A., Sánchez, L., & Aragall, F. 2014. Integrating Design for All in Living Labs. *Technology Innovation Management Review*, 4(5): 50–59.
http://timreview.ca/article/793

Hakkarainen, L., & Hyysalo, S. 2013. How Do We Keep the Living Laboratory Alive? Learning and Conflicts in Living Lab Collaboration. *Technology Innovation Management Review*, 3(12): 16–22.
http://timreview.ca/article/749

Hakkarainen, L., & Hyysalo, S. 2016. The Evolution of Intermediary Activities: Broadening the Concept of Facilitation in Living Labs. *Technology Innovation Management Review*, 6(1): 45–58.
http://timreview.ca/article/960

Higgins A., Klein S. 2011. Introduction to the Living Lab Approach. In Y. H. Tan, N. Björn-Andersen, S. Klein, & B. Rukanova (Eds), *Accelerating Global Supply Chains with IT-Innovation*: 31–36. Berlin: Springer, Heidelberg.
https://dx.doi.org/10.1007/978-3-642-15669-4_2

Joss, S. 2015. *Sustainable Cities: Governing for Urban Innovation*. Basingstoke, UK: Palgrave Macmillan.

Juujärvi, S., & Lund, V. 2016. Enhancing Early Innovation in an Urban Living Lab: Lessons from Espoo, Finland. *Technology Innovation Management Review*, 6(1): 17–26.
http://timreview.ca/article/957

Juujärvi, S., & Pesso, K. 2013. Actor Roles in an Urban Living Lab: What Can We Learn from Suurpelto, Finland? *Technology Innovation Management Review*, 3(11): 22–27.
http://timreview.ca/article/742

Karré, P. M., Vanhommerig, I., & van Bueren, E. 2015. De stad als lab voor sociale verandering. *Bestuurskunde*, 24(1): 1–11.
https://dx.doi.org/10.5553/Bk/092733872015024001001

Katzy, B. 2012. Designing Viable Business Models for Living Labs. *Technology Innovation Management Review*, 2(9): 19–24.
http://timreview.ca/article/604

Kieboom, M. 2013. Living Labs: Innoveren in een proeftuin vraagt om nieuwe manieren van tuinieren. *Kennisland*, August 20, 2013. Accessed August 2, 2016:
https://www.kl.nl/opinie/living-labs-innoveren-in-een-proeftuin-vraagt-om-nieuwe-manieren/

Kresin, F. 2009. Living Labs: Hoop of Hype? *Frankwatching*, March 31, 2009. Accessed August 1, 2016.
https://www.frankwatching.com/archive/2009/03/31/living-labs-hoop-of-hype/

Krueger, R., & Buckingham, S. 2009. Creative-City Scripts, Economic Development, and Sustainability. *Geographical Review*, 99(1): iii–xii.
https://dx.doi.org/10.1111/j.1931-0846.2009.tb00414.x

Kusiak, A. 2007. Innovation: The Living Laboratory Perspective. *Computer-Aided Design and Applications*, 4(6): 863–876.
http://dx.doi.org/10.1080/16864360.2007.10738518

Leminen, S., Westerlund, M., & Nyström, A.-G. 2012. Living Labs as Open-Innovation Networks. *Technology Innovation Management Review*, 2(9): 6–11.
http://timreview.ca/article/602

Leminen, S. 2013. Coordination and Participation in Living Lab Networks. *Technology Innovation Management Review*, 3(11): 5–14.
http://timreview.ca/article/740

Leminen, S. 2015. *Living Labs as Open Innovation Networks-Networks, Roles and Innovation Outcomes*. Doctoral dissertation, Aalto University, Finland.
https://aaltodoc.aalto.fi/handle/123456789/17899

Leminen, S. K., Nyström, A.-G., Westerlund, M., Kortelainen, M. J., & Zolkiewski, J. 2016. The Effect of Network Structure on Radical Innovation in Living Labs. *Journal of Business & Industrial Marketing*, 31(6): 737–757.
https://dx.doi.org/10.1108/JBIM-10-2012-0179

Leminen, S., Turunen, T., & Westerlund, M. 2015. The Grey Areas Between Open and Closed in Innovation Networks. *Technology Innovation Management Review*, 5(12): 6–18.
http://timreview.ca/article/948

Leminen, S., & Westerlund, M. 2017. Categorization of Innovation Tools in Living Labs. *Technology Innovation Management Review*, 7(1): 15–25.
http://timreview.ca/article/1046

Lemke, M. 2009. *Living Labs for User-Driven Open Innovation: An Overview of the Living Labs Methodology, Activities, Achievements*. Brussels: European Commission: New Infrastructure Paradigms and Experimental Facilities.
https://dx.doi.org/10.2759/34481

Lesnikowski, A., Ford, J., Biesbroek, R., Berrang-Ford, L., & Heymann, S. J. 2016. National-Level Progress on Adaptation. *Nature Climate Change*, 6(3): 261–264.
https://dx.doi.org/10.1038/nclimate2863

McCormick, K. 2016. *GUST Policy Brief: Urban Living Labs*. Governance of Urban Sustainability Transitions (GUST).
http://www.urbanlivinglabs.net/p/publications.html

Mikkelä, K. 2008. *Finnish Living Lab and User Driven Open Innovation Situation*. Paper presented at the Workshop on Living Labs, Halmstad University, Sweden, May 29–30, 2008.

Mulder, I. 2012. Living Labbing the Rotterdam Way: Co-Creation as an Enabler for Urban Innovation. *Technology Innovation Management Review*, 2(9): 39–43.
http://timreview.ca/article/607

Nevens, F., Frantzeskaki, N., Gorissen, L., & Loorbach, D. 2013. Urban Transition Labs: Co-Creating Transformative Action for Sustainable Cities. *Journal of Cleaner Production*, 50(1): 111–122.
https://dx.doi.org/10.1016/j.jclepro.2012.12.001

Niitamo, V. P., Kulkki, S., Eriksson, M., & Hribernik, K. A. 2006. State-of-the-Art and Good Practice in the Field of Living Labs. In Proceedings of the IEEE International Technology Management Conference (ICE) 2006: 1–8.
https://dx.doi.org/10.1109/ICE.2006.7477081

Niitamo, V.-P., Westerlund, M., & Leminen, S. 2012. A Small-Firm Perspective on the Benefits of Living Labs. *Technology Innovation Management Review*, 2(9): 44–49.
http://timreview.ca/article/608

Pallot, M., & Pawar, K. 2012. *A Holistic Model of User Experience for Living Lab Experiential Design*. Paper presented to the 18th International ICE Conference on Engineering, Technology and Innovation, Munich, Germany.
https://dx.doi.org/10.1109/ICE.2012.6297648

Perry, B. 2006. Science, society and the University: A Paradox of Values. *Social Epistemology*, 20(3-4): 201–219.
https://dx.doi.org/10.1080/02691720600879798

Pierson, J., & Lievens, B. 2005. Configuring Living Labs for a 'Thick' Understanding of Innovation. In *Proceedings of the Ethnographic Praxis in Industry Conference, 2005:* 114–127.
https://dx.doi.org/10.1111/j.1559-8918.2005.tb00012.x

Prahalad, C. K., & Ramaswamy, V. 2004. Co-Creation Experiences: The Next Practice in Value Creation. *Journal of Interactive Marketing,* 18(3): 5–14.
https://dx.doi.org/10.1002/dir.20015

Rits, O., Schuurman, D., & Ballon, P. 2015. Exploring the Benefits of Integrating Business Model Research within Living Lab Projects. *Technology Innovation Management Review,* 5(12): 19–27.
http://timreview.ca/article/949

Schaffers, H., Cordoba, M. G., Hongisto, P., Kallai, T., Merz, C., & Van Rensburg, J. 2007. Exploring Business Models for Open Innovation in Rural Living Labs. In *Proceedings of the IEEE International Technology Management Conference (ICE), 2007:* 1–8.

Schaffers, H., & Turkama, P. 2012. Living Labs for Cross-Border Systemic Innovation. *Technology Innovation Management Review,* 2(9): 25–30.
http://timreview.ca/article/605

Schaub, A. 2016. Amsterdam Living Lab. *Waag Society,* Accessed August 2, 2017:
https://www.waag.org/nl/project/amsterdam-living-lab

Schuurman, D., & De Marez, L. 2012. Structuring User Involvement in Panel-Based Living Labs. *Technology Innovation Management Review,* 2(9): 31–38.
http://timreview.ca/article/606

Schuurman, D., De Marez, L., & Ballon, P. 2013. Open Innovation Processes in Living Lab Innovation Systems: Insights from the LeYLab. *Technology Innovation Management Review,* 3(11): 28–36.
http://timreview.ca/article/743

Schuurman, D., De Marez, L., & Ballon, P. 2016. The Impact of Living Lab Methodology on Open Innovation Contributions and Outcomes. *Technology Innovation Management Review,* 1(6): 7–16.
http://timreview.ca/article/956

Schuurman, D., & Tõnurist, P. 2017. Innovation in the Public Sector: Exploring the Characteristics and Potential of Living Labs and Innovation Labs. *Technology Innovation Management Review,* 7(1): 7–14.
http://timreview.ca/article/1045

Schuurman, D., De Marez, L., & Ballon, P. 2016. The Impact of Living Lab Methodology on Open Innovation Contributions and Outcomes. *Technology Innovation Management Review,* 1(6): 7–16.
http://timreview.ca/article/956

Ståhlbröst, A. 2012. A Set of Key Principles to Assess the Impact of Living Labs. *International Journal of Product Development,* 17(1-2): 60–75.
https://dx.doi.org/10.1504/IJPD.2012.051154

Ståhlbröst, A. 2013. A Living Lab as a Service: Creating Value for Micro-enterprises through Collaboration and Innovation. *Technology Innovation Management Review,* 3(11): 37–42.
http://timreview.ca/article/744

Ståhlbröst, A., & Lassinantti, J. 2015. Leveraging Living Lab Innovation Processes through Crowdsourcing. *Technology Innovation Management Review,* 5(12): 28–36.
http://timreview.ca/article/950

Steen, K., & van Bueren, E. 2017. *Urban Living Labs: A Living Lab Way of Working.* AMS Research Report 2016–2017. Amsterdam: AMS Institute.
http://www.ams-institute.org/news/out-now-urban-living-labs-a-living-lab-way-of-working/

Tang, T., & Hämäläinen, M. 2012. *Living Lab Methods and Tools for Fostering Everyday Life Innovation.* Paper presented at the 18th International ICE Conference on Engineering, Technology and Innovation (ICE), Munich, Germany.
https://dx.doi.org/10.1109/ICE.2012.6297644

Tukiainen, T., Leminen, S., & Westerlund, M. 2015. Cities as Collaborative Innovation Platforms. *Technology Innovation Management Review,* 5(10): 16–23.
http://timreview.ca/article/933

U.S. Congress. 1969. National Environmental Policy Act of 1969. *Public Law,* 91(190), 1–5.

van Bijsterveldt, M. 2016. Living Labs. *Ruimtelijke Adaptatie.* Accessed August 2, 2016.
https://ruimtelijkeadaptatie.nl/living-labs/

van Bueren, E., & Broekhans, B. 2013. Individual Projects as Portals for Mainstreaming Niche Innovations. In R. L. Henn & A. J. Hofman (Eds.), *Constructing Green: The Social Structures of Sustainability:* 145–167. Cambridge, MA: MIT Press.
https://dx.doi.org/ 10.7551/mitpress/9780262019415.003.0007

van Bueren, E., & De Jong, J. 2007. Establishing Sustainability: Policy Successes and Failures. *Building Research & Information,* 35(5): 543–556.
http://dx.doi.org/10.1080/09613210701203874

van der Heijden, E. 2016. Urban Management Fieldlabs: Living Lab Description. *European Network of Living Labs.* Accessed August 2, 2016:
http://www.openlivinglabs.eu/livinglab/urban-management-fieldlabs

Veeckman, C., Schuurman, D., Leminen, S., & Westerlund, M. 2013. Linking Living Lab Characteristics and Their Outcomes: Towards a Conceptual Framework. *Technology Innovation Management Review,* 3(12): 6–15.
http://timreview.ca/article/748

Veeckman, C., & van der Graaf, S. 2015. The City as Living Laboratory: Empowering Citizens with the Citadel Toolkit. *Technology Innovation Management Review,* 5(3): 6–17.
http://timreview.ca/article/877

Veryzer, R. W., & de Mozota, B. B. 2005. The Impact of User-Oriented Design on New Product Development: An Examination of Fundamental Relationship. *The Journal of Product Innovation Management,* 22(2): 128–143.
http://dx.doi.org/10.1111/j.0737-6782.2005.00110.x

Vincent, S. 2016. Lab Builders: Some Thinking about the Next Generation of Lab Programmes. *Design for Europe,* June 21, 2016. Accessed August 2, 2016:
http://www.designforeurope.eu/news-opinion/lab-builders-some-thinking-about-next-generation-lab-programmes

Voytenko, Y., McCormick, K., Evans, J., & Schliwa, G. 2016. Urban Living Labs for Sustainability and Low Carbon Cities in Europe: Towards a Research Agenda. *Journal of Cleaner Production,* 123: 45–54.
https://doi.org/10.1016/j.jclepro.2015.08.053

Wallin, S. 2014. *APRILab: Guidelines to Define and Establish an Urban Living Lab.* Amsterdam: University of Amsterdam, Amsterdam Institute of Social Science Research.

Westerlund, M., & Leminen, S. 2011. Managing the Challenges of Becoming an Open Innovation Company: Experiences from Living Labs. *Technology Innovation Management Review,* 1(1): 19–25. http://timreview.ca/article/489

Wheeler, S. M., & Beatley, T. (Eds.). 2014. *Sustainable Urban Development Reader.* London: Routledge.

Keywords: living labs, urban living labs, definition, characteristics, cities, TIM Review

Appendices

Available online at: tinyurl.com/ycoua757

Appendix A. Overview of definitions of living labs in the TIM Review

Appendix B. Defining characteristics of (urban) living labs mentioned in living lab literature

14

Innovation Rhetoric and Reality: An Introduction to the TIM Review's Special Issue on Innovation and Entrepreneurship in Australia

Rowena Barrett

"Innovation as a term has become meaningless."

Scott Middleton
Chief Executive Officer
Terem Technologies (Australia)

In this introduction to the *Technology Innovation Management Review's* special issue on Innovation and Entrepreneurship in Australia, Guest Editor Rowena Barrett reflects on the perceptions of "innovation", both in terms of its evolving concepts and terminology (the rhetoric) and its frontline application (the reality). Prompted by the recent launch and ongoing implementation of Australia Government's National Innovation and Science Agenda's, this special issue focuses on insights into innovation and entrepreneurship from the Australian context.

Introduction

It is rumoured that the word "innovation" was barred from ministerial lexicons during Tony Abbott's terms as Prime Minister of Australia. Now, with Malcolm Turnbull as Australia's Prime Minister, by way of total contrast, it is difficult to find a ministerial utterance that does not connect innovation with jobs and growth. However, as the 2015 Australian Innovation System report makes clear, innovation is a complex phenomenon (Department of Industry, Innovation and Science, 2015).

Innovation exists along a number of axes ranging from radical (or disruptive) to incremental, first-in-organization to first-in-world, product to process, sector to sector, as well as over the lifecycle, and simply a change in focus over time (i.e., design thinking, open innovation). One danger is that innovation is simply reduced to a notion of change. As writers from Marx and Schumpeter to more recently Stan Metcalfe and academics at the Science Policy Research Unit (SPRU; http://www.sussex.ac.uk/spru/) at the University of Sussex have all pointed out, change is endemic to capitalism. Therefore any discussion of innovation must move beyond the obvious threat of tautology.

What Is Innovation – Other than Simply Change?

Innovation is a poorly defined concept. The meaning of the term varies considerably depending on the context in which it is used, and for what purpose. It is most widely assumed that innovation is about technology and scientific advancements. This view dominates analyses and discussions around policy and regulatory issues, such as intellectual property (IP) protection. This more narrow focus on technological innovation has, unsurprisingly, placed the policy spotlight on the technical skill requirements for innovation, such as skills in STEM (science, technology, engineering, and mathematics). As a result, some governments' policies for promoting growth in STEM skills have called for compulsory science education in schools and at universities, or for students to be introduced to skills such as coding from primary school (Australian Government, 2015; US Government Accountability Office, 2005). These skills are undoubtedly critical to a great deal of new innovation, especially as new technologies disrupt established systems of producing goods and services and challenge traditional business models, but innovation needs to be understood through a lens that is broader than simply technology.

The most generally accepted definition of innovation comes from the OECD's Oslo Manual (2005):

"The implementation of a new or significantly improved product (good or service), or process, a new marketing method, or a new organisational method in business practices, workplace organisation or external relation."

This definition captures innovations that are new or significant to the firm, as well as new to the world. It emphasizes innovation as being broader than simply technical breakthroughs and their application in industry. The definition draws attention to implementation of discoveries and highlights the importance of diffusion of innovation.

The definition includes any activity in an *innovation process* – from the conception of new ideas, inventions, and discoveries; to development and testing; to the production, marketing, and commercialization of those inventions within the ecosystem, not just within particular organizations alone. This also takes us far beyond a simple, technical, and research and development focused view of innovation. These activities in the innovation cycle occur in a dynamic and complex manner, and require the firm to undertake three distinctive sets of activities. Teece (2010) argues these are:

1. *Sensing:* the identification and assessment of new opportunities for growth and profit

2. *Seizing:* the mobilization of resources, skills, and capabilities to realize the opportunity and to capture opportunities for creating value

3. *Transforming:* ongoing efforts to improve and renew the original innovation to sustain the value creating opportunities such innovations present

The Government's *Australia Innovation System Report 2015* (Department of Industry, Innovation and Science, 2015) takes a systems approach to innovation, as the report's title suggests. In it, the argument goes that, without understanding how the components of the national innovation ecosystem interact, it is impossible to identify the causes and implications of innovation. The report defines an innovation system as follows:

"an open network of organisations that interact with each other and operate within framework conditions that regulate their activities and interactions. The three components of the innovation system – networks, innovation activities and framework conditions – collectively function to produce and diffuse innovations that have, in aggregate, economic, social and/or environmental value."

Networks includes geographic clusters, business associations, and supply chains. Framework conditions encompass a range of macro-economic, cultural, educational, and policy settings that nurture innovation. Innovation activities can include training, research and development, venture capital investment, and patenting activity.

As Salter and Alexy (2014) have commented, there is a whole industry of consultants and academics putting new words in front of the word innovation (for an extended discussion see Cunningham et al., 2016). Starting with a distinction between incremental and radical innovation (with the recent addition of the concept of disruption), what we have allied to this is the distinction between first-in-organization compared to first-in-world innovation. The concept widened to include service, technological, and organizational innovation. More recently, the concept of open innovation has gained credence with the allied concepts of customer or user-linked innovation. These latter two are also described as hidden innovation. However, this approach is not the only way in which the concept of innovation has expanded.

The above can be contrasted with the Ten Types of Innovation or "the building blocks of breakthroughs", expounded by Keeley and colleagues (2013). The Ten Types are broken down into three categories:

1. *Configuration:* profit model; network; structure; process

2. *Offering:* product performance; product system

3. *Experience:* service; channel; brand; customer engagement

Sophisticated innovation, it is suggested, uses many types of innovation, customized elegantly and orchestrated with care (Keeley et al., 2015). There are similarities and overlaps with the first list but these are not the only choices available. A Google search under "types of innovation", will yield any number of alternative lists and categorizations.

The problem is that the concept of innovation threatens to become, in Andrew Sayer's (1992) terms, a "chaotic concept". This means the concept, while having great political purchase, has little analytical value as it is packed with many differing and sometimes contradictory elements. It lumps together the unrelated and the inessential and divides the indivisible. However, all this notwithstanding, there has never been a more exciting time to launch a national innovation and science agenda.

Australia's National Innovation and Science Agenda

In December 2015, Australia's National Innovation and Science Agenda was launched with the subheading, "Welcome to the ideas boom" (Australian Government, 2015). A sound bite perhaps, but also a signal to Australians that the resources boom might not be so boomlike anymore! The National Innovation and Science Agenda makes clear that innovation and science are seen to be critical for Australia delivering new sources of growth, maintaining high-wage jobs and seizing the next wave of economic prosperity (Australian Government, 2015). Innovation is viewed as being important to every sector of the economy and is about new and existing businesses creating new products, processes, and business models.

Australia's National Innovation and Science Agenda has four key pillars: i) culture and capital; ii) collaboration; iii) talent and skills; and iv) government as an exemplar. To support the agenda, AUD$1.1 billion in funding was allocated to initiatives in these four areas, key aspects of which are outlined below:

1. *Culture and capital:* this pillar builds on the acknowledgement that Australia has a poor record in converting bright ideas into commercial realities – only 9% of Australian small and medium-sized enterprises brought a new idea to market in 2012–13 compared to 19% in the top 5 OECD countries (Australian Government, 2015). Under this pillar, new tax breaks are offered to help overcome what is seen to be a bias against businesses that take risks and innovate. In addition the Commonwealth Scientific and Industrial Research Organisation (CSIRO; csiro.au) and the Biomedical Translation Fund (tinyurl.com/hovnyc7) will co-invest in commercializing promising ideas.

2. *Collaboration:* this pillar is framed around the understanding that Australia's rate of collaboration between industry and academia is reported to be the lowest across OECD nations (PwC, 2015). The aim here is to encourage researchers and businesses to collaborate. The National Innovation and Science Agenda promotes partnership-based research for universities and will target investment to what is seen to be critical research infrastructure.

3. *Talent and skills:* this pillar is almost entirely focused on STEM skills. The first aim of funding initiatives here is to encourage more Australian students to study science, mathematics, and computing in schools. The second aim is to make it easier to attract more entrepreneurial and research talent to come to Australia from overseas.

4. *Government as exemplar:* under this pillar, the aim is to move government from its position of lagging the private sector in innovation. Funding here is all about helping government to be more innovative in how services are delivered as well as making it easier for startups and innovative small businesses to sell technology services to government.

As part of the National Innovation and Science Agenda's promise, the Government claims that it will make innovation central to all policies. To this end, a new body, Innovation and Science Australia (tinyurl.com/z6fgxo4) has been established alongside the Digital Transformation Office (dto.gov.au), a Digital Market Place (tinyurl.com/hxqmlja) and a Business Research and Innovation Initiative (tinyurl.com/j7wd822). The Data61 (www.csiro.au/en/Research/D61) data innovation group has been established to help develop new technology-based industries and transform existing ones.

The National Innovation and Science Agenda approach is based on the assumption that problems in innovation can be laid at the door of an unresponsive and over burdensome government/public sector, lack of support for startups, insufficient focus on STEM education (particularly for women), and a lack of encouragement for collaboration, particularly between industry and universities. However, there are indications that the problems might be somewhat more fundamental.

(Mis)management?

The recently published major study of Australian Leadership at Work (Gahan et al., 2016) suggests there may be more fundamental problems with Australian management. After surveying some 8000 people across 2703 organizations and 2561 workplaces, as well as interviewing people from the top to the bottom of organizations, the conclusion is "Australian leaders, on the whole, have not mastered the fundamentals of management" (Gahan et al., 2016). Gahan and colleagues (2016) identify seven significant gaps and weaknesses, which together are a cause for major concern:

1. Many Australian workplaces are underperforming.

2. Many Australian organizations do not get the basics right.

3. Few Australian organisations report high levels of innovation.

4. Many Australian leaders are not well-trained for the job.

5. Too many Australian organizations under-invest in leadership development, especially at the frontline.

6. Leadership in Australian organizations does not reflect wider social diversity.

7. Many senior leaders do not draw on strategic advice in making decisions about the future.

In many ways, these were the same criticisms of Australian management that were outlined in the 1995 Enterprising Nation report (the Karpin Report). Given the 20-year gap between these studies, the lack of progress is worrying.

Focusing specifically on innovation, and again following the findings of the Karpin Report (1995) as well as Green (2009), the Leadership at Work study suggests that lack of leadership for innovation remains a long-standing challenge. In the report, Gahan and colleagues (2016) conclude:

• Australian organizations struggle to turn knowledge and information inputs into innovation outputs.

• Small organizations struggle most with radical innovation. Public sector organizations score considerably higher on radical innovation.

• Short-term performance is more affected by incremental product and service improvements and is less susceptible to radical innovation

• Highly innovative organizations actively sense, internalize, and act upon new knowledge and changes in their operational environment.

• High-performance work practices relate positively to innovation.

• Employee management based on clear goals produces more innovative outcomes.

In a similar vein, the 2016 American Express CFO Future-Proofing Survey (American Express, 2016) suggests that Australian business may not be on the cusp of an ideas boom, with about 70% of Australian mid-sized businesses not significantly investing in innovation. This led the *Australian Financial Review* to speculate as to whether "the Government's ideas boom may be hot air" (Sherbon, 2016) when 40% of Chief Futures Officers said their organization had failed to bring in any new product or service innovation in the previous three years. Half of all CFOs believed their organizations were not evolving fast enough, and that innovation was an ad hoc rather than strategic activity.

Unsurprisingly, when OECD data was drawn upon, the American Express report concluded that Australia has one of the weakest levels of network and collaborative innovation compared with other OECD countries. This finding helps explain why Australia ranks 21st out of 32 countries for innovation in the OECD ranking (American Express, 2016). This position was acknowledged in the Australian Government's own Australian Innovation System Report. In a world economy increasingly dominated by global value chains, the relative isolation of Australian business from these structures (within which innovation and R&D are increasingly concentrated) is a matter of concern. This point was stressed in the Australian Innovation System Report (Department of Industry, Innovation and Science, 2014):

> *"Businesses that participate in global value chains have been argued to be more innovative, more engaged in research and development (R&D) and skills development, drive the highest productivity premium, and can support high unit labour costs... Participation in global value chains also drives a step change in business culture by challenging participants to upgrade their management, financing and technology, and encourages greater collaboration... Investing in research and innovation will be the key to maintaining a strong position in a global value chain as a price maker."*

However, as a number of commentators point out, Australia is relatively under-integrated into the world economy (see Drake-Brockman, 2014). More particularly, in the 2014 Australian Innovation System Report, it is argued that, according to the OECD's global value chain participation index, Australia's overall participation in global value chains is below the OECD median and well behind global value chain hub countries (Department of Industry, Innovation and Science, 2014).

In reviewing the results of the Global Innovation Index (GII), produced jointly by Cornell University, INSEAD, and the World Intellectual Property Organization, Cunningham and colleagues (2016) argue:

"In 2015, Australia ranked 17 overall out of the 141 countries for which a GII score is available, confirming that Australia has established a comparatively healthy environment for enterprise innovation. However, a further breakdown of this index reveals some important points of weakness. Significantly, Australia shows a notable disparity between innovation inputs (ranked 10th) and innovation outputs (ranked 24th), which include technology and knowledge (ranked 39th) and creative outputs (ranked 7th). This implies a lack of 'innovation efficiency'.

While the input measures are generally adequate, Australia performs significantly below its overall ranking on measures capturing the human capital inputs into innovation systems. On the output side, low scores are particularly prevalent in the areas of knowledge impact and knowledge diffusion. Overall, the GII indicates shortcomings in the capacity of Australian enterprise to generate and, more specifically, to bring innovations to application and diffusion. The low scores in business sophistication—especially in innovation linkages (ranked 38th) and knowledge absorption (ranked 63rd)—indicate that rather than lacking skills in general, Australia lacks capacity in using these skills and other inputs for innovation."

Research for the Australian Government suggest that although Australia has a relatively high (but declining) rate of small business formation, only a very small percentage of startups (3.2%) exhibit serious employment growth and yet this tiny minority accounts for 77% of total post-entry job creation by micro startups (Hendrickson et al., 2015, 2016).

There appears to be some fundamental issues that go wider and deeper than those identified in the National Innovation and Science Agenda. Since it was announced, there has been an Australian budget handed down, but it was not met with universal approval. Professor Roy Green, Dean of the University of Technology Sydney Business School, suggested "we have to look very hard to find the 'ideas boom' in this budget" (Green et al., 2016), arguing that only AUD$1 billion of the AUD$3 billion cut in research and innovation expenditure under the previous Abbott government was being restored. Or, as Professor Marek Kowalkiewicz and colleagues from Queensland University of Technology's PwC Chair of Digital Economy team also noted in their assessment of the 2016 Australian budget, it does not address those who face industry and professional disruption due to automation and the changing nature of work (Green et al., 2016).

John Bessant from Essex University has suggested that the Red Queen in *Alice in Wonderland* is a powerful metaphor for innovation in the modern world (Bessant, 2016). Remember, the Red Queen keeps changing rules, and the game, and this is perfectly normal in Wonderland – and reflects the reality of our own world, it seems. As Bessant argues, "simply recognizing that we need to change what we offer, and how we create and deliver it, isn't going to be enough. We've also got to have the ability to step back and reconfigure our approaches to doing so as the game shifts and the rules change beneath our feet" (Bessant, 2016).

Thankfully, the contributors to this special issue help provide a pathway through this chaos, which reflects both the rhetoric and reality of innovation. Through sharing insights gleaned from the Australian context, we hope the articles in this issue will benefit all players and supporters of the innovation game, with its ever-changing rules, wherever in the world it is being played.

About the Author

Rowena Barrett is Head of the School of Management at Queensland University of Technology (QUT) in Brisbane, Australia. She has played leading roles in organizing, managing, and administering research in universities in Australia and the United Kingdom. Rowena holds a PhD from the University of Melbourne, Australia, on industrial relations in smaller firms, and she has spent many years investigating the ways people in smaller firms are managed and how those firms are organized. She has a sizeable portfolio of research publications including books, book chapters, journal articles, as well practitioner papers and reports. Her research has been funded by university, industry, and government sources in Australia and the United Kingdom. She has extensive experience in teaching in the areas of management, human resource management, and industrial relations, while she has supervised over 10 PhDs to successful completion. She has also held numerous board and council member positions, both within and beyond academia. In 2015, Rowena chaired the ISPIM Innovation Summit "Changing the Innovation Landscape", which took place at QUT in Brisbane, Australia, from 6–9 December.

References

American Express. 2016. Game Plan for Growth: 2016 American Express CFO Future-Proofing Survey. *Chief Future Officer; American Express Australia.* Accessed April 15, 2016: http://www.chieffutureofficer.com/

Australian Government. 2015. *National Innovation and Science Agenda.* Canberra: Commonwealth of Australia, Department of the Prime Minister and Cabinet. http://www.innovation.gov.au/page/agenda

Bessant, J. 2016. Playing Chess with the Red Queen. *The Hype Innovation Blog,* June 8, 2016. Accessed June 15, 2016: http://blog.hypeinnovation.com/playing-chess-with-the-red-queen

Cunningham, S., Theilacker. M., Gahan, P., Callan, V., & Rainnie, A. Forthcoming in 2016. *Skills and Capabilities for Australian Enterprise Innovation.* Report for the Australian Council of Learned Academies (ACOLA).

Department of Industry, Science and Innovation. 2014. *Australian Innovation System Report 2014.* Canberra: Australian Government, Office of the Chief Economist. http://www.industry.gov.au/innovationreport

Department of Industry, Science and Innovation. 2015. *Australian Innovation System Report 2015.* Canberra: Australian Government, Office of the Chief Economist. http://www.industry.gov.au/innovationreport

Drake-Brockman, J. 2014. Advanced Manufacturing Global Value Chains and Policy Implications. In *Advanced Manufacturing: Beyond the Production Line:* 57–70. Melbourne: Committee for Economic Development of Australia (CEDA). http://apo.org.au/files/Resource/ceda_advanced_manufacting_2014_final.pdf

Gahan, P., Adamovic, M., Bevitt, A., Harley, B., Healy, J., Olsen, J. E., & Theilacker, M. 2016. *Leadership at Work: Do Australian Leaders Have What It Takes?* Melbourne: Centre for Workplace Leadership, University of Melbourne. http://sal.workplaceleadership.com.au

Green, R. 2009. *Management Matters in Australia: Just How Productive Are We? Findings from the Australian Management Practices and Productivity Global Benchmarking Project.* Canberra: Australian Government, Department of Innovation, Industry, Science and Research. http://www.industry.gov.au/industry/OtherReportsandStudies/Documents/ManagementMattersinAustraliaReport.pdf

Green, R., Webster, B., Kowalkiewicz, M., Barrett, R., & Shahiduzzaman, S. M. 2016. Budget Does Little to Help 'Transition' the Economy. *The Conversation,* May 3, 2016. Accessed June 15, 2016: http://theconversation.com/budget-does-little-to-help-transition-the-economy-57637

Hendrickson, L., Bucifal, S., Balaguer, A., & Hansell, D. 2015. *Employment Dynamics of Australian Entrepreneurship.* Research Paper 4/2015. Canberra: Australian Government, Office of the Chief Scientist. http://www.industry.gov.au/Office-of-the-Chief-Economist/Research-Papers/Pages/The-employment-dynamics-of-Australian-entrepreneurship.aspx

Hendrickson, L., Bucifal, S., Balaguer, A., & Hansell, D. 2016. Employment Dynamics of Australian Entrepreneurship: A Management Perspective. *Technology Innovation Management Review,* 6(6): 33–40. http://timreview.ca/article/995

Karpin, D. 1995. *Enterprising Nation: Renewing Australia's Managers to Meet the Challenge of the Asia-Pacific Century.* Canberra: Australian Government. http://trove.nla.gov.au/version/45053541

Keeley, L., Pikkel, R., Quinn, B., & Walters, H. 2013. *Ten Types of Innovation.* New Jersey: John Wiley & Sons.

OECD. 2005. *The Measurement of Scientific and Technological Activities: Proposed Guidelines for Collecting and Interpreting Innovation Data: Oslo Manual.* Paris: Organisation for Economic Co-operation and Development (OECD). http://www.oecd.org/sti/inno/oslomanualguidelinesforcollectingandinterpretinginnovationdata3rdedition.htm

PwC. 2015. *Innovate and Prosper: Ensuring Australia's Future Competitiveness through University–Industry Collaboration.* Sydney: PricewaterhouseCoopers (PwC). http://www.atn.edu.au/siteassets/publications/atninnovateprosper.pdf

Salter, A., & Alexy, O. 2014. The Nature of Innovation. In M. Dodgson, D. M. Gann, & N. Phillips (Eds), *The Oxford Handbook of Innovation Management:* 26–52. Oxford: Oxford University Press.

Sayer, A. 1992. *Method in Social Science. A Realist Approach.* Abingdon, England: Routledge.

Sherbon, J. 2016. Government's Ideas Boom May Be More Hot Air than Substance. *The Australian Financial Review,* May 31, 2016. Accessed June 15, 2016: http://www.afr.com/news/special-reports/culture-tech/governments-ideas-boom-may-be-more-hot-air-than-substance-20160530-gp71rn

Teece, D. J. 2010. Technological Innovation and the Theory of the Firm: The Role of Enterprise-Level Knowledge, Complementarities, and (Dynamic) Capabilities. In B. H. Hall & N. Rosenberg (Eds.), *Handbook of the Economics of Innovation,* 1: 679–730. http://dx.doi.org/10.1016/S0169-7218(10)01016-6

US Government Accountability Office. 2005. *Higher Education: Science, Technology, Engineering, and Mathematics Trends and the Role of Federal Programs.* Washington, DC: United States Government Accountability Office. http://www.gao.gov/products/GAO-06-702T

Keywords: innovation, entrepreneurship, Australia, National Innovation and Science Agenda

Accelerating Research Innovation by Adopting the Lean Startup Paradigm

Kaisa Still

> "*It is the unlikely partnership between academia, the public sector and private enterprise that allow us to navigate the path from discovery, to innovation, to transformation. The process, however, is often unwieldy, taking decades to go from primary discovery to a measurable impact on society.*"
>
> Greg Satell
> Innovation advisor and author
> In *Forbes*, Dec. 18, 2015

Converting scientific expertise into marketable products and services is playing an increasingly important role in the launching of new ventures, the growth of existing firms, and the creation of new jobs. In this article, we explore how the lean startup paradigm, which validates the market for a product with a business model that can sustain subsequent scaling, has led to a new process model to accelerate innovation. We then apply this paradigm to the context of research at universities and other research organizations. The article is based on the assumption that the organizational context matters, and it shows how a deeper understanding of the research context could enable an acceleration of the innovation process. We complement theoretical examples with a case example from VTT Technical Research Institute of Finland. Our findings show that many of the concepts from early-acceleration phases – and the lean startup paradigm – can also be relevant in innovation discussions within the research context. However, the phase of value-proposition discovery is less adequately addressed, and that of growth discovery, with its emphasis on building on a scalable, sustainable business does not seem to be addressed with the presented innovation approaches from the research context. Hence, the entrepreneurial activities at the research context differ from those in startups and internal startups in established organizations.

Introduction

Today, commercialization, or "going to market" (Friedman, 2002), is particularly challenging, especially given the pressure to do it quickly. Indeed, it has been claimed that the need for speed in development and speed in going to market have a tremendous impact on the dynamics of technology transfer and is the most important driver of change, which means that a firm's ability to develop and market products and technologies quickly is crucial (Amesse & Cohendet, 2001).

However, the context of this study is the research world, which is not oftentimes characterized as a dynamic environment with a high clock-speed, though it has been stated that "in the long run, only dynamic and innovative research environments survive" (Graversen, 2007). For example, some years ago, it was estimated that it takes three to six months to put a research team together; hence, this long-term perspective that is needed for several design iterations can only be achieved by government funding (Nunamaker, cited in Winter, 2010). A bit more than a third of all R&D activities in Europe continue to be funded by government (Eurostat, 2016). Reaping the benefits of product and service innovation – in a timely, successful manner, continues to be a challenge. In Europe, a so-called European paradox has been acknowledged, referring to the perceived failure of European countries to translate scientific advances into marketable innovations (European Commission, 1995). Furthermore, the European Commission has placed renewed emphasis on the conversion of Europe's scientific expertise into marketable products and services by seeking to use public sector intervention to stimulate the private sector and to remove bottlenecks that stop such ideas reaching the market (Eurostat, 2016).

Technology transfer – simply defined as "the movement of know-how, technical knowledge, or technology from one organizational setting to another" (Roessner, 2000) – from universities and other public research organizations is increasingly seen to play a significant role in contributing to new venture creation, the growth of existing firms, and new job creation. This role is oftentimes addressed under the umbrella concept of innovation, which for about 20 years has been viewed as a complex, interactive process that is essentially systemic in character (Autio, 1997), with technology transfer activities loosely regarded as "boundary spanners" or "brokers" between academia and industry (O'Kane et al., 2015).

In this article, we seek to transfer a technology innovation methodology from industry to academia. By applying the "lean startup" paradigm to research activities of universities (and other public research institutions), we seek to accelerate innovation in a research context. The article is structured as follows. First, we briefly examine the lean startup paradigm and share insights about innovation in the research context. Then, we introduce a new model of innovation acceleration developed specifically to address the European paradox of commercialization failure. Next, we align our innovation acceleration model with theories and methods from the research context and share practical insights from a case from VTT Technical Research Centre of Finland. We then discuss the findings, and we conclude by highlighting the implications for researchers, managers, and policy makers.

Lean Startup Paradigm

The lean startup paradigm envisions a new company arising based on a new product or service that will be embraced by a particular market (Maurya, 2012; Ries, 2011) because it solves the customer's urgent problem (Moogk, 2012). Simply put, the paramount goal of the startup management team is to find product/market fit with a business model that can scale before they run out of cash (Blank 2014). It has been stated that the uniqueness of this lean startup methodology consists of its ability to explicitly take into account the numerous uncertainties regarding the sustainability and suitability of a given solution towards a specific customer problem (Rasmussen & Tanev, 2015).

The ideas behind the lean startup were launched by the introduction of the customer development process by Steve Blank (2007). Subsequently, it has been refined by Blank to emphasize experimentation (Blank, 2013), with the focus on the adaptiveness and effectiveness of new startups in dealing with the scarcity of resources in their go-to-market efforts (Maurya, 2012). Around the same time, Erik Ries (2011) tied the lean startup concept to lean manufacturing and lean thinking by applying them to the context of entrepreneurship. He wrote the game-changing book *The Lean Startup: How Constant Innovation Creates Radically Successful Businesses,* which is based on five principles that guide the translation of a specific entrepreneurial vision into new products and solutions:

1. Entrepreneurs are everywhere.
2. Entrepreneurship is management.
3. Learning should be validated.
4. Follow a build-measure-learn approach.
5. Apply innovation accounting: a qualitative approach that allows entrepreneurs to see whether the innovation engine is working.

In the lean startup, the first part of the innovation process is to determine whether the product vision can be matched with a problem worth solving using a combination of qualitative customer observation and interviewing techniques (Maurya, 2012). If the product vision is validated with customer data (i.e., if there is a problem/solution fit), the startup rapidly builds prototypes or initial versions of its product that are complete enough to demonstrate the value it brings to the users. This is known as a minimum viable product (MVP) (Moogk, 2012). Once the problem is found and the MVP is built, the goal of a startup is to validate that there is an attractive enough market for the solution. In effect, the startup is looking for a viable business model. Throughout this process, learning can take place, supporting the question whether to pivot the original strategy or persevere (Ries, 2011).

Ries (2011) further states that the lean startup must avoid doctrines and rigid ideology. Some have labelled lean startup "a movement" (Rasmussen & Tanev, 2015) as well as "a collection of tools and techniques that can be employed by entrepreneurs to build their ventures faster and at lower cost" (Harms, 2015), both emphasizing that it tends to be viewed more as a practice-based methodology than a theory for innovation. However, Harms (2015) points out that innovation and technology management scholars may know the lean startup approach under the names of "disciplined entrepreneurship", "hypothesis-driven entrepreneurship", or "probe and learn".

Innovation in a Research Context

The importance of context
In a special issue focusing on the importance of context in entrepreneurial innovation, Autio and colleagues (2014) state: "All human action occurs in contexts: it is the context that regulates what individuals and teams get to see, what choices they are likely to make, what the outcomes of those choices are likely to be. For this reason, context must play a central role in our understanding of the origins, forms, micro-processes, functioning, and diverse outcomes of entrepreneurial activities." They go on to acknowledge and emphasize the tendency in the entrepreneurship literature to focus on the individual, the team, and the resulting venture while neglecting the context.

Still, earlier research has emphasized that the environment in which an individual works is likely to have a great influence on that individual's behaviour, with the main environmental factors being culture, policies, and routines (Kalar & Antoncic, 2015). The research streams that have concentrated on contexts explaining entrepreneurial innovation have explored nations, regions, and industries – and, according to these scholars, context matters because it shapes not only the opportunities that are available but also the dynamics that unfold (Garud et al., 2014). Hence, the research context as such has not been a dominant research stream, though universities and public research organizations have been presented as key players in knowledge ecosystems, advancing technological innovation within the system (Clarysse et al., 2014).

The boundaries between science and business are blurring (O'Kane et al., 2015), as evidenced by the introduction of concepts such as "academic capitalism", the "triple-helix", and the evolving entrepreneurial university. By extending the role of universities beyond simply producing new knowledge, but also disseminating this knowledge to industry and society, the entrepreneurial university is expected to provide a culture and suitable atmosphere for encouraging academics through activities that are more entrepreneurial in nature (Kalar & Antoncic, 2015). Still, the contrast is clearly acknowledged: the startup culture "encourages individual initiative and autonomy, and creates a shared esprit de corps that results in the passionate and relentless pursuit of opportunity. This is the antithesis of the process, procedures and rules that make up large companies" (Blank, 2014).

In his seminal work focusing on university-directed commercialization and entrepreneurial innovation at Stanford University, Nelson (2014) showed how the organizational context not only shaped the decision to engage in entrepreneurship but also the specific approach taken to the commercialization process. In Nelson's work, interviewees argued that the availability of specific funding sources influenced the specific commercialization focus and urgency. Furthermore, he finds it doubtful that individual initiatives such as faculty workshops on entrepreneurship, access to technical equipment, or a seed grant in the tens of thousands of dollars (all common approaches taken by universities in their attempts to spur entrepreneurship) would have changed the fundamental commercialization process and its impact.

Nelson (2014) hence suggested that understanding entrepreneurial behaviours may demand attention not only to individual-level characteristics but also to contextually-informed approaches to action. In addition, he advises that it might be good to embrace those aspects of the university context that mark it as distinct from the firm – for it might be these very features that plant and nurture the research that leads to future marketable technologies.

Understanding innovation in a research context
In the 1980s and 1990s, the focus of technology transfer was on competiveness and was based on a rather simple logic: universities and government labs make, industry takes (Bozeman, 2000). In introducing the concept of design science for the research context of information systems, the purpose was to produce systems that do not yet exist, to achieve better results, with an engineering emphasis (Nunamaker et al., 1991). The concept of "the last mile" was introduced to identify where the value to society is created: the last mile is where you make the lasting difference (Nunamaker et al., 2015; Winter, 2010). This is seen to proceed in three stages: i) proof-of-concept research to demonstrate the functional feasibility of a solution; ii) proof-of-value research to investigate whether a solution can create value across a variety of conditions; and iii) proof-of-use research to address complex issues of operational feasibility. This means that the solution is not really un-

derstood and cannot really be evaluated before it is actually implemented, and researchers cannot predict the impact of their research when only sitting at their desks (Nunamaker et al., 2015).

In the late 1990s, Autio (1997) discussed new technology-based firms emerging as spinoffs from universities and public research organizations. He saw some key differences between science-based firms (firms developing applications concerning physical phenomena or theoretical constructs) and engineering-based firms (firms expanding the scope of use of these applications), arguing that the dominant focus of science-based firms was on a technology-push mode of technology transfer, whereas the emphasis of engineering-based firms appeared to be on a market-pull mode. However, according to Autio, ultimately the most important economic impact of these spinoffs may well be as a catalyst delivered through technology interactions between the firms and their operating environment.

In the 2000s, the research context was addressed by introducing the term "public research institutions". In 2007, a paper suggested two modes of innovation: the science, technology, and innovation (STI) mode, which is based on the production and use of codified scientific and technical knowledge, and the doing, using, interacting (DUI) mode, which relies on informal processes of learning and experience-based know-how (Jensen et al., 2007). It also argued – just like Autio did 10 years earlier – that, in most areas, the results of scientific research are not directly useful for technological advances. Rather, they are more about providing guidance and clues for further development. Furthermore, it suggested that firms with an exclusive focus on developing their science and technology base could benefit from adopting practices and measures designed to promote informal learning by DUI. For public research institutions and universities, it was pointed out that, if they cannot foster sufficient commercialization and entrepreneurial skills among their academics and technology transfer officers, it may be appropriate to place more emphasis on licensing inventions. Hence, in technology transfer, patenting and subsequent licensing were still heavily favoured.

In the 2010s, the linkages between science, technology, and university spinoffs and universities were increasingly emphasized. In explaining their multi-stage, holistic model for creating university spinoffs, Pattnaik and Pandey (2014) argued that universities are moving from their traditional roles of research, teaching, and knowledge dissemination into a more advanced role of creating spinoffs and promoting academic entrepreneurship, with significant impact on regional development and economic growth. They highlighted the role of spinoffs as significant engines toward the commercialization of the technologies that were previously developed, but they also noted that university researchers tend to be more focused in technology development than in other equally important aspects of business. Pattnaik and Pandey's (2014) model consisted of four steps: i) competences and funding; ii) test and confirmation of results; iii) invention disclosure and patenting; and iv) creating spinoffs, leasing technology, and focusing on general economic and social value. More recently, Boh and colleagues (2016) stated that faculty and students are most heavily involved in the earliest phases of the technology commercialization process of the university spin-offs. These authors also identified six stages in the early technology commercialization process: i) idea generation; ii) the commercializing decision; iii) prototype generation and establishment of commercial and technical viability; iv) founding team formation; v) strategy and commercialization process determination; and vi) fundraising to sustain activities, with the aim of convincing investors that the new technology has commercial and technical viability (Boh et al., 2016).

On the research organization side – now addressed as public research organizations – Steinmo and Rasmussen (2016) concluded that they play a crucial role in R&D and innovation across a wide range of industries. On one hand, public research organizations are seen as valuable collaboration partners; firms that collaborate with public research organizations are more likely to develop innovations than other firms. On the other hand, most firms find it difficult to collaborate with public research organizations. Business organizations and public research organizations are seen to pursue different goals. They are therefore structurally different from each other in many ways such as, for example, in their incentive structures and management styles (Steinmo & Rasmussen, 2016). These differences often prevent firms from using public research organizations as sources of external information, and firms generally rate them very lowly as information sources and potential partners.

An Innovation Acceleration Model Inspired by Lean Startup

In many public research organizations, research projects have been undertaken to address the European paradox. Indeed, this has also been a starting point for the Accelerate project (www.accelerateproject.eu), which is part of the second instalment of the ITEA programme (itea3.org) to support innovative, industry-driven, pre-competitive R&D projects in the area of software-intensive systems and services. Overall, the project is focused on enabling: i) the commercialization of better products and services, ii) while being faster to the market, and iii) with sustainable business models. During the project, a new model of innovation acceleration has been developed to overcome some of the key limitations of current models of innovation. Most existing models represent some variations on the familiar pipeline-process architecture: they are not embedded in the strategy issues of company boards, and hence there is misalignment between the newly emerging theoretical models of innovation and innovation in practice (Berkhout et al., 2010).

Inspiration for the model of innovation acceleration has been found from the new kind of innovation thinking that has emerged from within the startup world. The principles of lean startup entrepreneurship have been extended beyond the startup context as the early-stage entrepreneurial challenges have been found to exist beyond the context encountered by typical high-technology ventures (Lockett et al., 2005). The "internal startup" concept, in which a company launches a separate (semi-)independent initiative to pursue a new innovation or idea (Mäkijärvi et al., 2016), has even been claimed so successful that "an internal startup is an ideal environment to nurture innovation and entrepreneurship in large companies" (Edison et al., 2016).

The resulting innovation acceleration model has been defined with four phases (customer discovery, solution discovery, value proposition discovery, and growth discovery), moving from the initial idea to the scalable, fast-growing, and sustainable business (Figure 1). The customer discovery phase aims at producing an initial concept: a vision for a new business with committed people. The solution discovery phase aims at a high-value concept with user acceptance and resources to move forward. In the value proposition discovery phase, a validated and desired solution is produced

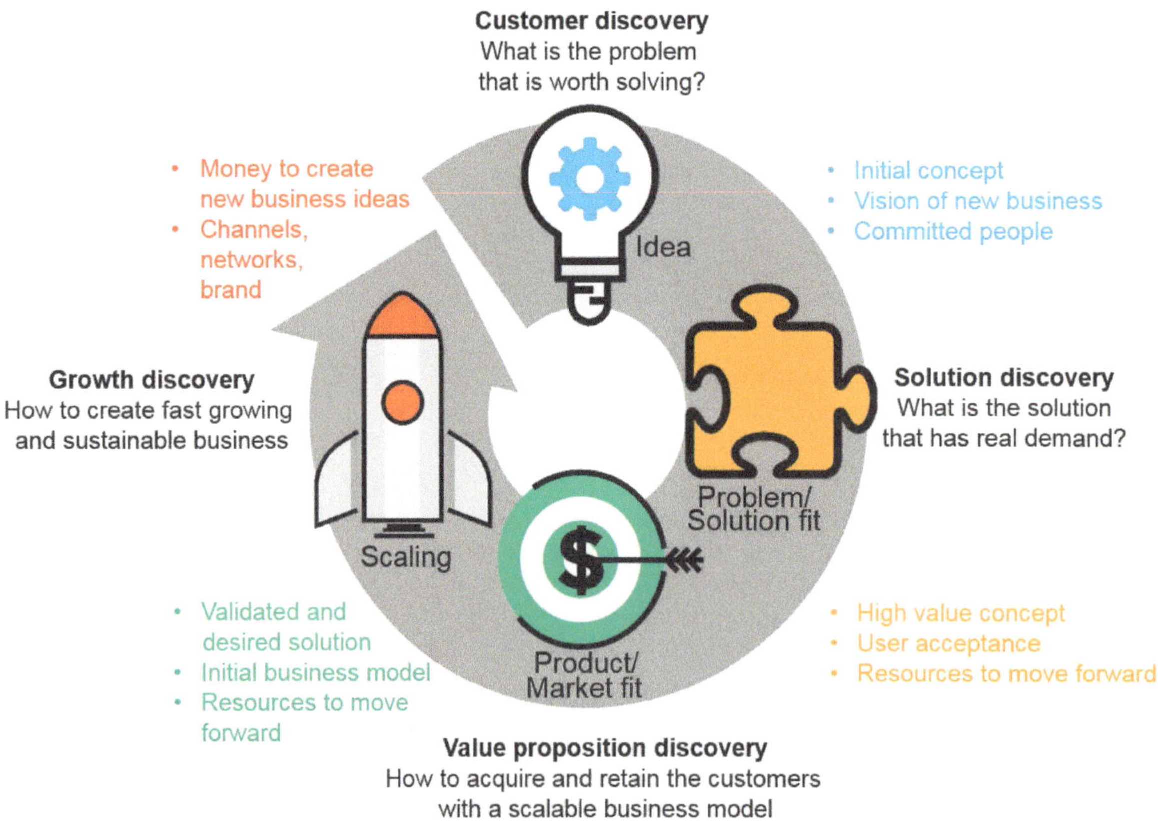

Figure 1. The innovation acceleration model

with an initial business model and resources to move forward. The growth discovery phase is about scaling and creating a sustainable business, which then is expected to result in money to create new business ideas, as well as channels, networks, and brand. Hence, it is a model of continuous refinement, in which especially the two last phases are go-to-market activities as the entity is operating in the market with its value proposition, going toward scaling.

The new process model has been validated by working with Finnish startups that could easily place themselves within the continuum of the process elements (Wallin et al., 2015). It has been successfully used by some of startups and internal startups of the Accelerate project. It has been also used in European startup masterclasses (Mohout, 2014).

Applying the Innovation Acceleration Model to the Research Context

This article is based on the premise that, in today's fast-paced world faced with global competition, it is imperative for many stakeholders to explore the opportunities for accelerating innovation in the research context. Hence, we set out to answer the following research question: *How can the innovation acceleration model be applied in the context of research?*

As described above, it is well known that the research context (referring to universities and research organizations, especially public research organizations) differs from the context of business. Overall, it has been stated that traditional research projects focus on exploration while startups focus on exploitation (Nelson, 2014). Furthermore, in academia, it has been stated that ideas are generated for their own sake, and their implementation is considered less important or even irrelevant (Winter, 2010). For the purpose of this article, we have selected and presented some perspectives that would be relevant in highlighting the differences in innovation acceleration. Still, in this so-called non-commercial environment, speed has been proven to matter (Lockett et al., 2005).

Our research is less about the entrepreneurial individuals – although we understand that innovation in all of its phases is done by individuals – and more about the organizational context. We agree, therefore, with the hypothesis that organizational context plays a role in shaping how participants approach technology commercialization (Autio et al., 2014; Nelson, 2014). We seek to explore the extension of the possibilities of innovation acceleration beyond standalone startups and internal startups.

To explore the importance of context for innovation, and innovation acceleration in particular, we analyzed a selection of different approaches. These approaches, representing more than 20 years of innovation research in a research context, were categorized and mapped onto the schematic representation of the innovation acceleration methodology, which, as already presented, is very much based on the lean startup paradigm.

Context of a case study

After aligning our innovation acceleration framework with theories and methods from the research context, we sought to derive complementary practical insights by applying it within VTT Technical Research Centre of Finland Ltd (vttresearch.com). VTT is a leading research and technology company in the Nordic countries, and it offers research and innovation services to both private and public partners. In 2015, it had a net turnover of 185M€, received 48 new patent applications, and created 21 spinoffs.

Recently, VTT has been re-focusing its strategy by focusing on growth: with scientific and operational excellence, it wants to bring its own innovations to the market, simultaneously supporting Finnish companies and society on the whole. The new strategy journey supports the analysis of the existing innovation and technology transfer processes such as the one made for the purposes of this article. There are a number of improvement projects under way, for example related to intellectual property rights, sales and spinoffs, service model development, as well as a wider implementation of the lean startup methodology. It should be noted that spinoffs are not new to VTT: Autio's (1997) early research into technology-based new business creation included 29 VTT spinoffs.

As just one example of VTT's innovation activities, the Innovative Business from Emerging Technologies (iBet) program encourages internal startups and internal entrepreneurship within VTT. It explores future opportunities, concentrating on renewing the technology basis and leading to societal, industrial, and technological impacts. VTT is looking for great ideas that can be incubated, refined, and transformed into inventions, further developing them into innovations aiming at marketable products, solutions, and services. The iBet programme proceeds as follows:

1. For selected ideas, team members are invited to spend 3–5 days to prepare a pitch.

2. Teams with successful pitches are given one month for further development.

3. Approved plans are given funding for one year, with an option for further continuation.

From 2016 to 2017, 168 ideas were submitted to the iBet process. Of these, 20 ideas were selected, 13 of which were granted funding for one year.

Findings

With the aim of understanding the applicability and use of the lean startup paradigm within the research context, the research question was explored using both the theoretical approaches as well as one concrete case. The findings are presented in Table 1, which shows how the four phases of the innovation acceleration model are addressed.

The major finding of the analysis is that the front-end of the model is addressed by all of the approaches. However, the "go-to-market" elements corresponding to the two last phases of the innovation model are not widely addressed. The value proposition discovery phase was oftentimes presented as something that another entity than the university or research organization, such as a spinoff, would take care of. The phase of growth discovery, with its emphasis on building on scalable, sustainable business does not seem to be addressed with the presented innovation approaches from the research context.

The first phase of customer discovery encourages entrepreneurs (or entrepreneurial teams) to create an initial concept and vision of new business with selected committed people. This phase responds to the tensions of market pull versus technology push with developing applications and technology interactions (Autio, 1997) and the doing-using-interacting model (Jensen et al., 2007). This phase also emphasizes the competences and funding from the creating of spinoff model (Pattnaik & Pandey, 2014) as well as idea generation, the commercialization decision, and founding team formation of the early technology commercialization process at universities (Boh et al., 2016).

The second phase is about solution discovery, when the goal is to create a high-value concept, with user acceptance and resources to move forward. The proof-of-concept and proof-of value (Nunamaker et al., 1991) and the continuation of doing-using-interacting (Jensen et al., 2007) can be placed into this phase. Also, the test and confirmation (Pattnaik & Pandey, 2014), and the prototype generation and establishment of commercial and technical viability as well as invention disclosure and patenting (Boh et al., 2016) correspond to this phase.

In the third phase of value proposition discovery, developing a validated and desired solution is key, with corresponding efforts to create the initial business model, and again, with resources to move on. The proof-of use (Nunamaker et al., 1991) correlates to the validated solution. Whether to move on with a spinoff or with leasing technology (Pattnaik & Pandey, 2014) can be seen as creating an initial business model – however, again, taking the innovation process outside of the university or research organization setting, and hence forcing the innovation acceleration cycle to start from the beginning. The strategy and determination of the commercialization process as well as fundraising to sustain activities (Boh et al., 2016) also can be seen to be part of this third phase. However, whether all of these activities create a high-value concept is to be determined, as many of the research context approaches do not mention user acceptance or real contact and feedback from customers, which is imperative in the lean startup paradigm.

The fourth phase, growth discovery, aims at creating scalable, sustainable businesses. Other than with the hint that spinoffs or licensing agreements are the tools for taking the innovation to the market, it does not seem to be addressed with the innovation approaches discussed in the research context. Channels, networks, and brands are not mentioned in the descriptions of these approaches, nor is the business sustainability focusing on securing the money to create a new business.

VTT's approach to supporting entrepreneurship with the iBet program also reflects the phases of the acceleration methodology, and shows how the lean startup paradigm is increasingly being applied to the research context. Idea selection corresponds to the idea phase, development of plans corresponds to the problem/solution fit, and continued funding allows for exploring toward the product/market fit. Again, the words "scaling", "brand", and "growth" are note mentioned in the iBet program, and hence the growth discovery phase does not appear to be addressed.

Table 1. Mapping the innovation in research context to the model for innovation acceleration

<table>
<tr><th rowspan="2">Innovation in Research Context</th><th colspan="4">Innovation Acceleration Model</th></tr>
<tr><th>Customer Discovery
What is the problem that is worth solving?</th><th>Solution Discovery
What is the solution that has real demand?</th><th>Value Proposition Discovery
How to acquire and retain the customers with a scalable business model</th><th>Growth Discovery
How to create fast growing and sustainable business</th></tr>
<tr><td>Nunamaker et al. (1991)
Nunamaker et al. (2015)</td><td>–</td><td>• Proof-of-concept
• Proof-of-value</td><td>• Proof-of-use</td><td>–</td></tr>
<tr><td>Autio et al. (1997)</td><td colspan="2">• Developing applications
• Market pull
• Technology interactions</td><td>–</td><td>–</td></tr>
<tr><td>Jensen et al. (2007)</td><td colspan="2">• Doing-Using-Interacting (DUI model)</td><td>–</td><td>–</td></tr>
<tr><td>Pattnaik & Pandey (2014)</td><td>• Competences and funding</td><td>• Testing and confirmation of results
• Invention disclosure and patenting</td><td colspan="2">• Creating spinoffs
• Leasing technology</td></tr>
<tr><td>Boh et al. (2016)</td><td>• Idea generation
• Commercializatio n decision
• Founding team formation</td><td>• Prototype generation and establishment of commercial and technical viability
• Strategy and commercialization process determination</td><td>• Fundraising to sustain activities, with the aim of convincing investors</td><td>–</td></tr>
<tr><td>VTT Case: iBET</td><td>• Idea selection</td><td>• Development of selected plans</td><td>• Continued funding for selected plans</td><td>–</td></tr>
</table>

Discussion

Universities and research organizations, identified as non-commercial environments, oftentimes with commercialization and go-to-market activities going through technology transfer offices, have been described with their specific innovation activities, characteristics, and processes, hence separating them from companies and businesses. For example, new technology-based firms have been described as growth oriented, taking risks to pursue growth, growing or perishing with their technology (Autio, 1997), and none of those descriptions were found in the research innovation literature.

Still, it can be concluded that many of the innovation-related concepts derived from the research context can be fitted into the acceleration methodology. However, it is also evident that the context does make a difference, and that the research context has been approaching innovation from a different perspective. The descriptive exploration of more than 20 years of innovation activities within a research environment indicated the impact of context on entrepreneurial activities (Autio et al., 2014; Clarysse et al., 2014; Garud et al., 2014; Nelson, 2014). Hence, our results from analyzing innovation approaches from a research context with the model for innovation acceleration bring evidence to the notion that entrepreneurial innovation is profoundly affected by its context (Autio et al., 2014).

Some of the principles of the lean startup approach (Ries, 2011) could be found in the analyzed innovation approaches. For example, the doing, using, interacting (DUI) model as well as informal learning (Jensen et al., 2007) correspond very well to the overall idea of validated learning. The research context is seen to be moving from exploration to exploitation (Nelson, 2014) – hence, going toward the goal of a successful business, which is at the core of the lean startup paradigm, and indeed, Pattnaik and Pandey (2014) already mention spinoffs as the means for that. However, with this creation of spinoffs as well as with leasing technology, Pattnaik and Pandey (2014) move the innovation process away from the university or research organization setting. Hence, research organizations are seen to be valuable partners in innovation (Steinmo & Rasmussen, 2016), perhaps with the implicit connotation that they are not actual innovators and entrepreneurs, which conflicts with the first lean startup principle: "entrepreneurs are everywhere" (Ries, 2011). In addition, the emphasis on the user (or market) is not dominant in the research context approaches. For example, the multi-stage, holistic model of creating university spinoffs (Pattnaik & Pandey, 2014) does not seem to focus on user needs. Also, discussions of management or innovation accounting were not directly addressed in the selected approaches.

The practical example from VTT, the iBet program, can be seen as one way of going beyond the traditional innovation process of a research organization. Although it also showed that the growth discovery phase was notably not included, it did include some emphasis on go-to-market activities. Furthermore, the fact that this entrepreneurship focused program was attractive to the personnel at VTT is interesting and encouraging: with the total number of VTT researchers at about 1450, this translates to submission from more than 11 percent of them (assuming that each researcher only submitted one idea). This finding reflects that the research context (at least in Finland) is changing, and entrepreneurship is increasingly seen to be part of it.

Limitations of the study

Innovation continues to be a multi-disciplinary topic with a large amount of literature related to its processes, characteristics, impact, etc. This article does not pretend to be comprehensive in its quest to analyze how the lean startup paradigm explicitly presented with an innovation acceleration methodology can be applied in the context of research. Rather, the article is intended as more of an exploratory discussion that starts from an existing model for innovation acceleration, and its applicability and validity in the research context. Hence, the number of research approaches analyzed was limited, and the approach was aligned with a single case example from VTT. In particular, the major finding of the "go to market" elements corresponding to the two last phases of the innovation model but especially on the growth discovery/scaling being largely absent from the research context could be explored further, with more examples as well and on the basis of a more thorough theoretical review.

Managerial implications

There are many different actors in the research context who are involved in the innovation process. Beyond research personnel, new business development people, intellectual property specialists, and early sales professionals, accelerating innovation is important to all. All of us must know the terminology and the process steps that are generally used to describe innovation within our specific contexts. However, we should not feel limited by the semantic framework, especially if the organization wants to go beyond traditional plans and development activities, and really starts "doing, using, and interacting".

As the early phases of innovation are increasingly emphasized in research contexts, research organizations should start addressing this gap by adopting lean startup practices. The adaptation of competence development, hiring practices, and strategic partnering are also possible methods in developing go-to-market capabilities by taking into account the different early stages of the research process context. With the new capabilities, the issue of speed can also be addressed.

Looking into the demand side of technology and innovation is also encouraged. Going beyond the traditional technology transfer is emphasized by Blank (2014), who states that companies manage innovation by building innovation internally, buying it, or partnering with resources outside of the company. It should be noted that he heavily emphasizes the acquisition of startups as the corporate innovation strategy, which is also supported by the evidence of growth strategies of companies such as Google. A question worth exploring in more detail hence becomes: will startups become the de-facto standard for technology transfer from research to the corporate world?

Policy implications

Innovation policy has not been immune to the paradigm shift the lean startup: it sees the promise of successful businesses that will contribute to wealth, competitiveness, and quality of life. Not only are the results of lean startups on the wish list, but so are the ingredients. For example, Finland wants to become an international leader in experimental culture, at both national and regional levels as well as in organizations (kokeilevasuomi.fi/en/frontpage). Accordingly, the Finnish government has continued to develop research funding programs that focus on and support experimentation.

This article highlights the interdependencies of the research context's innovation process with the surrounding innovation ecosystem, and the impacts created by such activities. This aspect was also emphasized by Autio and colleagues (2014): "the connections and ties across these specific contexts are crucial for entrepreneurial innovation and future policy development deeds to recognize these inter-dependencies and the possible synergies and conflicts between them." For example, the innovation policy has direct links to customer needs; the speed of knowledge transfer is partly controlled by the scientific publishing processes.

At the same time, the impact of context on an individual's innovation activities is emphasized. For example, to put it bluntly, it may be detrimental to expect the same kind of entrepreneurship from scientists at research organizations as from individuals working at startups.

Overall, there is a clear call for clarity of roles and responsibilities of various ecosystem players and for addressing the dynamics of such systems. Also, the value of having these players coming from various contexts and still co-creating together should be further studied.

Conclusion

In this article, we conducted a limited literature review to highlight that the research context was seen to have its own innovation characteristics and processes for technology transfer. We then explored how the lean startup paradigm explicitly presented as an innovation acceleration model matched the special context of research. The concepts from research innovation were complemented with a practical example: the iBet case at VTT Technical Research Institute of Finland.

The findings show that many of the concepts of the early acceleration phases can also be found in the innovation discussions within the research context. The phases for going-to-market receive less attention: i) the value proposition discovery phase was addressed sometimes, and even then, it was oftentimes presented that another entity than the university or research organization such as a spinoff would take care of it; and ii) the phase of growth discovery, with its emphasis on building on scalable, sustainable business does not seem to be addressed with the presented innovation approaches from the research context.

Hence, the entrepreneurial activities at the research context differ from those in startups and internal startups, which supports the impact of context presented, for example, by Nelson (2014) and Autio and colleagues (2014). This gap should be discussed and addressed at research organizations, for example with practical tools for competence development, hiring, and the selection of strategic partners. Furthermore, when looking into the demand-side of the technologies and innovation, there seems to be an inclination to emphasize startups as means of technology transfer. If this continues, it challenges not only the culture and processes but also the outputs of the research context. In addition, policy makers should take this into account when clarifying the roles, responsibilities, value-creation activities, and dynamics within the innovation ecosystem.

About the Author

Kaisa Still is a Senior Scientist at VTT Technical Research Centre of Finland. She has extensive experience of innovation management gained within a research organization and a university, in a startup and in growth companies, as well as in a business incubator. Supporting collaboration, co-creation and innovation with technology continues to be at the core of her interests. Her current work concentrates on platforms and innovation ecosystems, accelerating innovation activities, and digital opportunities. Combined with the policy perspective, her work extends to private and public organizations, in regional and global contexts.

References

Amesse, F., & Cohendet, P. 2001. Technology Transfer Revisited from the Perspective of the Knowledge-Based Economy. *Research Policy,* 30(9): 1459–1478.
https://doi.org/10.1016/S0048-7333(01)00162-7

Autio, E. 1997. New, Technology-Based Firms in Innovation Networks Symplectic and Generative Impacts. *Research Policy,* 26(3): 263–281.
https://doi.org/10.1016/S0048-7333(96)00906-7

Autio, E., Kenney, M., Mustar, P., Siegel, D., & Wright, M. 2014. Entrepreneurial Innovation: The Importance of Context. *Research Policy,* 43(7): 1097–1108.
https://doi.org/10.1016/j.respol.2014.01.015

Berkhout, G., Hartmann, D., & Trott, P. 2010. Connecting Technological Capabilities with Market Needs Using a Cyclic Innovation Model. *R&D Management,* 40(5): 474–490.
http://dx.doi.org/10.1111/j.1467-9310.2010.00618.x

Blank, S. 2007. *The Four Steps to Epiphany: Successful Strategies for Products that Win.* Raleigh, NC: Lulu Enterprises.

Blank, S., 2013. Why the Lean Start-up Changes Everything. *Harvard Business Review,* 91(5): 635–72.

Blank, S. 2014. Corporate Acquisitions of Startups – Why Do They Fail? *Forbes,* April 22, 2014. Accessed April 21, 2017:
https://www.forbes.com/sites/steveblank/2014/04/22/corporate-acquisitions-of-startups-why-do-they-fail/

Boh, W. F., De-Haan, U., & Strom, R. 2016. University Technology Transfer through Entrepreneurship: Faculty and Students in Spinoffs. *The Journal of Technology Transfer,* 41(4): 661–669.
http://dx.doi.org/10.1007/s10961-015-9399-6

Bozeman, B. 2000. Technology Transfer and Public Policy: A Review of Research and Theory. *Research Policy,* 29(4–5): 627–655.
https://doi.org/10.1016/S0048-7333(99)00093-1

Clarysse, B., Wright, M., Bruneel, J., & Mahajan, A. 2014. Creating Value in Ecosystems: Crossing the Chasm between Knowledge and Business Ecosystems. *Research Policy,* 43(7): 1165–1176.
https://doi.org/10.1016/j.respol.2014.04.014

Edison, H., Wang, X., & Abrahamsson, P. 2016. *Product Innovation through Internal Startup in Large Software Companies: A Case Study.* Paper presented at the 42nd Euromicro Conference on Software Engineering and Advanced Applications (SEAA), Limassol, Cyprus, August 31 – September 2, 2016.
https://doi.org/10.1109/SEAA.2016.36

Eurostat. 2016. R&D Expenditure. *Eurostat: Statistics Explained.* Accessed April 21, 2017:
http://ec.europa.eu/eurostat/statistics-explained/index.php/R_%26_D_expenditure

Friedman, L. G. 2002. *Go To Market Strategy: Advanced Techniques and Tools for Selling More Products, To More Customers, More Profitably.* Oxford, UK: Butterworth-Heinemann, Elsevier Science.

Garud, R., Gehman, J., & Giuliani, A. P. 2014. Contextualizing Entrepreneurial Innovation: A Narrative Perspective. *Research Policy,* 43(7): 1177–1188.
https://doi.org/10.1016/j.respol.2014.04.015

Graversen, E. B. 2005. Dynamic Research Environments: A Development Model. *The International Journal of Human Resource Management,* 16(8): 1498–1511.
http://dx.doi.org/10.1080/09585190500220754

Harms, R. 2015. Self-Regulated Learning, Team Learning and Project Performance in Entrepreneurship Education: Learning in a Lean Startup Environment. *Technological Forecasting & Social Change,* 100: 21–28.
https://doi.org/10.1016/j.techfore.2015.02.007

Jensen, M. B., Johnson, B., Lorenz, E., & Lundvall, B. Å. 2007. Forms of Knowledge and Modes of Innovation. *Research Policy,* 36(5): 680–693.
https://doi.org/10.1016/j.respol.2007.01.006

Kalar, B., & Antoncic, B. 2015. The Entrepreneurial University, Academic Activities and Technology and Knowledge Transfer in Four European Countries. *Technovation,* 36–37: 1–11.
https://doi.org/10.1016/j.technovation.2014.11.002

Lockett, A., Siegel, D., Wright, M., & Ensley, M. D. 2005. The Creation of Spin-Off Firms at Public Research Institutions: Managerial and Policy Implications. *Research Policy,* 34(7): 981–993.
https://doi.org/10.1016/j.respol.2005.05.010

Märijärvi, J., Hokkanen, L., Komssi, M., Kiljander, H., Xu, Y., Raatikainen, M., Seppänen, P., Heininen, J., Koivulahti-Ojala, M., Helenius, M., & Järvinen, J. 2016. *The Cookbook for Successful Internal Startups.* Helsinki: Digile, N4S.
http://www.n4s.fi/wordpress/wp-content/uploads/2016/04/the_cookbook_for_successful_Internal_startups.pdf

Maurya, A. 2012. *Running Lean: Iterate from Plan A to a Plan That Works,* 2nd Ed. Sebastopol, CA: O'Reilly Media Inc.

Mohout, O. 2014. *Startup Master Class 1 – Genesis: Idea Stage.* Antwerp: Antwerp Management School.
http://www.slideshare.net/omohout/genesis-43234459

Moogk, D. R. 2012. Minimum Viable Product and the Importance of Experimentation in Technology Startups. *Technology Innovation Management Review,* 2(3): 23–26.
https://timreview.ca/article/535

Nelson, A. J. 2014. From the Ivory Tower to the Startup Garage: Organizational Context and Commercialization Processes. *Research Policy,* 43(7): 1144–1156.
https://doi.org/10.1016/j.respol.2014.04.011

Nunamaker, J. F., Chen, M., & Purdin, T. D. M. 1991. Systems Development in Information Systems Research. *Journal of Management Information Systems,* 7(3): 89–106.
http://dx.doi.org/10.1080/07421222.1990.11517898

Nunamaker, J. F. Jr., Briggs, R. O, Derrick, D. C., & Schwabe, G. 2015. The Last Research Mile: Achieving Both Rigor and Relevance in Information Systems Research. *Journal of Management Information Systems,* 32(3): 10–47.
http://dx.doi.org/10.1080/07421222.2015.1094961

O'Kane, C., Mangematin, V., Geoghecan, W., & Fitzgerald, C. 2015. University Technology Transfer Offices: The Search for Identity to Build Legitimacy. *Research Policy,* 44(2): 421–437.
https://doi.org/10.1016/j.respol.2014.08.003

Pattnaik, P. N., & Pandey, S. C. 2014. University Spinoffs: What, Why, and How? *Technology Innovation Management Review,* 4(12): 44–50.
https://timreview.ca/article/857

Rasmussen, E. S., & Tanev, S. 2015. The Emergence of the Lean Global Startup as a New Type of Firm. *Technology Innovation Management Review,* 5(11): 12–19.
https://timreview.ca/article/941

Ries, E. 2011. *The Lean Startup.* New York: Crown Business.

Roessner, J. D. 2000. Technology Transfer. In C. Hill (Ed). *Science and Technology Policy in the US: A Time of Change.* London: Longman.

Steinmo, M., & Rasmussen, E. 2016. How Firms Collaborate with Public Research Organizations: The Evolution of Proximity Dimensions in Successful Innovation Projects. *Journal of Business Research,* 69(3): 1250–1259.
https://doi.org/10.1016/j.jbusres.2015.09.006

Wallin, A., Still, K., & Komi, M. 2015. Start-Up Entrepreneurs' Key Concerns on Path of Entrepreneurial Innovation. In *Proceedings of the XXVI ISPIM Conference – Shaping the Frontiers of Innovation Management,* Budapest, Hungary, June 14–17, 2015.

Winter, R. 2010. Interview with Jay F. Nunamaker, Jr. on "Toward a Broader Vision of IS Research". *Business & Information Systems Engineering,* 2(5): 321–329.
http://dx.doi.org/10.1007/s12599-010-0122-8

Keywords: innovation, innovation acceleration, research, research organization, innovation paradox, lean startup, context, VTT, commercialization

16

Is There a Lean Future for Global Startups?

Stoyan Tanev

> *In real life, unlike in Shakespeare, the sweetness of the rose depends upon the name it bears. Things are not only what they are. They are, in very important respects, what they seem to be.*
>
> Hubert Humphrey (1911–1978)
> 38th Vice President of the United States

This article integrates insights from the latest research on the lean startup entrepreneurial method, born-global firms, and global startups. It contributes to the clarification of terminology referring to the global aspects of startups, summarizes insights from previous literature focusing on global startups, and further substantiates the articulation of the need for considering the lean global startup as a new type of firm. The main message is that the lessons learned from the emergence of lean startup entrepreneurship offer a basis for promoting a similar lean phase in technology-based global startup research and practice. The analysis should benefit both researchers and practitioners in technology entrepreneurship, international entrepreneurship, and global innovation management.

Introduction

This article examines some of the developments following the publication of my earlier article in this journal (Tanev, 2012), in which I summarized the characteristics of born-global firms in a context relevant to new technology startups. I defined a born-global firm as a new venture that acts to satisfy a global niche from day one by searching for and accessing resources that could help its global reach. The definition focuses on new ventures that are international or global by design and not by emergence.

Since the publication of my 2012 article, there have been two major developments that could be both related to the context of born-global technology startups. The first one was the spread of the lean startup movement (Ries, 2012) as a systematic articulation of a set of entrepreneurial practices inspired by the customer development process suggested earlier by Steve Blank (2007). Blank's approach was described in greater detail in the *Startup Owner's Manual* (Blank & Dorf, 2012) as a step-by-step process for managing the search for a new business model, providing entrepreneurs with a practical path from idea to a scalable business model. It could be considered as an enhancement of Moore's (1991) technology adoption lifecycle approach, because it offers a more systematic way of dealing with the entrepreneurial challenges of "crossing the chasm" between the early adopters and the first viable market niche. Eisenmann, Ries, and Dillard (2012) defined a lean startup as a firm that follows a hypothesis-driven approach to the evaluation of an entrepreneurial opportunity and the development of a new product for a specific market niche. The lean startup approach has also attracted the attention of entrepreneurship researchers trying to position its key insights in the context of existing management and innovation theories (Frederiksen & Brem, 2017; Stolze et al., 2014).

The second development is the further maturation of international entrepreneurship research. I could refer to several recent research papers (Alcácer et al., 2016; Cavusgil & Knight, 2015; Coviello, 2015; Sarasvathy et al., 2014; Weerawardena et al., 2015; Yang & Gabrielsson, 2017) as examples of key developments in the field. The paper by Coviello (2015) is particularly relevant for the context of new technology firms because it points out the relevance of born-global startup over born-global firm definitions and highlights the research relevance of the global intentions of startups at their inception instead of the characteristics of young firms that have undergone a successful internationalization. In an interview with the author as part of the current special issue, Coviello has also provided additional comments on the international entrepreneurship research maturation process and the possibility of considering the lean global startup as a new type of firm (Coviello & Tanev, 2017).

The remainder of this article is structured as follows. First, I discuss some of the insights by Coviello (2015) that could be highly relevant for the context of new technology firms engaged in a global business path from their inception. Next, I identify the key lessons learned from research focusing on born-global startups. Finally, I offer additional arguments that could better highlight the benefits of considering lean global startups as a new type of organization/firm.

Born-Global Firms or Global Startups?

Coviello (2015) provides an explanation of the problem of using the term "born global" in the context of startups. One of Coviello's key points is that, in order "to study a 'new' international venture, it is important to investigate (for example) their decisions or actions when they are, in fact, still new." In other words, the identification of a firm as "born global" is not a post factum conclusion that is the result of looking at how successful a firm was in its internationalization efforts. In other words, there is a difference between firms that were truly born with the intent to serve multiple foreign markets and firms that simply happen to export early. The difference in firms' growth intentions at their very founding will definitely lead to the emergence of different internationalization paths resulting in qualitatively different growth modes. Studying the different growth modes without taking into account the difference in firms' intentions at their founding will not help the proper articulation of business design principles focusing on establishing new global startups. In addition, the call for a closer attention on firms' intentions at their founding should not neglect the fact that there could have been many activities contributing to the global orientation of the firm even before its formal founding. In many cases, the founders incorporate a new firm after having worked for years on their product and partnership development.

The second important point made by Coviello is that there is a much better term referring to the globalness of startups: "global startup". This term was introduced by Oviatt and McDougall (1994) almost at the same time as the term "born global" and has a different meaning. Indeed, using the two terms interchangeably is confusing. Oviatt and McDougall have specifically discussed the characteristics of global startups, including the drivers for the emergence of such firms as a "powerful economic engine" (Oviatt et al., 1995). In their view, a global startup is just one specific type of international new venture that seeks to derive competitive advantage by coordinating many organizational activities – not just export sales – across many countries from its very inception. The other three types of international new ventures are export/import startups, multinational traders, and geographically-focused startups. They differ from global startups by serving a smaller number of international markets (export/import startups and geographically-focused startups) or coordinating a smaller range of (mostly logistical) activities across different countries (multinational trader companies). The focus on two global startup criteria – many coordinated activities across countries and many international markets – is an important point because it emphasizes a key difference with the definition of born-global firms as "entrepreneurial startups that, from or near their founding, seek to derive a substantial proportion of their revenue from the sale of products in international markets" (Knight & Cavusgil, 2004) or "young companies that derive a significant portion of their revenue from international sales" (Cavusgil & Knight, 2015). The broader focus of the international new venture definition includes the possibility for the coordination of activities other than export, such as importing, offshore R&D, joint ventures, or production subsidiaries (Coviello, 2015). The broader international new venture definition of global startups fits much better with today's context of new technology firms that could and usually benefit from upstream activities across national borders before, or in parallel to, reaching sales capacity.

An additional aspect of the above discussion is the need to account for the difference between "international" and "global". Coviello points out that "Oviatt and McDougall (1994) specifically chose to use the term 'international new venture' (encompassing four types of firms) because many of the firms they observed did not have a global focus. Rather, most competed primarily in their regional markets or in a relatively limited number of countries" (Coviello, 2015). This statement implies that the key criterion for qualifying a new firm as a global startup focuses on the quantitative measure corresponding to the multiplicity of the international markets served and not so much on the geographical and cultural distance between these multiple markets. Such implication opens two interesting questions.

The first question is whether the tangibility of this measure contradicts the intentional character of the early-stage activities of startups. On the one hand, following Coviello's advice, we should focus on the actual *intentions* of the startups to serve multiple foreign markets at their inception and, on the other hand, on the multiplicity of international markets *served* by the startups and their various activities *coordinated* across different countries. The problem here is in the difference between *intending to serve* and *actually serving, intending to coordinate,* and *actually coordinating.* How many startups could actually serve multiple international markets and coordinate many activities across multiple countries at their very inception?

The second interesting question is whether the meaning of "global" should be restricted to refer to the multiplicity of international markets and coordinated activities across multiple countries without any additional requirements for the geographical, psychic, and cultural distance between these countries. Should the export of a Danish firm to Germany, or a Canadian firm to the Unites States, be considered as part of their global market reach? It is not by accident that Madsen and Servais (1997) define born globals as "firms that seek to derive significant advantages from the use of resources from or the sale of outputs to multiple countries/continents right from their legal birth." The use of continents as a potential global location on an equal basis with countries indicates the tendency to associate the meaning of "global" with locations at significant geographical, psychic, and cultural distances. Another expression of this tendency is the recent definition of born-global firms suggested by Andersson, Evers, and Griot (2013): "business organizations that have a global mindset from inception and aim to derive significant competitive advantages from the use of resources and the sale of outputs in multiple countries spanning the three economic trading blocs of NAFTA, EU and Asia-Pacific." This definition refers to the initial intentions of the firms with respect to both resources and markets, and it emphasizes a multi-continent perspective on globalness that is inclusive of the context of startups.

At the end of the present section, one could reiterate the judgement by Lopez and colleagues (2009) that the born-global literature has been lacking a precise definition of what a born-global firm is, and some existing definitions are tautological. In this sense, the call for more terminological clarity and for a new more mature phase of international entrepreneurship research made by Coviello (see her interview in this special issue: Coviello & Tanev, 2017) is timely and highly relevant. One specific point could be particularly emphasized: "using the terms INV [international new venture] and BG [born global] synonymously and/or interchangeably is inaccurate, as is any reference to firms as 'INVs/BGs'" (Coviello, 2015). Rather, it is important to recognize that, by definition, these organizational forms differ. For example, a study of firms labeled as born global will examine firms that, very close to birth, actively export to global markets. If the study captures multiple and global value chain activities very close to birth, the term "global startup" from Oviatt and McDougall's (1994) typology of international new ventures is more appropriate" (Coviello, 2015). At the same time, the overall spirit of the definitions seems to predominantly reflect the context of younger small firms and not of actual startups. Actual startups deal with the challenges of maximizing the value of resources acquired ex ante from factor markets; younger small firms deal with the challenges of capturing value by competing on product markets (Schmidt & Keil, 2013). This was one of the key reasons for proposing the integration of the lean startup and the born-global or global startup paradigms. The next section will summarize some of the insights about global startups such as articulated by Oviatt, McDougall, and Loper as early as 1995, and by others after them.

The Emergence of the Global Startup Phenomenon

In their paper, Oviatt and colleagues (1995) do not seem to overemphasize their initial definition. It is in a footnote that they mention that a global startup is a type of international new venture that coordinates many organizational activities across many countries. In the main text, they add that "such firms seem to have aggressive growth objectives in that they rapidly exploit technological advantages, acquire foreign technologies, and follow clients into foreign lands" (Oviatt et al., 1995). According to them, the emerging phenomenon of rapid internationalization has become a reality due to specific technological and competitive forces. In addition, the factors enabling early internationalization have affected the formation of new ventures and made the usual slowly staged internationalization efforts risky for many firms.

Oviatt and colleagues (1995) summarized the forces that drive the emergence of new global startups by suggesting that entrepreneurs, investors, and corporate executives exploring venturing options should consider the following six conditions when determining whether a new business should be a global or a domestic startup.

1. *The best human resources are dispersed among various countries.* Certain locations in the world offer access to unique workforce skills. Global startups could profit from actual presence in such locations by procuring the best resources at lower prices.

2. *Foreign financing could be easier or more suitable.* Startups having difficulties in raising financing could benefit from pursuing funding outside their own countries of origin. According to Oviatt and colleagues (1995) the international sourcing of funds may be one of the most important forces driving startups to internationalize because of the inherent benefits of overseas funding. "Foreign investors will want their venture to move rapidly into their own markets, and founders will certainly consider moving into a country from which they have received funding. In some cases, a prime motivation of foreign investors may be to get new technology into their own home market." (Oviatt et al., 1995).

3. *The target customers require the new venture to be international.* The increasingly global nature of demand in many markets is one of the main forces encouraging the formation of global startups.

4. *Rapid worldwide communications allow for quick competitive responses.* The pervasive availability of rapid communications and transportation capabilities worldwide has enabled almost instantaneous access to information about virtually any market. Entrepreneurs in foreign countries can learn quickly about potential business opportunities in any other country. In addition, there is a variety of unprotected market niches at different global locations. Many startups became global because quite often "the best domestic defense might be a superior international offense" (Oviatt et al., 1995).

5. *Worldwide sales are required to support the venture.* Many industries cannot afford the comfort of remaining regional or national. Worldwide sales might be needed to justify large R&D expenses and to effectively address the target market. A global startup enjoys more potential customers than a domestic venture selling the same product. Some startups could actually be forced to target foreign customers because of the insufficiency of revenue in their home countries. Oviatt and colleagues (1995) call such firms "reluctant global startups". They do not have a global strategy from the start but need to be large to be effective, and being international is a vehicle for doing that.

6. *Domestic inertia will be crippling if internationalization is postponed.* If a new venture starts out domestically, its policies and procedures are driven by the logic of the domestic market. A newly emerging international market opportunity will most likely lead to the need of disruptive operational changes that could be met with resistance by its employees. It may take great efforts and a long time to overcome such inertia. An initial international orientation facilitates the adoption of operational standards and product architectures that permit easy internationalization (Oviatt et al., 1995). In brief, a global startup benefits in the long run from being international at inception.

In addition to summarizing the key factors enabling early internationalization, Oviatt and colleagues (1995) have identified seven characteristics that are commonly associated with the survival and growth of global startups, including:

1. *Existence of a global vision from inception.* The founders must be able to compellingly communicate a global vision to everyone else associated with the venture.

2. *Founders or top managers are internationally experienced.* Understanding the logic of cross-border business conduct is absolutely necessary.

3. *The entrepreneurs have strong international business networks.* New ventures are resource poor and usually depend on a supportive network of business associates. "Having a network of international alliances to access vital resources rather than owning those resources outright is the increasingly preferred way of conducting international business." (Oviatt et al., 1995).

4. *Exploitation of preemptive technology or marketing approach.* Most successful global startups begin by selling a unique product or service in leading markets. The way for them to overcome the advantages of indigenous firms is to be first to market with a distinctively valuable product or service.

5. *Possession or privileged access to a unique intangible asset.* Marketing a distinctively valuable product is preemptive only if its distinctiveness is sustained through a competitive advantage. In most cases, the competitive advantage is some type of unique special knowledge that only the startup has. Global startups should recruit, train, and manage their human resources very effectively and use them to continuously innovate.

6. *There is a close link between product or service extensions.* Continuous innovation ensures that a firm's advantage is a moving target that is more difficult for competitors to hit. Yet, innovations subsequent to the founding of the startup must be incremental due to the venture's limited experience and resources. Successful global startups seem to follow their initial product or service with extensions that are closely linked to the unique assets from which they derived their original competitive advantage.

7. *The organization is closely coordinated worldwide.* All the above characteristics would require a strong top management team to coordinate R&D, procurement, production, marketing, distribution, sales, and other activities in several worldwide locations that need to be closely coordinated. Such close coordination implies several things, including: i) a close interaction among top managers aligned with a strong commitment to the main goals of the new venture; ii) efficient communication of the global vision throughout the organization to ensure every employee's commitment to that vision; iii) top managers should enjoy and endure travel: international business requires face-to-face meetings to establish and nurture the network of international business associates; and iv) the technological communication infrastructure of a successful global startup must be more sophisticated than that of the usual domestic startup of equal size.

Meyer and Xia (2012) offer additional insights about global startups by focusing on the experience of British global entrepreneurs. They emphasize that the pursuit of global markets is only one way to exploit global opportunities. Many ambitious entrepreneurs, especially in high-tech ventures, push further and create business models that explore and exploit resources in multiple countries. The actuation of the opportunity behind these business models involves the need to identify, attract, transform, and apply resources in different countries, but it also requires a focus on the implementation of the business idea from day one. Meyer and Xia (2012) also provide examples of British high-tech startups that would not have existed if they did not develop business models by tapping into resources all around the world. According to them, there are three globalization trends that stand out in creating opportunities for global entrepreneurs: global value chain fine-slicing, global communities of practice, and global communications technologies.

The global value chain fine-slicing trend has emerged within the dominant practices of most multinational companies that tend to locate different tasks in different countries and selectively outsource some of these tasks. Examples of such tasks are manufacturing, back-office services, collaborative research, and acquisition of innovations from independent technology firms through licensing or specific contractual arrangements. According to Meyer and Xia (2012), the emerging granularity of the localization of these tasks creates opportunities for entrepreneurs to tap into a value chain by developing a specialized component or service targeted at major players in an industry. The emergence of such opportunities requires a certain degree of modularity within the dominant product design and architecture that would allow the pursuit of a displacement innovation (Christensen et al., 2004).Displacement is a specific type of innovation that could take place at a point of modularity by targeting the mainstream market. A new globally-minded startup could specialize by focusing on one particular component of a product or service and positioning it competitively in a global value chain by taking part of the market from well-established multinational incumbents. Such startups cannot win unless their products can interface with the established product architecture at points of well-defined modularity. This becomes possible when the interactions across the interfaces between different product components or modules are well understood and predictable. Interestingly, displacements could also enable low-end disruption by new startups that emerge by assembling value chain components in new ways to offer new customer benefits (Christensen et al., 2004). A good example of a company that was able to do that is Dell – it took advantage of the modularity of personal computers and developed its low-end disruptive business model. The example demonstrates the opportunity of articulating global startup design principles by incorporating insights based on theories of disruption (Gans, 2016).

The global communities of practice trend has emerged within many fields of specialization where close communities have evolved between experts who, despite geographic distances, meet frequently, exchange ideas and best practices, and establish informal rules of conduct in the community. For example, academic and industry-based researchers participate in conferences worldwide where they exchange ideas on new technologies and establish actual global networks. Such conferences allow researchers to create personal relationships with experts in multiple countries and global locations. Entrepreneurs can build on these ties when creating their own ventures, hiring, contracting, or simply accessing the best people, resources, tools, and assets across multiple geographically distant locations.

The emergence of *the global communications technologies trend* was already discussed above. The point is that many entrepreneurs have learned to employ existing communication technologies and many open source tools to combine occasional intensive face-to-face meetings with more frequent, but shorter and practically costless virtual meetings. The interaction between members of a geographically dispersed team can easily become an everyday routine.

The reason for summarizing the three trends above was to emphasize that global entrepreneurs take advantage of these opportunities and design their business models by combining and exploiting talent and resources from all around the world. The vastly reduced set-up costs of communication technology systems enables new startups to make efficient interpersonal exchanges across large distances. The access to talent and resources worldwide allows startups to become global from the start by integrating complementary technological products developed by partners from multiple global locations, coordinating resources, working with distributors, and seeking additional support from investors. Most of the entrepreneurs who have succeeded in making a global move are mature individuals who have achieved professional success before either in business or in academia. According to Meyer and Xia (2012), the commonality between them is that they share a global mindset, deep industry experience, ability to build networks by connecting with the best in their industry and technological domain around the world, persistence in pursuing global business opportunities, and readiness to adopt a lean startup-like vision of the business given that it is not always that the first attempt that is the most successful. The latest research offers even stronger arguments that the reality of the present information age suggests the adoption of a new techno-economic paradigm for the emerging structures and processes in international business in general. The reality of the new paradigm includes the changing nature of the competitive advantages of places (a new understanding of locations), the competitive advantages and strategies of firms (a new understanding of ownership), and the governance structure of international business networks (a new understanding of internationalization) (Alcácer et al., 2016). The implications of the emergence of this paradigm should be explored further in the context of new technology-based global startups.

The Emergence of the Lean Global Startup Paradigm

This section will focus on some of the issues mentioned in the introduction, and it will offer additional argumentation that could substantiate the introduction of the lean global startup as a new type of firm. It will therefore indirectly address Coviello's pessimism: "At any rate, I don't consider a lean global startup to be a new form of organization or growth mode. Lean is a way of operating. In the same way that we study organization structure, orientation etc., we should study the influence of lean (or effectual logic) as young firms internationalize" (Coviello & Tanev, 2017). It would be good to start with a reminder of the lean startup definitions as suggested by its "Godfathers": Steve Blank (2013) and Eric Ries (2011). The definitions of the lean startup emphasize several points:

- A lean startup is not a smaller version of a large company. Whereas well established companies focus on executing a business model, a startup struggles to articulate and establish one. In this sense, "a lean startup is a temporary organization designed to search for a repeatable and scalable business model" (Blank, 2013).

- A lean startup is "a human institution designed to create new products and services under conditions of extreme uncertainty" (Ries, 2012).

- Blank and Ries pioneered three key concepts to characterize the essence of lean startups: customer development, minimum viable product (MVP), and pivot. The MVP is a product consisting of a minimum set of features that is used, *first,* as a tactic to reduce wasted engineering hours and, *second,* as a way of getting the product in the hands of early and visionary customers as soon as possible. Pivot is a term used to describe a major change in direction of a startup while staying grounded in learning.

- A lean startup, as a temporary organization/institution, deals with the challenges and uncertainties of transforming the ex-ante value of resources acquired in factor markets into ex-post product market success of a newly created company. This transformation is associated with a business transition from a temporary startup mode of operation of a customer development team searching for a business model, to a sustaining operational mode of a newly established company based on functional departments focusing on the execution of the business model (Blank, 2013).

- In parallel to the articulation of a viable business model, a lean startup is striving to build the minimum viable ecosystem (MVE) of partners that would enable the new company to demonstrate its potential to create value on the product market. The concept of MVE was introduced by Adner (2012) as one of the principles for the construction of a successful business ecosystem. The MVP and MVE concepts fit very well together. The link between MVP and MVE could be expressed as follows: establishing a viable business means using an MVP as part of a customer development approach to transforming a lean startup into a new company with a well-configured MVE. Such an understanding of new business creation resembles the definition of creation suggested by Deleuze (1998) – the act of making configurations.

The reason to provide a more detailed definition of the lean startup was to emphasize several points, as follows. First, the lean startup is a special type of startup; it is not just the early or immature stage of any startup. There are startups that are not lean startups. If we focus on the technology business area, we could say that the other type – regular or non-lean startups – usually focuses on intensive R&D and product development until they come up with a product that is then in need of a market (i.e., technology and product development dominate over marketing). Such were most of the startups in the 1990s. Such are many of today's startups that do not really reach a viable business stage in the form of a well-established new company. The reason for Blank and Ries to introduce and insist on promoting the lean startup concept was exactly to differentiate the (lean) startup described above from the regular ones. According to their lean philosophy, "lean" is about reducing waste, and the greatest waste for a startup is to develop a product that nobody wants.

There is a danger of considering a lean startup just as an initial phase of a new startup that is preliminary to its future viable business stage. One can, of course, do that at the cost of missing the point and never reaching a viable business stage. The reason for the existence of such danger is that everything in the lean startup and in the viable business stage of the future new company is different – the type of product that is being developed, the type of customers being targeted, the type of employees that need to be hired, and the type of business goals being set. This fact was realized as early as the 1990s through the promotion of the *Crossing the Chasm* approach articulated by Geoffrey Moore (1991). There is however a key difference in the messages of the lean startup and crossing-the-chasm approaches. The main focus of the crossing-the-chasm approach was to emphasize that: i) the early customers (i.e., enthusiasts, innovators, and early adopters) are not a startup's ultimate customer target market segment: the late adopters; ii) the technological solution that impresses the early adopters is not the whole product that is going to impress the late adopters – the startup should focus on interacting with customers to help the development of the whole product; iii) the initial marketing approach to early adopters is not the way to deal with later adopters.

The lean startup approach could be seen as a valuable development of Moore's crossing-the-chasm approach. The lean startup "new development" consists of shifting the above messages into another key: i) the lean startup is not the company that the entrepreneur is trying to build; ii) the product of the lean startup is not its product but the business model; and iii) the focus of product development should be on using customer feedback to evolve the MVP into an awesome product that is going to impress the first customer target niche (i.e., product development is based on customer development). In this sense, the lean startup concept emphasizes two key differences: lean startups versus other (non-lean) startups and lean startups versus the companies that are going to be built on them. The claim that a lean startup is a temporary organization amplifies the emphasis of the second difference. According to Blank (2007), "The idea of not having a functional organization until the organization has found a proven business model is one of the hardest things for new startups to grasp" (Blank & Dorf, 2012). The current knowledge in global startup research and practice could greatly benefit from a similar enhancement based on the lean startup concept.

Second, it is clear that global startups have been studied before as a special type of international new venture and should not be considered as something new in both business practice and research. The purpose of the section of this article dedicated to global startups was to illustrate this fact. The global startup literature, however, does not seem to address the ongoing adoption, popularity, and relevance of the lean startup approach. The lean startup literature does not speak about the opportunities of a global start either. The lean global startup concept was introduced with the intention of linking the aspects of lean and global (Lemminger et al. 2014; Tanev et al., 2015; Rasmussen & Tanev, 2015). Almost at the same time, Steve Blank (2014) followed-up with a valuable insight: Startups have to be born global or die local because most countries do not have sufficient population to support a potential scale up with their local market (i.e., growth-oriented startups ultimately need to become global players from day one). In the definition of lean global startup, "lean" refers to global and employs a lean startup logic that is similar to the one described above. A lean global startup adopts the lean approach as part of its global move and not simply as part of its product and customer development strategies focusing on addressing a first local market niche. In a typical lean global startup, the target market niche would be at a global location and all local marketing efforts would be in parallel to the global one or for the sake of experimentation and learning. It is clear, however, that there could be global startups that are not lean startups. In this sense, it would be highly beneficial to consider the lean global startup as a unique firm type.

Third, the global startup definition suggested by Oviatt and McDougall's (1994) and Oviatt, McDougall, and Loper (1995) does not appear to be good enough to address the context of actual technology startups that have adopted a global marketing or global resourcing strategy from their inception. It misses some of the key points of the lean global startup context such as the multi-continent aspects of globalness and the evolutionary nature of a startup's transition to a newly established company. It appears to be too much attribute-based and not so much circumstance-based, to use the Christensen's (2006) terminology. More importantly, it does not seem to have taken into account the difference between the ex ante resource focus of early-stage startups and the ex post product market focus of established young companies. According to Doz and Wilson (2012), this is a typical problem for the managers of global organizations – judging the value or success of a global location by using ex post performance measurements of value. At the same time, the focus on the lean global startup context is a direct expression of Coviello's concern that international new ventures or global startups should be studied at their very early stages, when their initial intentions are most clearly articulated. In this sense, the introduction of the lean global startup concept appears to be quite useful in emphasizing the lean aspects of reaching a global presence. It is inclusive for both "finger-push" firms, with the potential to serve global markets, and for other startups that have committed to a more evolutionary lean global path from their start. Previous research offered examples of these two types of global startup scenarios and labelled them lean and global startups and lean-to-global startups (Borseman et al., 2016; Rasmussen & Tanev, 2015; Tanev et al., 2015). However, the meaning of "global" in the lean global startup definition has to be refined with respect to the global startup definition provided by Oviatt and McDougall (1994) by reducing the requirement for the quantitative intensity of the global resource coordination or market presence at the time of inception of a startup. For example, the simultaneous operation of a startup at two globally distant locations (for example, in two different continents) should be considered to be global enough for the sake of born global startup research. Establishing a new business in a foreign country enables the actualization of special innovative capabilities (Jones & Coviello, 2005; Weerawardena et al., 2015; Zijdemans & Tanev, 2014;). Once demonstrated in one specific country, these capabilities could be more easily replicated in other countries. A softer requirement for the quantitative intensity of the global upstream and downstream engagement of startups will allow the development of analytical and practical frameworks that would help globally-driven startups to pursue their global business path more systematically in the same way the lean startup approach does.

Conclusion

This article has two main contributions. The first one is taking into account some of the key insights from most recent research on born-global firms and, more specifically, the need for clarification and refinement of existing definitions by focusing on startups that have engaged into a global path from their very inception (Coveillo, 2015). On the one hand, it adopts the idea that existing born-global startup definitions provide a better fit to the startup context and summarizes some of the key lessons learned from born-global startup research (Meyer & Xia, 2012; Oviatt et al., 1995). On the other hand, it offers a critical reflection on these insights and suggests that existing definitions remain grounded in the context of younger firms and not so much on actual startups. The second contribution is the summary of the characteristics of lean startups in a way that could help further substantiate the claim about the benefits of considering lean global startups as a new type of organization/firm. The analysis suggests that the lessons learned from the emergence of lean startup entrepreneurship offer a basis for promoting a similar lean phase in technology-based global startup research and practice. Last but not least, international entrepreneurship research could greatly benefit from adopting a global instantiation of the lean startup paradigm in addressing the actual context of new technology firms dedicated to operating in a global context from their very inception.

About the Author

Stoyan Tanev is an Associate Professor in the Department of Technology and Innovation, Faculty of Engineering, Southern Denmark University (SDU) in Odense. Dr. Tanev is leading the Technology Entrepreneurship stream of the Master Program of Product Development and Innovation at SDU. He is also an Adjunct Research Professor in the Sprott School of Business at Carleton University in Ottawa, Canada, where he is associated with the Technology Innovation Management Program. He has a MSc and a PhD in Physics jointly from the University Pierre and Marie Curie, Paris, France, and the University of Sofia, Bulgaria, a PhD in Theology from the University of Sofia, Bulgaria, an MEng in Technology Innovation Management from Carleton University, Canada, and an MA from the University of Sherbrooke, Canada. He has multidisciplinary research interests with a focus on the fields of global technology entrepreneurship, technology innovation management, business model design, and value co-creation. Dr. Tanev is Senior IEEE member, as well as member of the editorial boards of the *Technology Innovation Management Review*, the *International Journal of Actor-Network Theory*, and *Technological Innovation.*

References

Adner, R. 2012. *The Wide Lens. What Successful Innovators See that Others Miss.* New York: Portfolio/Penguin.

Alcácer, J., Cantwell, J., & Piscitello, L. 2016. Internationalization in the Information Age: A New Era for Places, Firms, and International Business Networks? *Journal of International Business Studies,* 47(5): 499–512.
http://dx.doi.org/10.1057/jibs.2016.22

Andersson, S., Evers, N., & Griot, C. 2013. Local and International Networks in Small Firm Internationalization: Cases from the Rhône-Alpes Medical Technology Regional Cluster. *Entrepreneurship & Regional Development,* 25(9–10): 867–888.
http://dx.doi.org/10.1080/08985626.2013.847975

Blank, S. 2007. *The Four Steps to the Epiphany – Successful Strategies for Products that Win.* Raleigh, NC: Lulu Enterprises.

Blank, S. 2013. Why the Lean Start-Up Changes Everything. *Harvard Business Review,* 91(5): 63–72.

Blank, S. 2014. Born Global or Die Local – Building a Regional Startup Playbook. *SteveBlank.com,* October 31, 2014. Accessed May 1, 2017: http://steveblank.com/2014/10/31/born-global-or-die-local-building-a-regional-startup-playbook/

Blank, S., & Dorf, B. 2012. *The Startup Owner's Manual: The Step-by-Step Guide for Building a Great Company.* Pescadero, CA: K&S Ranch Incorporated.

Borseman, M., Tanev, S., & Weiss, M. 2016. Lost in the Canvases: Managing Uncertainty in Lean Global Startups. In *Proceedings of ISPIM Innovation Forum: Charting the Future of Innovation Management,* March 13–16, 2016, Boston, USA.

Cavusgil, S., & Knight, G. 2015. The Born Global Firm: An Entrepreneurial and Capabilities Perspective on Early and Rapid Internationalization. *Journal of International Business Studies,* 46, 3–16.
http://dx.doi.org/10.1057/jibs.2014.62

Christensen, C., Roth, E., & Anthony, S. 2004. *Seeing What's Next: Using the Theories of Innovation to Predict Industry Change.* Boston, MA: Harvard Business Review Press.

Christensen, C. 2006. The Ongoing Process of Building a Theory of Disruption. *Journal of Product Innovation Management,* 23(1): 39–55.
http://dx.doi.org/10.1111/j.1540-5885.2005.00180.x

Coviello, N. 2015. Re-Thinking Research on Born Globals. *Journal of International Business Studies,* 46(1): 17–26.
http://dx.doi.org/10.1057/jibs.2014.59

Coviello, N., & Tanev, S. 2017. Initiating a New Research Phase in the Field of International Entrepreneurship: An Interview with Professor Nicole Coviello. *Technology Innovation Management Review,* 7(5): 52–56.
http://timreview.ca/article/1077

Deleuze, G. 1998. Qu'est-ce que l'acte de création? *Trafic,* 27: 133–142.

Doz, Y., & Wilson, K. 2012. *Managing Global Innovation: Frameworks for Integrating Capabilities Around the World.* Boston, MA: Harvard Business Review Press.

Eisenmann, T., Ries, E., & Dillard, S. 2011. *Hypothesis-Driven Entrepreneurship: The Lean Startup.* Harvard Business School Entrepreneurial Management Case No. 812–095.
https://papers.ssrn.com/sol3/papers.cfm?abstract_id=2037237

Frederiksen, D., & Brem, A. 2017. How do Entrepreneurs Think They Create Value? A Scientific Reflection of Eric Ries' Lean Startup Approach. *International Entrepreneurship and Management Journal,* 13(1): 169–189.
http://dx.doi.org/10.1007/s11365-016-0411-x

Gans, J. 2016. *The Disruption Dilemma.* Cambridge, MA: The MIT Press.

Jones, M. V., & Coviello, N. E. 2005. Internationalisation: Conceptualising an Entrepreneurial Process of Behaviour in Time. *Journal of International Business Studies,* 36(3): 284–303.
http://dx.doi.org/10.1057/palgrave.jibs.8400138

Jones, M., Coviello, N., & Tang, Y. 2011. International Entrepreneurship Research (1989–2009): A Domain Ontology and Thematic Analysis. *Journal of Business Venturing*, 26(6): 632–659. http://dx.doi.org/10.1016/j.jbusvent.2011.04.001

Knight, G., & Cavusgil, S. 2004. Innovation, Organizational Capabilities, and the Born-Global Firm. *Journal of International Business Studies*, 35: 124–141. http://dx.doi.org/10.1057/palgrave.jibs.8400071

Lemminger, R., Svendsen, L., Zijdemans, E., Rasmussen, E., & Tanev, S. 2014. Lean and Global Technology Start-ups: Linking the Two Research Streams. In *Proceedings of the ISPIM Americas Innovation Forum*, October 4–8, 2014, Montreal, Canada.

Lopez, L., Kundu, S., & Ciravegna, L. 2009. Born Global or Born Regional? Evidence from an Exploratory Study in the Costa Rican Software Industry. *Journal of International Business Studies*, 40(7): 1228–1238. http://dx.doi.org/10.1057/jibs.2008.69

Madsen, T. K, & Servais, P. 1997. The Internationalization of Born Globals: An Evolutionary Process? *International Business Review*, 6(6): 561–583. https://doi.org/10.1016/S0969-5931(97)00032-2

Meyer, K., & Xia, H. 2012. British Entrepreneurs, Global Visions. *Business Strategy Review*, 23(2): 52–57. http://dx.doi.org/10.1111/j.1467-8616.2012.00846.x

Moore, G. A. 1991. *Crossing the Chasm: Marketing and Selling High-Tech Products to Mainstream Customers*. New York: Harper Business.

Oviatt, B. M., & McDougall, P. P. 1994. Toward a Theory of International New Ventures. *Journal of International Business Studies*, 25(1): 45–64. http://dx.doi.org/10.1057/palgrave.jibs.8490193

Oviatt, B. M., McDougall, P. P., & Loper, M. 1995. Global Start-Ups: Entrepreneurs on a Worldwide Stage. *The Academy of Management Executive*, 9(2): 30–44. http://www.jstor.org/stable/4165256

Rasmussen, E. S., & Tanev, S. 2015. The Emergence of the Lean Global Startup as a New Type of Firm. *Technology Innovation Management Review*, 5(11): 12–19. http://timreview.ca/article/941

Ries, E. 2011. *The Lean Startup: How Today's Entrepreneurs Use Continuous Innovation to Create Radically Successful Businesses*. New York: Crown Business.

Sarasvathy, S., Kumar, K., York, J. G., & Bhagavatula, S. 2014. An Effectual Approach to International Entrepreneurship: Overlaps, Challenges, and Provocative Possibilities. *Entrepreneurship Theory and Practice*, 38(1): 71–93. http://dx.doi.org/10.1111/etap.12088

Schmidt, J., & Keil, T. 2013. What Makes a Resource Valuable? Identifying the Drivers of Firm-idiosyncratic Resource Value. *Academy of Management Review*, 38(2): 206–228. http://dx.doi.org/10.5465/amr.2010.0404

Stolze, A., Arnsfeld, T., Kelly, L., & Lüdtke, C. 2014. *The Lean Startup Status Quo: Deconstructing the Lean Start-Up Movement to Assess its Validity as a Strategic Planning Tool for Entrepreneurs*. Working Paper No. 3. Faculty of Business Management and Social Sciences. Osnabrück, Germany: Osnabrück University of Applied Sciences.

Tanev, S. 2012. Global from the Start: The Characteristics of Born-Global Firms in the Technology Sector. *Technology Innovation Management Review*, 2(3): 5–8. http://timreview.ca/article/532

Tanev, S., Rasmussen, E. S., Zijdemans, E., Lemminger, R., & Limkilde, L. 2015. Lean and Global Technology Start-Ups: Linking the Two Research Streams. *International Journal of Innovation Management*, 19(3): 1540008. http://dx.doi.org/10.1142/S1363919615400083

Weerawardena, J., Mort, G. S., Salunke, S., Knight, G., & Liesch, P. W. 2015. The Role of the Market Sub-System and the Socio-Technical Sub-System in Innovation and Firm Performance: A Dynamic Capabilities Approach. *Journal of the Academy of Marketing Science*, 43(2): 221–239. http://dx.doi.org/10.1007/s11747-014-0382-9

Yang, M., & Gabrielsson, P. 2017. Entrepreneurial Marketing of International High-Tech Business-to-Business New Ventures: A Decision-Making Process Perspective. *Industrial Marketing Management*. http://dx.doi.org/10.1016/j.indmarman.2017.01.007

Zijdemans, E., & Tanev, S. 2014. Conceptualizing Innovation in Born-Global Firms. *Technology Innovation Management Review*, 4(9): 5–10. http://timreview.ca/article/826

Keywords: international entrepreneurship, lean startup, born global, global startup, lean global startup, international new venture, technology startup

17

The Evolution of Intermediary Activities: Broadening the Concept of Facilitation in Living Labs

Louna Hakkarainen and Sampsa Hyysalo

> *“It is hardly possible to overrate the value... of placing human beings in contact with persons dissimilar to themselves, and with modes of thought and action unlike those with which they are familiar. ... Such communication has always been... one of the primary sources of progress.”*
>
> John Stuart Mill (1806–1873)
> In *Principles of Political Economy*

Innovation intermediaries play an important role in open innovation endeavours. In living lab projects, where different professional identities and organizational cultures are at play, intermediary actors facilitate learning between stakeholders and manage tensions and conflicts of interest. The current living lab literature recognizes the importance and multifacetedness of these actors, but does not shed light on the work they do at a more practical level. Our study seeks to capture the variety and evolution of work tasks of user-side innovation intermediaries during and after a four-year technology project in a living lab. The study explores how these mediating actors tackle the everyday challenges of a living lab project. This article is grounded on a longitudinal qualitative case study of a innovation process for a floor monitoring system for elderly care – the "smart floor".

Introduction

Living labs are real-life experimentation environments in which new products and services are given shape through collaborative efforts of users and developers. They aim to extend co-design and open innovation activities from mere concept design and ideation to design-in-use, which is often requisite for co-realizing the true value points of new technologies and services (Botero & Hyysalo, 2013; Hartswood et al., 2002; Hillgren et al., 2011; Hyysalo, 2010; Leminen et al., 2015; Voss et al., 2009).

The success of such real-life collaboration, which aims to promote learning between different stakeholders, hinges on how the co-design process has been orchestrated, facilitated, and managed. In discussions about living labs notions such as “quadruple helix” and “public–private–people partnerships” flag the issue prominently. However, research on collaboration dynamics in living labs remains nascent, and it seems that often the complex knowledge exchange tends to be taken for granted, overlooked, or simplified beyond what, for instance, the kind of guidance practitioners would benefit from the most.

This article on intermediation work in a living lab project is based on a longitudinal qualitative study of a four-year (2005–2009) living lab project that took place in four units of a large public nursing home in Finland. The data allows us to describe and analyze how the user-side innovation intermediaries facilitated learning between developers and users during a long-term co-design project. We focus on the intermediation work done by three living lab project workers, whose educational background was in nursing and elderly care. After the four-year living lab project, the developer company hired the key project worker as a customer care specialist. This made it possible to extend the scope of our research to a total of eight years and to include the after-market launch period, when the locally tailored product was "generified" to serve a widening clientele (Hyysalo, 2010; Pollock & Williams, 2008).

In order to address the variety of intermediation work in the case, we have turned to research on innovation intermediaries. Innovation intermediaries have been central in social learning processes in technological innovation (Stewart & Hyysalo, 2008; Williams et al., 2005). In innovation studies, these mediating actors have been studied for some time. Howells (2006) describes an innovation intermediary as "[a]n organization or body [or an individual] that acts an agent or broker in any aspect of the innovation process between two or more parties".

For a long time, research around the topic focused on supply-side actors, such as industry associations and knowledge-intensive business services, but lately, work has been done to highlight the significance of innovation intermediaries in the user-side activities and processes of social learning: "The highly visible supply-side intermediaries [...] and the easily identifiable middle-ground agencies [...] tend to overshadow the often more informal yet just as crucial intermediaries at the user-end of the supply-use relation. Intermediate users, local experts and 'tailors' facilitate, configure and broker systems, usages and knowledge about systems and their deployments, helping users to domesticate them and suppliers to respond to actual, realised uses." (Stewart & Hyysalo, 2008). Our present study focuses on the role of public sector user-side innovation intermediaries in a collaborative innovation process.

Theoretical Framework

Our understanding of living labs relies on findings from science and technology studies – especially around social learning (Hyysalo, 2009; Williams et al., 2005) and domestication of technology (Berger et al., 2006; Silverstone et al., 1992; Sørensen, 1996).

The social learning in technological innovation approach (Williams et al., 2005) grew out of research on the social shaping of innovation (MacKenzie & Wajcman, 1999; Williams & Edge, 1996). The concept of social learning places particular emphasis on the activity of the users during the appropriation of new technology and highlights the importance of simultaneously studying processes of design, implementation, and use.

Social learning refers especially to two simultaneous, complementary, and intertwined processes: innofusion (Fleck, 1988) and domestication of technology (Sørensen, 1996). Innofusion (innovation that takes place during diffusion) refers to "processes of technological design, trial and exploration, in which user needs and requirements are discovered and incorporated in the course of the struggle to get the technology to work in useful ways, at the point of application" (Fleck, 1988). The concept of domestication has its origins in cultural consumption studies, and it refers to the work users go through in "fitting [technologies] into the pre-existing heterogeneous network of machines, systems, routines and culture" (Sørensen, 1996).

From these perspectives, we see living labs as a co-design infrastructures in which users' creativity around technology use and their efforts to fit technology to cultural, organizational, and material contexts become resources for product development. However, the potential of this kind of collaboration does not realize automatically, which is why we focus on the crucial work done by innovation intermediaries in living lab networks.

Innovation intermediaries

Stewart and Hyysalo (2008) define user-side innovation intermediaries as organizations or individuals that "attempt to configure the users, the context, the technology and the 'content', *but they do not, and cannot define and control use or the technology*". They are thus actors who seek to influence users and developers, but do not have final say over how the technology is eventually used (this is what users and managers at user organizations do) nor do they hold decision-making power, or necessary skills, to alter the form of the technology at the developer end.

In their seminal studies, Howells (2006) and Bessant and Rush (1995) have listed functions and bridging activities of innovation intermediaries (Box 1). Shortcomings of these kinds of listings are that they leave aside the common types of engagements that these actors are involved in during their "bridging activities".

Stewart and Hyysalo (2008) have attempted to move from a mere ordered list of functions to an analytically ordered set of concepts that describe how intermediaries act and what are the different facets of their work in innovation. They have recognized three user-side innovation intermediary roles with respect to social learning: facilitating, configuring, and brokering.

Box 1. Functions and activities of innovation intermediaries

Intermediary functions (Howells, 2006)
1. Foresight and diagnostics
2. Scanning and information processing
3. Knowledge processing and (re)combination
4. Gatekeeping and brokering
5. Testing and validation
6. Accreditation
7. Validation and regulation resources; organizational development
8. Protecting the results
9. Commercialization
10. Evaluation of outcomes

Bridging activities (Bessant & Rush, 1995)
1. Articulation of needs; selection of options
2. Identification of needs; selection training
3. Creation of business cases
4. Communications; development
5. Education; links to external info
6. Project management; managing external resources; organizational development

Facilitating means providing opportunities to other *people,* by educating, gathering and distributing resources, influencing regulations, developing the local rules, and creating "spaces" for others to act. Configuring means material and symbolic alteration of *technology,* adjusting its form and content (often in minor ways), as well as how it is interpreted and used. Brokering refers to the establishing, nurturing, adjusting, and altering of *connections* between different actors. This work on connections is not just neutral bridging, but is often selective and occasionally self-serving to the position of the intermediary actor itself.

Intermediation work in living labs

In recent years, living labs also have been analyzed as innovation intermediaries (e.g., Almirall & Wareham, 2011; Baltes & Gard, 2010; Katzy et al., 2013). Almirall and Wareham (2011) define living labs as "[...] open innovation intermediaries that seek to mediate between users, research, public and private organisations, advance our concept of technology transfer by incorporating not only the user based experimentation, but also by engaging firms and public organisations in a process of learning and the creation of pre-commercial demand."

Some attempts have been made to shed light on the interaction dynamics inside living labs on a more detailed level. Such research has focused on communities of practice and boundary objects (Johansson & Snis, 2011), living lab actors' roles and role patterns (Nyström et al., 2014; Box 2), living lab networks' modes of coordination and participation (Leminen, 2013), functions and roles of public open innovation intermediaries (Bakici et al., 2013), strategic capabilities of living labs (Katzy et al., 2013), paradoxical tensions in living labs (Leminen et al., 2015), complexity in the stakeholder interactions (Pade-Khene et al., 2013), and possibilities of social and cognitive translation between stakeholders (Svensson & Ebbesson, 2010). Part of this work has been attempts to also identify the roles of intermediary actors in living labs (Heikkinen et al., 2007; Nyström et al., 2014; see Box 2).

Although helpful in gaining a sense of what functions actors perform in collaborative innovation, empirically derived listings and classifications bear close similarity to previous empirically derived listings of innovation intermediaries such as those of Howells (2006) or Bessant and Rush (1995) (see Box 1).

Gregor (2002) has characterized such listings as "naming theory", the most rudimentary form of theory within a research domain, a stepping stone on which more analytically ordered typologies and gradually more explanatory theory building can take place. One of the steps needed to move beyond naming and answering simple "what" questions is to conduct empirical studies that expose the situatedness and context-specific aspects of the innovation process and can shed light on "how" questions. This is important also for gaining practical sense of what works (Gregor, 2002; Woolrych et al., 2011)

Thus, with regard to actor roles in living labs, further work is called for, particularly in two respects. First, there is a need to empirically gain better specificity in what kinds of engagements the roles relate to. The current lists of actor roles by Nyström and colleagues (2014) have been derived from multiple projects and multiple different actors and beg for further clarification, as do the contents of the different roles. Furthermore, only some of the roles are present in different projects and, at that, different *phases* of projects. Existing analysis of processes of intermediation in or by living labs address the systemic or organizational level, but fail to describe in detail how individuals tackle the challenges posed by everyday life in living labs.

Box 2. Identified actor roles

Previously identified actor roles (Heikkinen et al., 2007)
1. Webber: Acts as the initiator; decides on potential actors
2. Instigator: Influences actors' decision-making processes
3. Gatekeeper: Possesses resources
4. Advocate: Background role; distributes information externally
5. Producer: Contributes to the development process
6. Planner: Participates in development processes; input in the form of intangible resources
7. Accessory provider: Self-motivated to promote its products, services, and expertise

Newly identified roles (specific to living labs) (Nyström et al., 2014)
8. Coordinator: Coordinates a group of participants
9. Builder: Establishes and promotes the emergence of close relationships between various participants in the living lab
10. Messenger: Forwards and disseminates information in the living lab network
11. Facilitator: Offers resources for the use of the network
12. Orchestrator: Guides and supports the network's activities and continuation; tries to establish trust in the network to boost collaboration to further the living lab's goals
13. Integrator: Integrates heterogeneous knowledge, development ideas, technologies, or outputs of different living lab actors into a functional entity
14. Informant: Brings users' knowledge, understanding, and opinions to the living lab
15. Tester: Tests innovation in (customers') real-life environments (e.g., hospitals, student restaurants, and classrooms)
16. Contributor: Collaborates intensively with the other actors in the network to develop new products, services, processes, or technologies
17. Co-creator: The user co-designs a service, product, or process together with the company's R&D team and the other living lab actors.

Second, although the more detailed empirical examination of roles and their prevalence in actual living lab projects is in order, the research on actor roles in living labs would also benefit from seeking to move beyond mere naming towards better understanding of the interrelations of different roles, as was done with innovation intermediaries previously (Stewart & Hyysalo, 2008). Our focus on living lab facilitators happens to reside within the broader notion of innovation intermediary, and hence we shall examine whether our previously developed typology of configuring, brokering, and facilitating would be fit for further organizing the findings in the present article.

Research Approach

Our work enriches the previous research by focusing on the innovation intermediaries' work on the level of tasks and activities. We map the evolution of the intermediation work during and after the living lab project, covering almost eight years' time on the biography of the maturing artefact.

The study continues an analysis started in licentiate study by Hakkarainen (2013) and continued during the follow-up phase of study (Hakkarainen & Hyysalo, 2013; Hyysalo & Hakkarainen, 2014). The living lab project workers documented nearly all the collaboration meetings held with different assemblies over the course of the four-year project. In addition to memos, the data included project reports, plans, and marketing material – altogether 151 different documents related to the development and use of the "smart floor", which we describe later in the article. The overall number of qualitative in-depth interviews is 21: 16 during the living lab project and five after it. Four of the latter interviews were conducted with the developer company's sales manager and customer care specialist (who was previously a living lab project worker), and one was conducted with the customer care specialist alone. The last interview was conducted after the both interviewees had quit working for the company.

The units of analysis are *intermediary activities and tasks* of the living lab project personnel. By task, we mean an organized set of actions that can be either a one-time effort or a repeated pattern in the practices of the mediating personnel – in any case, a set of actions that formed a mutually recognized whole by both the mediating personnel and their colleagues (Strauss, 1993).

The coded tasks were ordered chronologically and reorganized under higher-level activities. The result of the analysis were 31 different tasks, which were categorized under 13 activities. The results were organized in a matrix (see Table 4) that shows how the activities and tasks evolved over time in different phases of the innovation process.

The smart floor innovation process has been divided in four phases (Figure 1). The division is based on empirical work done by Van de Ven and colleagues (1999) on innovation journeys and by Pollock and Williams (2008) on biographies of artefacts as well as process dynamics observed in the study by Hakkarainen (2013). Each transition represents significant changes in the innovation network as well as in the smart floor artefact.

In the final step of the analysis, we structured the tasks according to facilitating, configuring, and brokering (Stewart & Hyysalo, 2008) to see if there are changes in the broader-level orientation of the intermediaries in the course of the innovation project.

Case Study: A Smart Floor System

The origins of smart floor system are in the Helsinki University of Technology (now Aalto University), where the motion-tracking technique behind it was discovered in the early 1990s. Years later, a group of researchers and students created the first version of the smart floor – a simple floor monitoring system – and a company was founded around it in 2005. The idea for creating a gerontechnological device originally came from the user side: a well networked, innovation-oriented nursing home manager became aware of the discovery and encouraged the engineers to advance the technique into a system for elderly care.

The technology was next developed in an enabler-driven living lab (Leminen et al., 2012), which was established in 2006 as part of Helsinki Living Lab, an early member of the European Network of Living Labs. The lab focused on a large public nursing home. The public-sector actors were the initiators of the collaboration and were also responsible for applying funding and hiring of the project personnel that acted as innovation intermediaries. The nursing home manager later became the head of the innovation undertaking, wherein the smart floor was one of the four sub-projects. The main stakeholders of the project are presented in the Figure 2. The number of project workers varied between two and three fulltime workers in different stages of the project.

The smart floor system – the outcome of the collaboration – consists of a sensor foil, which is installed under the flooring material; a user interface, which is accessed on a computer situated in the office; and cell phones, which the nurses carry with them during their work shifts. The movements of the residents generate alerts, which the nurses receive through the cell phones. The system can inform the nurses about, for example, a situation where a frail elderly person is getting out of bed, entering or leaving the room, entering the toilet, or occupying the toilet for an unusually long time. The alarms are tailored individually to each person.

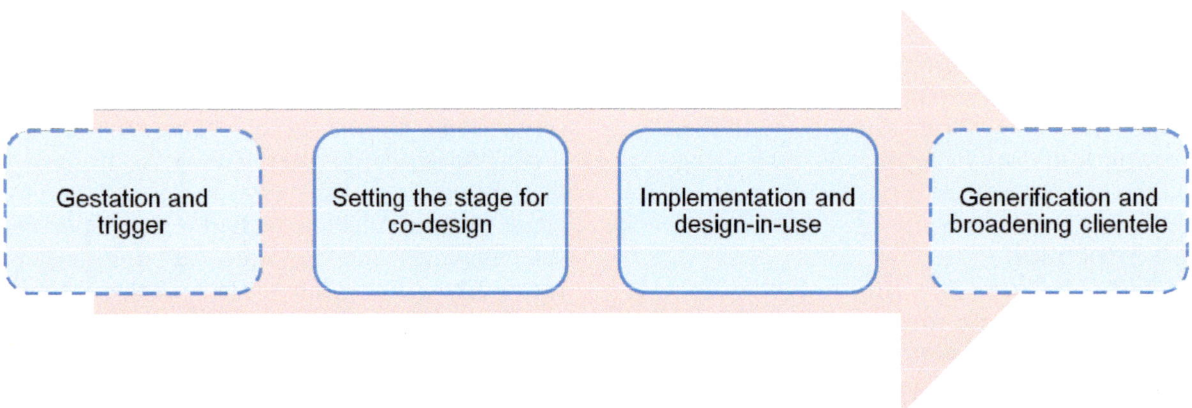

Figure 1. Phases of the smart floor innovation process

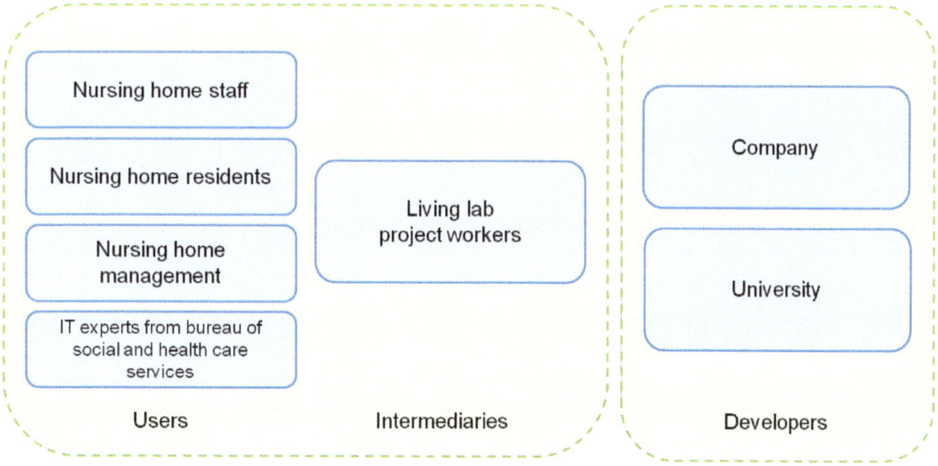

Figure 2. Stakeholders in the smart floor living lab

Setting the stage for co-design

Technology development was not the purpose of the collaboration project from the beginning. The initial plan was to explore ways to efficiently utilize the smart floor technology in the everyday life of the nursing home. However, due to the immaturity of the product, the focus of the collaboration changed to technology development.

The project workers had background in care work and, during the first months of the project, they participated in regular care duties in the units. This meant that the project workers had a profound understanding of the users, their work practices, and the context of use. However, they were not familiar with formal co-design or participatory design methods.

The collaboration started officially with a workshop in which the intermediaries, developers, and care workers defined the first user requirements for the system. After this, the information exchanges took place mostly in regular meetings. The project workers could organize the collaboration as they saw best, and the goals and methods were reassessed regularly and adjusted to the needs of the project.

The project was formally divided in two sub-projects: the main purpose of the first part was to test the smart floor in two rooms and to develop it further, especially by fixing technical bugs and getting rid of false alarms, so that the second part, a larger-scale implementation, was possible. The project workers had significant responsibility in diagnosing and weeding out technical problems.

From the beginning, the engineers and the nursing home staff and management – project workers included – had strongly differing understandings about the maturity of the product and each other's roles in the collaboration. The company was in a hurry to launch their product, but from the users' perspective, the smart floor was not even ready for the test implementation. The client – as represented by nursing home staff and project workers – was frustrated with the functioning of the system and severity of its bugs; they saw the engineers as arrogant and indifferent to the welfare of the residents and nursing home staff. The developers, for their part, saw the users' requests as unreasonable and unrealistically scheduled. The goal of the company was to create a generic product instead of a tailored system, and they were sceptical about the representativeness of the client's demands.

Finally, the nursing home management and project workers refused to proceed with the implementation unless their demands were met. At the end of 2007, two out of three members of the living lab project staff – including the project manager and project co-ordinator – resigned, as did technology company's CEO, bringing the whole undertaking to the verge of collapse.

A summary of intermediary activities and tasks in the first phase is presented in Table 1.

Implementation and design-in-use

Changes in staff eased the tensions, and the collaboration continued, after the developers, two project workers (one newly hired), and management of the nursing home found common ground prior to the implementa-

Table 1. Intermediary activities and tasks in the stage-setting phase

Activity	Task	
Technical tinkering	Diagnosing and fixing bugs with the engineers	C
	Taking part in the installation and testing	C
Co-designing	Defining preliminary user requirements with the users	C
	Formulating project plan and choosing methods of collaboration	C/B
	Documenting the co-design process	F
User research	Studying the users, their work, and context of use	F
Advocating	Communicating the user perspective to the developers	B
	Pressuring the developers to realize users' wishes	B

C = Configuration; F = Facilitation; B = Brokering

tion phase. At the end of 2007, the smart floor was installed in two rooms as a pilot and then rolled out to three other units (each with around 20 residents), where the sensor foil was installed in all the rooms and public spaces.

The hiring of a new project worker was pivotal for the new consensus. At this point, the project management had better understanding of the requirements of the intermediary position. This time, they were looking for an independent and innovative negotiator, someone who would be technology-oriented and able to change perspectives when needed. In a delicate situation, the project workers needed to convince different stakeholders of each other's good intentions, recognize shared interests, and react quickly to changing circumstances. Nevertheless, they had to be practical enough to push through the demanding implementation phase and support the care workers by taking part in the regular care duties.

The implementation phase invoked a new kind of division between the living lab project stakeholders: many of the end-users – the nursing home staff – reacted negatively to the smart floor. The nursing staff was unwilling to study new things alongside their normal workload or to change their work routines. Their job was demanding enough on its own. In addition, the nurses saw themselves as caregivers, not machinists, and were generally reserved about complex gerontechnological devices. Many care workers boycotted the project and the system, for example, by not carrying cell phones with them during their shift and continuing to work as they used to. Pushing forward with the rollout of the system required developers, project workers, and nursing home management to ally themselves against the care personnel, among who many were reluctant to put the system to use let alone participate in its improvement and to make the use of the system. Attendance at the feedback meetings was made obligatory for the nurses.

During the implementation, the strict discipline was counterbalanced by the devotion of the project workers, who were also care professionals by education. They spent time in the living lab units on a daily basis and helped the nurses in the implementation of the system, even occasionally assisting them with normal care duties. The weekly (later monthly) feedback meetings provided the care personnel an opportunity to speak out, comment on the system, and express new development ideas. The project workers and the nurses discussed how the system had been utilized, what its benefits were, and how it affected the care practices and the elderly people. This feedback was complemented by observing the smart floor's daily use, which the project workers valued as the most important way to collect information for the improvement of the system. Their background as care workers helped them to make sense of the daily work in the units, which was needed because the burden of developing the system further was placed on their shoulders. The project workers observed use, identified problems and solutions with the engineers, and thought of ways to utilize different functionalities and properties of the system with the care personnel. Another important area was how the system should be used in order to produce optimal results: for example, how to determine the right mix of alarms for each resident, how the system affects elderly people in the long term, and what should be done when a nurse receives overlapping alarms. They also had to think about the challenges that the living lab project created, for example, what practical actions to take when the system does not work the way it is supposed to.

In addition, the project workers were active in planning, organizing, and executing effectiveness research of the impact of the smart floor on, for example, resident safety and nursing work. The work was done primarily for the client (the City of Helsinki), but the results were highly valuable for the company as well. Later in the project, the project workers were also active in showcasing the system and the project to numerous potential customers from all over the world.

A summary of intermediary activities and tasks in the design-in-use phase is presented in Table 2.

After the living lab project: Generification and broadening the clientele

In the course of the living lab project, the startup company had merged with an established electronics company. When the living lab project was coming to an end, the company hired, as a customer care specialist, the key project worker – the one that had started in the middle of the project and who managed to turn the confrontation into fruitful cooperation.

After the market launch of the product, the clientele of the company grew, and new contextual problems arose, for example, in new buildings where the concrete was more humid and disrupted the normal functioning of the system. There were also minor differences in work practices at different institutions, which required some changes to the system.

From the onset, the company adopted a tailoring strategy, which meant that the system was customized to each customer organization's needs. After a while, this strategy was found to be unviable, and a more generic product was needed. Hence, the company sought to repackage its offering as a more standard product and servicing, where the customer care services, that previously were offered freely, were billed separately.

The customer care specialist organized user training and took care of the customer concerns, but she also continued to participate in the R&D activities by collecting user feedback, ideating improvements in the system, and networking with potential partners. She acted as a link between the customers and the company, and for this reason she had a very realistic understanding of the customers' reactions, concerns, and preferences. Her technical know-how, which had accumulated during the living lab project, allowed her to participate actively in the technical installation, testing, and problem solving in new client organizations. She also had credibility and the ability to consult management of the client organizations in renewing their care practices in order to get the biggest benefit out of the system.

Committing the client organizations to the use of the system remained as one of the biggest challenges for the company. The use of a complex system such as the smart floor can easily degenerate in new client organizations, because the end users and mid-level managers are usually not the ones making the purchasing decision.

The customer care specialist also participated in the marketing and sales negotiations. Because of a shared professional identity, she was able to ally herself with the client organization and even make some critical comments if the sales manager's pitch was too direct.

In 2013, the company was sold once more and the sales manager was laid off. At this point, the customer care specialist also decided to resign, because she was expected to assume the sales manager's responsibilities in addition to her existing responsibilities. By the start of 2016, the smart floor had become a stable product in the market and it has been installed in over 2000 apartments, mostly in northern Europe.

A summary of intermediary activities and tasks in the design-in-use phase is presented in Table 3.

Evolution of Intermediary Activities

The mapping of the responsibilities of the project personnel shows how intermediary activities and tasks are spread out through the course of the innovation process, and how they continue and change along with the project (see Table 4). Above all, it reveals the diversity of responsibilities undertaken by the intermediary actors.

The most intensive engagement took place at the implementation and design-in-use phase, during which the

Table 2. Intermediary activities and tasks in the design-in-use phase

Activity	Task	
Technical tinkering	Diagnosing and fixing bugs with the engineers	C
	Documenting technical problems and false alarms with the users	C/F
Co-designing	Formulating project plan and choosing methods of collaboration	C/B
	Documenting the co-design process	F
	Collecting, filtering, and transferring end users' ideas to the developers	B
	Coming up with development ideas and evaluating them with the users	C/F
User research	Studying the users, their work, and context of use	F
	Observing use and spotting usability problems	F
Advocating	Communicating the user perspective to the developers	B
	Pressuring the developers to realize users' wishes	B
Developing work practices	Developing new work practices which the system supports	C
User training	Creating and carrying out a training program for the users	F
Implementing	Making and carrying out an implementation plan	C
	Supporting users during the implementation phase	F
Developing uses	Discovering optimal ways to use the system with the users	C/F
	Defining codes of conduct for problematic situations with the users	C
	Encouraging the users to actively discover new ways to utilize the system	F
Studying effectiveness	Planning and carrying out studies to assess the effectiveness of the system	F/B
	Documenting the benefits of the system with the users	F/B
	Evaluating how the system affects the residents with the users	F/B
Negotiating	Recognizing and mediating interests of different stakeholder groups	B
	Pushing the end users and mid-level managers to use the system	B
	Building trust with the users	B
Marketing and sales	Demonstrating the system to potential customers	B
Customer service	Receiving and resolving customer concerns	B

C = Configuration; F = Facilitation; B = Brokering

Table 3. Intermediary activities and tasks after the living lab phase

Activity	Task	
Technical tinkering	Diagnosing and fixing bugs with the engineers	C
	Taking part in the installation and testing	C
Co-designing	Collecting, filtering and transferring end users' ideas to the developers	B
	Coming up with development ideas and evaluating them with the users	C/F
User research	Observing use and spotting usability problems	F
Advocating	Communicating the user perspective to the developers	B
Developing work practices	Developing new work practices which the system supports	C
User training	Creating and carrying out a training program for the users	F
	Assessing the need for user training	F
	Monitoring the use of the system	C/F
Negotiating	Pushing the end users and mid-level managers to use the system	B
	Building trust with the users	B
Networking	Negotiating finance and partners for the R&D activities	B
Marketing and sales	Demonstrating the system to potential customers	B
	Taking part in sales negotiations	B
Customer service	Receiving and resolving customer concerns	B

C = Configuration; F = Facilitation; B = Brokering

Table 4. Evolution of intermediary tasks and activities

Activity	Task	I	II	III
Technical tinkering	Diagnosing and fixing bugs with the engineers	C	C	C
	Taking part in the installation and testing	C		C
	Documenting technical problems and false alarms with the users		C/F	
Co-designing	Defining preliminary user requirements with the users	C		
	Formulating project plan and choosing methods of collaboration	C/B	C/B	
	Documenting the co-design process	F	F	
	Collecting, filtering and transferring end users' ideas to the developers		B	B
	Coming up with development ideas and evaluating them with the users		C/F	C/F
User research	Studying the users, their work, and context of use	F	F	
	Observing use and spotting usability problems		F	F
Advocating	Communicating the user perspective to the developers	B	B	B
	Pressuring the developers to realize users' wishes	B	B	
Developing work practices	Developing new work practices which the system supports		C	C
User training	Creating and carrying out a training program for the users		F	F
	Assessing the need for user training			F
Carrying out implementation	Making and carrying out an implementation plan		C	
	Supporting users during the implementation phase		F	
	Monitoring the use of the system			C/F
Developing uses	Discovering optimal ways to use the system with the users		C/F	
	Defining codes of conduct for problematic situations with the users		C	
	Encouraging the users to actively discover new ways to utilize the system		F	
Studying effectiveness	Planning and carrying out studies to assess the effectiveness of the system		F/B	
	Documenting the benefits of the system with the users		F/B	
	Evaluating how the system affects the residents with the users		F/B	
Negotiating	Recognizing and mediating interests of different stakeholder groups		B	
	Pushing the end users and mid-level managers to use the system		B	B
	Building trust with the users		B	B
Networking	Negotiating finance and partners for the R&D activities			B
Marketing and sales	Demonstrating the system to potential customers		B	B
	Taking part in sales negotiations			B
Customer service	Receiving and resolving customer concerns		B	B

I = Setting the stage for co-design; II = Implementation and design-in-use; III = After the living lab project

C = Configuration; F = Facilitation; B = Brokering

largest number of tasks were performed. The case history underscores, however, that despite fewer tasks in other phases, they are equally crucial for success: effective collaboration in the design-in-use phase requires great effort, and achieving the goal of a profitable, widely applicable technology after the living lab phase was equally crucial for the innovation projects' success.

With respect to our analysis considering facilitation, configuring, and brokering, we can see three patterns emerging: i) all three engagements are quite evenly distributed in the first part of the living lab project; ii) the design-in-use phase is dominated by facilitation and brokering; and iii) brokering played the most important role after the project.

The three types of engagement do indeed appear to characterize the tasks of living lab intermediaries – none of these more abstracted roles appear redundant or absent. They underscore how the common way to denote such people as living lab "facilitators" seems to be a misleading way to characterize what such people do as innovation intermediaries: this role comprises only one third of their engagements and is strongest only in the design-in-use phase of collaborative innovation in living lab. Without a longitudinal perspective that reaches beyond the design-in-use phase, the illusion of the centrality of facilitation would prevail in our data as well.

Conclusions

Our study shows that the nature of intermediation in living lab projects cannot be reduced to facilitation. Intermediation work in a living lab project consists of a range of tasks, including configuring of technology and use practices, brokering contacts and interactions between different actors, as well as facilitating their work, learning, and interactions. Furthermore, the content and form of intermediary work evolves in the course of successful living lab project. Altogether, we recognized the intermediaries participating in 13 different intermediary activities and 31 tasks. Engagements that are typically thought of as "facilitating" comprise only a third of what these mediating personnel need to handle and comprise the most common form of engagement only in the phase after implementation, when design-in-use efforts are most active.

Previous research has approached the topic of intermediation in living labs mostly through cross-case comparisons of multiple organizations participating in multiple projects and networks (e.g., Heikkinen et al., 2007; Nyström et al., 2012). Because of this approach, the granularity of the findings has remained coarse and has resulted in "naming theory" of identifying lists of "actor roles". Following Gregor's (2002) framework for theory development, this is the most rudimentary form of theory in a given area that merely answers "what" questions. In the present article, we have shown how moving to longitudinal in-depth case studies of particular projects conducted in living labs helps to reveal process descriptions and answer "how" questions: both how living lab projects are shaped over time and how actor roles play out. This approach offers a richer understanding of the tasks and actions of particular actors as well as how they evolve over the course of an innovation project, allowing us to further connect living lab actor roles to wider theoretical development within innovation studies on innovation intermediaries (Bessant & Rush, 1995; Howells, 2006; Stewart & Hyysalo, 2008), as well as in-depth process studies on innovation (e.g., Hyysalo, 2010; Van de Ven et al., 1999; Williams & Edge, 1996; Williams et al., 2005).

Considering the pivotal role that the intermediary actors play in open innovation processes, such as those using living labs, we are surprised how under-researched the topic is to date. Recent living lab research has actively focused on the network composition and different methods that are used in living labs, but we want to highlight the importance of focusing, in detail, on the active engagements between different stakeholder groups and between people and technology.

The complexity of the intermediary work also reveals important practical insights for living labs: in a real-life context with multiple stakeholders, the direction of the innovation and challenges the project has to face are very difficult to predict. Thus, the capability of intermediaries to adjust their role and actions to changing circumstances is essential. This view holds implications for the recruitment of employees to living lab projects and for the management of living lab activities. Intermediaries hired in a living lab project need to engage in technical configuration and substance issues of the user domain, and not only in the brokering and facilitating tasks. Our study also lends support to the findings by Nyström and colleagues (2014) regarding the need for role ambidexterity, temporality, and multiplicity – an actor's capability to flexibly change, create, adjust, and adapt to roles with respect to the evolving network structure as well the ability to hold multiple roles at the same time.

Acknowledgements

This research was supported by Academy of Finland project grants: 289520 Getting collaborative design done and 288402 Transition intermediaries.

About the Authors

Louna Hakkarainen, Lic.Soc.Sc., is a doctoral candidate in the School of Art, Design and Architecture of Aalto University in Helsinki, Finland. She holds a licentiate degree from the University of Helsinki's Faculty of Social Sciences. Her research focuses on social shaping of technology, living lab collaboration, and facilitation.

Sampsa Hyysalo is an Associate Professor in Co-Design at the Aalto University School of Art, Design and Architecture and a Senior Researcher at the Aalto University School of Economics in Helsinki, Finland. Sampsa's research and teaching focus on user involvement in innovation and the co-evolution of technologies, practices and organizations. He received his PhD in Behavioral Sciences from the University of Helsinki and holds a Docentship in Information Systems, specialising in user-centred design.

References

Almirall, E., & Wareham, J. 2011. Living Labs: Arbiters of Mid- and Ground-Level Innovation. *Technology Analysis & Strategic Management*, 23(1): 87–102.
http://dx.doi.org/10.1007/978-3-642-15417-1_13

Baltes, G., & Gard, J. 2010. *Living Labs as Intermediary in Open Innovation: On the Role of Entrepreneurial Support.* Paper presented at the 16th International Conference on Concurrent Enterprising, Lugano, Switzerland, 2010.

Bakici, T., Almirall, E., & Wareham, J. 2013. The Role of Public Open Innovation Intermediaries in Local Government and the Public Sector. *Technology Analysis & Strategic Management*, 25(3): 311–327.
http://dx.doi.org/10.1080/09537325.2013.764983

Berger, T., Hartmann M., Punie Y., & Ward, K. (Eds.). 2006. Domestication of Media and Technology. Berkshire, UK: Open University Press.

Bessant, J., & Rush, H. 1995. Building Bridges for Innovation: The Role of Consultants in Technology Transfer. *Research Policy*, 24(1): 97–114.
http://dx.doi.org/10.1016/0048-7333(93)00751-E

Botero, A., & Hyysalo, S. 2013. Ageing Together: Steps towards Evolutionary Co-Design in Everyday Practices. *CoDesign*, 9(1): 37–54.
http://dx.doi.org/10.1080/15710882.2012.760608

Fleck, J. 1988. *Innofusion or Diffusation? The Nature of Technological Development in Robotics.* Edinburgh PICT Working Paper No. 7. Edinburgh, UK: Edinburgh University.

Gregor, S. 2002. A Theory of Theories in Information Systems. In S. Gregor & D. Hart (Eds.), *Information Systems Foundations: Building the Theoretical Base:* 1–20. Canberra, Australia: Australian National University.

Hakkarainen, L. 2013. *Paremman teknologian perässä – Turvalattia ja käyttäjälähtöisyyden Lupaus* (In Search of Better Technology – Safety Floor and the Promise of User Centrism). Licentiate thesis: University of Helsinki, Finland.

Hakkarainen, L., & Hyysalo, S. 2013. How Do We Keep the Living Laboratory Alive? Learning and Conflicts in Living Lab Collaboration. *Technology Innovation Management Review*, 3(12): 16–22.
http://timreview.ca/article/749

Hartswood, M., Procter, R., Slack, R., Voß, A., Büscher, M., Rouncefield, M., & Rouchy, P. 2002. Co-Realisation: Towards a Principled Synthesis of Ethnometodology and Participatory Design. *Scandinavian Journal of Information Systems*, 14(2): 9–30.
http://dx.doi.org/10.1007/978-1-84628-901-9_3

Heikkinen, M.T., Mainela, T., Still, J., & Tähtinen, J. 2007. Roles for Managing in Mobile Service Development Nets. *Industrial Marketing Management*, 36(7): 909–925.
http://dx.doi.org/10.1016/j.indmarman.2007.05.014

Hillgren, P.-A., Seravalli, A., & Emilson, A. 2011. Prototyping and Infrastructuring in Design for Social Innovation. *CoDesign*, 7(3–4): 169–183.
http://dx.doi.org/10.1080/15710882.2011.630474

Howells, J. 2006. Intermediation and the Role of Intermediaries in Innovation. *Research Policy*, 35(5): 715–728.
http://dx.doi.org/10.1016/j.respol.2006.03.005

Hyysalo, S. 2010. Health Technology Development and Use: From Practice-Bound Imaginations to Evolving Impacts. New York: Routledge.

Hyysalo, S. 2009. Learning for Learning Economy and Social Learning. *Research Policy*, 38(5): 726–735.
http://dx.doi.org/10.1016/j.respol.2009.01.003

Hyysalo, S., & Hakkarainen, L. 2014. What Difference Does a Living Lab Make? Comparing Two Health Technology Innovation Projects. *CoDesign*, 10(3-4): 191–208.
http://dx.doi.org/10.1080/15710882.2014.983936

Johansson, L.-O., & Lundh Snis, U. 2011. The Dynamics of Interaction: Exploring a Living Lab Innovation Process from a Community of Practice Perspective. In *Proceedings of Pacific Asia Conference on Information Systems (PACIS) 2011*, Paper 85.

Katzy, B., Turgut, E., Holzmann, T., & Sailer, K. 2013. Innovation Intermediaries: A Process View on Open Innovation Coordination. *Technology Analysis & Strategic Management*, 25(3): 295–309.
http://dx.doi.org/10.1080/09537325.2013.764982

Leminen, S., Westerlund, M., & Nyström, A.-G. 2012. Living Labs as Open Innovation Networks. *Technology Innovation Management Review, 2*(9): 6–11.
http://timreview.ca/article/602

Leminen, S. 2013. Coordination and Participation in Living Lab Networks. *Technology Innovation Management Review,* 3(11): 5–14.
http://timreview.ca/article/740

Leminen, S., DeFillippi, R., & Westerlund, M. 2015. *Paradoxical Tensions in Living Labs.* Paper presented at the XXVI ISPIM Conference – Shaping the Frontiers of Innovation Management, Budapest, Hungary, June 14–17, 2015.

MacKenzie, D., & Wajcman, J. (Eds.) 1999. *The Social Shaping of Technology,* 2nd Ed. Buckingham, UK: Open University Press.

Nyström, A.-G., Leminen, S., Westerlund, M., & Kortelainen, M. 2014. Actor Roles and Role Patterns Influencing Innovation in Living Labs. *Industrial Marketing Management,* 43(3): 483–495.
http://dx.doi.org/10.1016/j.indmarman.2013.12.016

Pade-Khene, C., Luton, R., Jordaan, T., Hildbrand, S., Gerwel Proches, C., Sitshaluza, A., Dominy, J., Ntshinga., W., & Moloto, N. 2013. Complexity of Stakeholder Interaction in Applied Research. *Ecology and Society,* 18(2): 13.
http://dx.doi.org/10.5751/ES-05405-180213

Pollock, N., & Williams, R. 2008. *Software and Organizations: The Biography of the Packaged Enterprize System or How SAP Conquered the World.* London: Routledge.

Silverstone, R., & Hirsch, E. 1992. *Consuming Technologies: Media and Information in Domestic Spaces.* London: Routledge.

Sørensen, K.H. 1996. *Learning Technology, Constructing Culture. Socio-Technical Change as Social Learning.* STS Working Paper no. 18/96. Trondheim, Norway: University of Trondheim, Centre for Technology and Society.

Stewart, J., & Hyysalo, S. 2008. Intermediaries, Users and Social Learning in Technological Innovation. *International Journal of Innovation Management,* 12(3): 295–325.
http://dx.doi.org/10.1142/S1363919608002035

Strauss, A. 1993. *Continual Permutations of Action.* New York: Aldine de Gruyter.

Svensson, J., & Ebbesson, E. 2010. Facilitating Social and Cognitive Translation in Innovation Networks. In *Proceedings of the Mediterranean Conference on Information Systems (MCIS) 2010,* Paper 85.

Van de Ven, A., Polley, D. E., Garud, R., & Venkataraman, S. 1999. *The Innovation Journey.* Oxford: Oxford University Press.

Voss, A., Hartswood, M., Procter, R., Rouncefield, M., Slack, R., & Büscher, M. (Eds.). 2009. *Configuring User-Designer Relations Interdisciplinary Perspectives.* London: Springer.
http://dx.doi.org/10.1007/978-1-84628-925-5

Williams, R., & Edge, D. 1996. The Social Shaping of Technology. *Research Policy,* 25(6): 856–899.
http://dx.doi.org/10.1016/0048-7333(96)00885-2

Williams, R., Slack, R., & Stewart, J. 2005. *Social Learning in Technological Innovation – Experimenting with Information and Communication Technologies.* Cheltenham,UK: Edgar Elgar Publishing.

Woolrych, A., Hornbæk, K., Frøkjær, E., & Cockton, G. 2011. Ingredients and Meals Rather than Recipes: A Proposal for Research That Does Not Treat Usability Evaluation Methods as Indivisible Wholes. *International Journal of Human-Computer Interaction,* 27(10): 940–970.
http://dx.doi.org/10.1080/10447318.2011.555314

Keywords: living lab, innovation intermediaries, facilitation, elderly care, co-design, health technology

18

Overcoming Barriers to Experimentation in Business-to-Business Living Labs

Ruben D'Hauwers, Aron-Levi Herregodts, Annabel Georges, Lynn Coorevits, Dimitri Schuurman, Olivier Rits, and Pieter Ballon

> "*There are three principal means of acquiring knowledge: observation of nature, reflection, and experimentation. Observation collects facts; reflection combines them; experimentation verifies the result of that combination.*"
>
> Denis Diderot (1713–1784)
> Philosopher, art critic, and writer

Business-to-business (B2B) living lab projects have been mentioned in different areas of academic research, but the innovation management literature requires deeper analysis of their potential opportunities and challenges. Real-life experimentation is a key requirement for living labs as it enables deeper insights in the potential success of innovations. However, the literature has not provided insights on how living lab projects can implement real-life experimentation in B2B innovation projects and does not describe appropriate conditions for experimentation in these settings. In this study, we identified three main barriers preventing real-life experimentation in B2B living lab projects: the technological complexity, the need for integration, and the difficulty in identifying testers. The barriers are discussed in detailed and potential solutions are provided to help overcome these barriers and stimulate the adoption of real-life experimentation in B2B innovation projects.

Introduction

Providers of "living labs as a service" – who offer services such as designing the idea-generation processes, planning or carrying out real-world tests of innovations, and assessing pre-market launches (Ståhlbröst, 2013) – are confronted with an ever-increasing demand for B2B-oriented projects. B2B companies focus on transactions between companies, whereas business-to-consumer (B2C) companies sell their products directly to the end user (Chauhan & Anbalagan, 2014). Both B2B and B2C innovation projects are confronted with a range of uncertainties throughout their development process, but much of the focus of the living lab literature has been on B2C projects.

In this article, we draw upon experiences dating back to 2005 with the establishment of iLab.o, the predecessor of iMinds Living Labs, which is now imec.livinglabs (imec-int.com/en/livinglabs). The organization's first projects were situated in a B2C context (see Schuurman [2015] for a detailed historical overview). However, as iLab.o evolved into a living-lab-as-a-service offering and started to attract more and more utilizers, we witnessed an inflow of B2B projects (see Schuurman et al. [2016] for an overview of the projects). While putting the proof-tested methods used by open innovation research in B2C projects into practice in a B2B environment, we discovered that the application of real-life experimentation in B2B-oriented living lab projects poses particular methodological as well as practical challenges. Given that real-life testing with potential users of the innovation is one of the main characteristics of living labs, and the literature on B2B living lab projects is scant, we aim to contribute to the academic literature by analyzing opportunities for real-life testing in eight case studies of B2B living labs. Through a cross-case analysis, we identify the main barriers to B2B experimentation and their respective potential solutions.

Living Labs and Real-Life Experimentation

In the context of living labs, the innovation process has evolved from a single-inventor perspective towards a collaborative development of two or more actors. In these collaborative efforts, the crucial role of co-creation has to be emphasized (Bogers et al., 2010; Schuurman et al., 2015). As Schuurman (2015) describes, a living lab is "a tool for distributed innovation that drives co-creation between the different actors involved, while providing the user with a central role". Indeed, organizations want to utilize co-creation in order to tap into the knowledge of (end) users (Kristensson et al., 2008).

Følstad (2008) argues that, in order for users to provide valuable contributions to the innovation at hand, they need to be able to experiment with the innovation and ideally do so in a real-life context. Real-life experimentation is seen as a defining characteristic of living labs (Schuurman, 2015). Coorevits and Schuurman (2015) argue that innovation is unpredictable because of contextual factors, influencing the product usage during this real-life experimentation (Sein et al., 2011) and therefore the testing of products built in the front-end of design is crucial. Forlizzi and Ford (2000) also stress the importance of the context-of-use, which influences the interaction of the user with the innovation. Therefore, it is of utmost importance to provide users with ample opportunities to experiment with the innovation, at least in a familiar and preferably real-life context. It is here that living labs are different compared to other innovation methods (Niitamo et al., 2006; Schuurman & De Marez, 2012; Coorevits, 2015). Testing not only provides context-specific insights on the development and acceptance of the innovation, but also informs researchers and practitioners about the conditions of technology acceptance and the impact of the innovation on the society and on its environment (Frissen & Van Lieshout, 2004).

Towards B2B (B2B) Living Labs

Since 2009, imec.livinglabs (formerly known as iLab.o and iMinds Living Labs) has offered "living labs as a service" to reach its mission of facilitating digital innovation in Flanders, Belgium. The service offering of the imec.livinglabs is focused on confronting potential end users with innovations by small and medium-sized enterprises (SMEs) through co-creation and real-life experimentation. In order to succeed in this facilitation, a key asset of the imec.livinglabs organization is its B2C-focused panel of potential test users. The majority of these projects are based on bilateral agreements, with a project usually lasting three to six months. Table 1 shows the recent increase of B2B projects relative to B2C projects in the portfolio of the imec.livinglabs. Over the course of the organization's first five years (2009–2014), B2B projects accounted for less than 20% of all cases. Many innovations of SMEs in Flanders are in the B2B market, and while these innovations previously did not take into account the needs of business users, a shift in the market could be observed. imec.livinglabs reacted to this evolution by integrating business model expertise into its offering (see Rits et al., 2015). The positioning of living labs in the B2B market of imec.livinglabs proved successful as evidenced by the absolute increase in B2B projects and the shift in the proportion of B2B projects where more than half of recent projects were B2B oriented.

Table 1. Increasing proportion of B2B living lab cases at imec.livinglabs from 2009 to 2016

	B2B	B2C	Total	% B2B
First 5 years (2009–2014)	4	18	22	18%
Next 2 Years (2014–2016)	21	19	40	53%
Overall (2009–2016)	**25**	**37**	**62**	**40%**

This shift from B2C-oriented projects to B2B-oriented projects is important to investigate because of the different characteristics and needs of these two settings. In general, compared to B2C markets, B2B markets have a limited number of customers that generate the largest part of the revenue (Sheth et al., 2000). Thus, when compared to B2C, B2B transaction values tend to be larger and purchase cycles tend to be longer (Brennan et al., 2007; Griffin, 2001). Also, the markets feature different methods of interaction between the business and the client, with B2B traditionally favouring face-to-face interactions (Di Fiore, 2016). Thus, in new product development, a relatively small set of potential B2B customers can exert significant influence over a firm's innovation (Bonner & Walker, 2004). Moreover, the decision makers might not be the actual users of the innovation, which impact the open innovation process significantly. Given that top managers may play a crucial role in driving innovations (Tellis et al., 2009) and may determine the direction of the innovation, a different role needs to be attributed to different types of users. Abrell and colleagues (2016) argue that customers making the purchasing decisions can provide knowledge about short-term changes in market needs, whereas users working directly with the products provide long-term guidance for digital innovation. Looking at innovation, Castro (2015) states that B2B firms focus on internal processes and capabilities and product-mix innovations, whereas B2C firms innovate on the brand, presence, and customer experience. This difference affects which methodologies can be applied in this context. The application of real-life experimentation in B2B-oriented living lab projects poses methodological as well as practical challenges and implications for organizations offering living labs as a service, which have not been explored in the living lab literature.

The academic importance attributed to real-life testing and experimentation reinforces the need to assess such an approach in B2B-oriented living-labs-as-a-service projects. Although some authors (Ballon et al., 2005; Almirall et al., 2012) explicitly mention B2B living labs, no clear insights are provided on the application of real-life experimentation in these distinctive environments. Ballon and colleagues (2005), for example, make note of considerable differences in experimenting with innovations between B2B and B2C test and experimentation platforms (TEPs). However, they offer no guidelines on this matter. Further, Almirall, Lee, and Wareham (2012) report that most cases in the Catalan Living Labs are B2B projects. They also compare the general methodological approaches of four living lab intermediaries in terms of the act of user involvement, the interpretation of real-life contexts, and the public–private–partnerships. However, the specific methodological differences between B2C and B2B contexts are not discussed, nor are any guidelines provided in relation to that distinction.

Case Study: Eight B2B Living Labs

We used an exploratory action research approach (Davison et al., 2004). We selected eight cases that were executed by imec.livinglabs as part of their living lab as a service, which is tailored towards SMEs. To ensure reliability, relevance, and comparability, the cases were selected according following criteria: i) the living lab projects had to be completely finished, ii) the cases must have been carried out between 2012 and 2016, and iii) the cases must be of a B2B nature. A case study approach was selected due to the absence of a clear supporting theory (on B2B living labs) and the exploratory nature of the study, in which key variables and their relationship are under investigation (Eisenhardt, 1989; Yin, 2009).

We defined a five-point scale (Table 2) to measure the extent of user involvement in a B2B living lab context.

We followed Følstad (2008) to differentiate between a familiar, semi-real context (Level 3) and a real-life context (Level 4). We defined Level 3 as testing without interacting with the entire ecosystem the product usually would operate in and it is thus not integrated with other processes. The familiar context can serve as an alternative to the real world by allowing greater balance between the threat of low ecological validity related to test labs and the uncontrollable aspect of field studies. In B2B environments, a familiar context might be a pilot or prototype environment wherein the real-life context is simulated as much as possible. Researchers often opt for the familiar context so they can maintain control over a selection of elements they want to investigate, such as pre-defined task execution to determine the learnability of an application.

The testing in Level 4 goes one step further: users interact with the innovation in a real-life setting. The entire ecosystem is involved and integration is included as well. In the context of B2C-oriented living labs, (end)users are confronted with technology in their everyday lives. In this situation, researchers cannot control the users' actions and the external elements influencing their behaviour. The real-life aspect of the test environment should provide the researcher with "unexpected" outcomes to improve the innovation (Sauer, 2013). As described by Almirall and colleagues (2012), "Real-life contexts are much more than a more realistic scenario for validating proposals; they form an arena where new meanings can emerge, tacit knowledge can be captured, and the whole ecosystem can be validated."

Table 2. Five levels of user involvement in a B2B-oriented living lab

Level	Extent of User Involvement
0	No user involvement
1	Users describe their needs in an exploratory manner
2	Users provide feedback after observing an innovation but do not interact with it
3	Users test an innovation in a familiar, semi-real context
4	Users test an innovation in their real-life context

Table 3 summarizes our eight B2B living lab cases, five of which featured real-life testing (Level 4) and one of which featured testing in a familiar environment (Level 3). Table 3 also indicates that, in four of the cases, a real-life test was not performed within the scope of the actual imec.livinglabs project. In these cases, the entrepreneur performed the real-life testing on their own. Here, the role of imec.livinglabs was limited to identifying potential testing cases and coaching the entrepreneur on the execution of the real-life test, and potentially also assisting with the analysis of the results.

Through a cross-case study, we identified three main B2B-specific barriers to real-life testing – process integration, technological complexity, and tester identification – as shown in Figure 1. Each barrier is described in greater detail in the subsections that follow, along with proposed solutions for overcoming these barriers.

Barrier 1: Process integration
When setting up a field study (e.g., in cases 4, 6, and 8), integration was required between the innovation and the existing processes in the companies. If integration is required, the company needs to make a larger commitment to adapt existing processes in the firm, and the IT department of the company will need to be included in the project, which leads to higher project

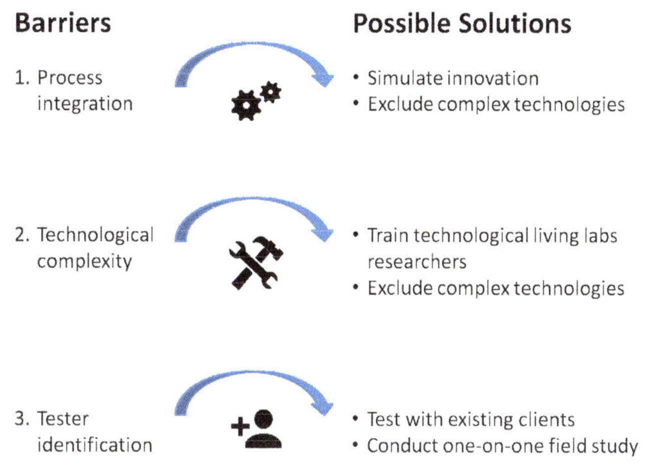

Figure 1. Three barriers to experimentation in B2B living labs and possible solutions

complexity. Nevertheless, in case 4, a proxy technology assessment was set up to simulate the technology through an alternative, simpler solution that could circumvent the difficult integration with existing procedures. A proxy technology assessment takes into account the context influencing the interaction of the user with the innovation in the front end of design and thus can provide an alternative to a field study early in the innovation process (Coorevits & Schuurman, 2015).

Table 3. Descriptions of the assessed cases of B2B living labs, their extent of user involvement, and the main testing entity

Case	Case Description	Extent of User Involvement	Testing Entity
1	A technology aiming to improve case management of processes in companies	Level 4	Entrepreneur
2	A startup focused on digital signage of content applications	Level 4	Entrepreneur
3	Improving the sourcing process by improving the information exchange between requestors, collaborators, and buyers	Level 4	imec.livinglabs
4	Coaching managers and employees on change	Level 3	imec.livinglabs
5	A solution to build, share, and manage knowledge in companies	Level 0	None
6	An Internet of Things (IoT) platform in a B2B context enabling companies to prototype IoT solutions	Level 4	Entrepreneur
7	A solution to track goods in the circular economy	Level 0	None
8	An integration layer for issue-tracking software	Level 4	Entrepreneur

In cases where real-life testing proves to be difficult due integration with other processes, we argue that simulations of the innovation, such as proxy technology assessments, can help overcome this barrier to experimentation.

Barrier 2: Technological complexity
In cases 6 and 8, the technology was highly complex, as the target market involved IT professionals in different organizations. The user researchers did not have a deep background or expertise on these innovations, which made it difficult to understand the technical needs of the users. For that reason, it was too difficult to test the concept in a field study because the user researchers would encounter difficulties in being the translator between the designer and user. The complexity in both cases was linked to the need for integration, thus the barrier of technological complexity and the barrier of process integration (possibly) go hand in hand, but this aspect needs further research.

We argue that, to overcome the barrier to technological complexity, complicated technologies should either be excluded from testing in from B2B living labs or technical experts should be trained to perform experimentation in technologically complicated environments.

Barrier 3: Tester identification
The identification and selection of testers proved to be challenging in cases 3 and 6, and it prevented the inclusion of field studies in those projects. The difficulties in identifying B2B testers arose due to a smaller pool of potential testers. Thus, the recruitment of testers may be more resource intensive in B2B projects than in B2C projects.

This barrier can be overcome by utilizing existing clients of the entrepreneur, which might make the process of identifying testing entities more efficient. A living lab project can also be a starting point for another research project focusing on the field study in a one-on-one relationship between two research partners (the entrepreneur and their potential client), as was the shown in case 6. Alternatively, the living lab can coach the entrepreneur to perform the field study themselves.

Conclusion

In this study, we identified and proposed solutions to three specific barriers hindering experimentation in B2B living labs: i) process integration between the existing company processes and the innovation, ii) technological complexity of the innovation, and iii) limitations on the identification and selection of relevant testers. These identified barriers require careful consideration and operationalization of living labs in the context of B2B projects.

Next to overcoming these three barriers, we can first try to avoid them with a more rigid selection of B2B projects that are suitable for living labs. A living lab could, for example, solely accept B2B projects with ready-to-test user interfaces and exclude B2B projects focused on process integration. This living lab self-criticism on the potential of methodologies and formats for B2B projects deserves its own discussion.

Another avenue for further exploration is to identify a positioning of the living labs in cases where real-life tests are performed by the entrepreneurs themselves. The entrepreneur potentially lacks the expertise and experience to perform a real-life test and might not focus on the user aspects of the innovation. Therefore, the providers of living-labs-as-a-service can position themselves as coaches rather than actual implementers of the real-life tests.

Potentially, living labs can also explore the potential of a B2B-focused panel similar to the B2C panel utilized by imec.livinglabs. This approach would potentially improve the identification and selection of testers for B2B innovations.

In conclusion, we believe that overcoming the identified contextual barriers through the different solutions we proposed – and others to be identified in future research – real-life experimentation in B2B living labs can prove to be highly beneficial to the development of B2B innovations.

About the Authors

Ruben D'Hauwers is a Researcher at imec – SMIT – VUB in Belgium. He holds a master's degrees in Business Engineering (2011) and Innovation and Entrepreneurship (2013). He has also worked in the field of business development in two different organizations. He joined imec in 2014 and focuses on business-to-business research, business modelling, and on capturing and validating the assumptions throughout the innovation process.

Aron-Levi Herregodts is a User Expert at imec.livinglabs and an affiliated researcher at imec – MICT – Ghent University in Belgium. He holds master's degrees in Communication Sciences (2013) and Complementary Business Economics (2014). As a user expert, his role is to translate user behaviour, needs, and wants into tangible recommendations to provide structure to the innovation process of both SMEs and large organizations. His main research interests include open innovation, user innovation, organizational learning, intermediary activities, and user-centric design and methodologies. He is also preparing a PhD on the configuration of intermediary user-oriented activities with innovation-relevant actors for distinct types of entrepreneurs and innovations.

Annabel Georges is a Junior Researcher in the research group at imec – MICT – Ghent University in Belgium. She holds a master's degree in Communication Sciences from Ghent University, with a specialization in New Media and Society. Her main interests are domestication theory, field tests, and improvements to living lab practices.

Lynn Coorevits is a Senior User Researcher for imec – MICT – Ghent University in Belgium, where she focuses on tools and techniques for open and user innovation, such as sensors and design thinking. Her current research focuses on the adoption and attrition of wearables as well as optimization of context integration in living lab projects. She works on several SME living lab projects ranging from the financial to social industry. She holds master's degrees in Psychology and in Marketing Analysis from Ghent University and has 9 years of experience in innovation research and consultancy.

Dimitri Schuurman is the Team Lead in User Research at imec.livinglabs and a Senior Researcher at imec – MICT – Ghent University in Belgium. He holds a PhD and a Master's degree in Communication Sciences from Ghent University. Together with his imec colleagues, Dimitri developed a specific living lab offering targeted at entrepreneurs in which he has managed over 100 innovation projects. Dimitri is responsible for the methodology and academic valorization of these living lab projects and coordinates a dynamic team of living lab researchers. His main interests and research topics are situated in the domains of open innovation, user innovation, and innovation management. His PhD thesis was entitled *Bridging the Gap between Open and User Innovation? Exploring the Value of Living Labs as a Means to Structure User Contribution and Manage Distributed Innovation.*

Olivier Rits holds a master's degree in Applied Physics (Engineering) from the Ghent University in Belgium. Previously, Olivier worked as a business developer for Alcatel-Lucent, where he was involved with many different technologies. Within SMIT, Olivier is leading the business modelling practice in the context of SME living lab innovation projects. With a team of researchers, he is organizing business model workshops and stakeholder interviews to apply business model methodologies to link the user feedback and insights to the business model aspects of the SME's strategy. Olivier is also involved in business modelling efforts in research projects dealing with ICT in general, with a specific interest in M2M and IoT and mobile telecommunications.

Pieter Ballon is the Academic Lead of imec.livinglabs, the International Secretary of the European Network of Living Labs, and Director of the research group imec-SMIT at Vrije Universiteit Brussel in Belgium. He specializes in business modelling, open innovation, and the mobile telecommunications industry. Formerly, he was Senior Consultant and Team Leader at TNO. From 2006–2007, he was the coordinator of the cross issue on business models of the Wireless World Initiative (WWI), which united five integrated projects in the European Union's 6th Framework Programme. Pieter holds a PhD in Communication Sciences from Vrije Universiteit Brussel and an MA in Modern History from Katholieke Universiteit Leuven.

References

Abrell, T., Pihlajamaa, M., Kanto, L., vom Brocke, J., & Uebernickel, F. 2016. The Role of Users and Customers in Digital Innovation: Insights from B2B Manufacturing Firms. *Information & Management,* 53(3): 324–335. http://dx.doi.org/10.1016/j.im.2015.12.005

Almirall, E., Lee, M., & Wareham, J. 2012. Mapping Living Labs in the Landscape of Innovation Methodologies. *Technology Innovation Management Review,* 2(9): 12.

Ballon, P., Pierson, J., & Delaere, S. 2005. *Test and Experimentation Platforms for Broadband Innovation: Examining European Practice.* Paper presented at the 16th European Regional Conference by the International Telecommunications Society, September 4–6 Porto, Portugal,.

Bogers, M., Afuah, A., & Bastian, B. 2010. Users as Innovators: A Review, Critique, and Future Research Directions. *Journal of Management,* 36(4): 857–875. http://dx.doi.org/10.1177/0149206309353944

Bonner, J. M., & Walker, O. C. 2004. Selecting Influential Business-to-Business Customers in New Product Development: Relational Embeddedness and Knowledge Heterogeneity Considerations. *Journal of Product Innovation Management,* 21(3): 155–169. http://dx.doi.org/10.1111/j.0737-6782.2004.00067.x

Castro, V. 2015. *Innovation Importance Across Dimensions for Business-to-Business and Business-to-Consumer Firms: A Case Study-Based Analysis.* Wharton Research Scholars, 132. http://repository.upenn.edu/wharton_research_scholars/132/

Coorevits, L., & Schuurman, D. 2015. *Increasing Relevance of Living Lab Outcomes through Proxy Technology Assessments.* Paper presented at the Open and User Innovation Conference, July 13–15, Lisbon, Portugal.

Davison, R., Martinsons, M. G., & Kock, N. 2004. Principles of Canonical Action Research. *Information Systems Journal,* 14(1): 65–86. http://dx.doi.org/10.1111/j.1365-2575.2004.00162.x

Eisenhardt, K. M. 1989. Building Theories from Case Study Research. *Academy of Management Review,* 14(4): 532–550. http://dx.doi.org/10.5465/AMR.1989.4308385

Følstad, A. 2008. Living Labs for Innovation and Development of Information and Communication Technology: A Literature Review. *eJOV: The Electronic Journal for Virtual Organizations,* 10: 99–131.

Forlizzi, J., & Ford, S. 2000. The Building Blocks of Experience: An Early Framework for Interaction Designers. In *Proceedings of the 3rd Conference on Designing Designing Interactive Systems:* 419–423. New York: ACM. http://dx.doi.org/10.1145/347642.347800

Frissen, V., & van Lieshout, M. 2004. *To User-Centred Innovation Processes: The Role of Living Labs.* Delft, Netherlands: TNO-ICT.

Griffin A. 2001. Product Development Cycle Time for B2B Products. *Industrial Marketing Management,* 31(4): 291–304. http://dx.doi.org/10.1016/S0019-8501(01)00162-6

Kristensson, P., Matthing, J., & Johansson, N. 2008. Key Strategies for the Successful Involvement of Customers in the Co-Creation of New Technology-Based Services. *International Journal of Service Industry Management,* 19(4): 474–491. http://dx.doi.org/10.1108/09564230810891914

Niitamo, V., Kulkki, S., Eriksson, M., & Hribernik, K. A. 2006. State-of-the-Art and Good Practice in the Field of Living Labs. In *Proceedings of the 12th International Conference on Concurrent Enterprising: Innovative Products and Services through Collaborative Networks:* 341–348. Milan: IEEE. http://dx.doi.org/10.1109/ICE.2006.7477081

Rits, O., Schuurman, D., & Ballon, P. 2015. Exploring the Benefits of Integrating Business Model Research within Living Lab Projects. *Technology Innovation Management Review,* 5(12): 19–27. http://timreview.ca/article/949

Sauer, S. C. 2013. *User Innovativeness in Living Laboratories: Everyday User Improvisations with ICTs as a Source of Innovation.* Thesis No. 13-256. Enschede, Netherlands: University of Twente.

Schuurman, D., & De Marez, L. 2012. Structuring User Involvement in Panel-Based Living Labs. *Technology Innovation Management Review,* 2(9): 31–38. http://timreview.ca/article/606

Schuurman, D. 2015. *Bridging the Gap between Open and User Innovation? Exploring the Value of Living Labs as a Means to Structure User Contribution and Manage Distributed Innovation.* Doctoral dissertation. Ghent University and Vrije Universiteit Brussel VUB.

Schuurman, D., De Marez, L., & Ballon, P. 2016. The Impact of Living Lab Methodology on Open Innovation Contributions and Outcomes. *Technology Innovation Management Review,* 1(6): 7–16. http://timreview.ca/article/956

Sein, M. K., Henfridsson, O., Rossi, M., & Lindgren, R. 2011. Action Design Research. *MIS Quarterly,* 35(1): 37–56.

Sheth, J. N., Sisodia, R. S., & Sharma, A. 2000. The Antecedents and Consequences of Customer-Centric Marketing. *Journal of the Academy of Marketing Science,* 28(1): 55–66. http://dx.doi.org/10.1177/0092070300281006

Ståhlbröst, A. 2013. A Living Lab as a Service: Creating Value for Micro-Enterprises through Collaboration and Innovation. *Technology Innovation Management Review,* 3(11): 37–42. http://timreview.ca/article/744

Tellis, G. J., Prabhu, J. C., & Chandy, R. K. 2009. Radical Innovation across Nations: The Preeminence of Corporate Culture. *Journal of Marketing,* 73(1): 3–23. http://dx.doi.org/10.1509/jmkg.73.1.3

Yin, R. K. 2009. *Case Study Research: Design and Methods.* Thousand Oaks, California: SAGE Publications, Inc.

Keywords: living labs, B2B, experimentation, user research, testing

19

Factors Affecting the Attrition of Test Users During Living Lab Field Trials

Annabel Georges, Dimitri Schuurman, and Koen Vervoort

"*Motivation is the art of getting people to do what you want them to do because they want to do it.*"

Dwight D. Eisenhower (1890–1969)
34th President of the United States

Next to active user involvement and a multi-method approach, a third major principle within living lab research consists of capturing the real-life context in which an innovation is used by end users. Field trials are a method to study the interaction of test users with an innovation in the context of use. However, when conducting field trials, there are several reasons why users stop participating in research activities, a phenomenon labelled as attrition. In this article, we elaborate on drop-outs during field trials by analyzing three post-trial surveys of living lab field trials. Our results show that several factors related to the innovation, as well as related to the field trial setup, play a role in attrition, including the lack of added value of the innovation and the extent to which the innovation satisfies the needs and time restrictions of test users. Based on our findings, we provide practical guidelines for managers to reduce attrition during field trials.

Introduction

Within living lab research, end users are involved actively to develop an innovation that is adapted to their needs and wants. A living lab environment is defined as "a user-driven open innovation ecosystem based on a business–citizens–government partnership which enables users to take an active part in the research, development and innovation process" (European Commission, 2009). In addition to this active user involvement, a multi-method approach and real-life interventions make up the three central characteristics of the living lab approach (Schuurman, 2015). Although questions have been raised about the extent to which living labs are capable of achieving the necessary levels of user engagement and keeping in mind that their interests are sometimes overlooked (Dutilleul et al., 2010), users are generally seen as very important actors.

A living lab study by Ebbesson and Eriksson (2013), in the context of an online platform to gather input from end users, showed good support for the end users during the startup phase of the projects, but also showed an increasing number of users dropping out or lowering their activity level. When studying the motivations of end users participating in open innovation processes, Ståhlbröst and Bergvall-Kåreborn (2011, 2013) found a close relationship between motivational factors and the values achieved, and thus that most voluntary contributors are satisfied when learning new things. Intrinsic motivations such as learning, being entertained, and stimulating curiosity are seen as the most important motivators to participate in an innovation intermediary context. Baccarne, Logghe, Veeckman, and Schuurman (2013) also found that the main motivator to participate in living lab research is intrinsic in nature, but for repeated participation, material incentives become more important as motivators. They also argue that the motivations to participate tend to differ according to the research step.

With this study, we wanted to dig deeper into the reasons why people participate or drop out during living lab research. Because there seem to be differences between research techniques (e.g., surveys, field trials, co-creation workshops) (Baccarne et al., 2013), we decided to focus on one research step in particular: field trials. Field trials can be defined as "tests of technical and other aspects of a new technology, product or service in a limited, but real-life environment" (Ballon et al., 2005).

They also link up with the "real-life intervention" characteristic of living lab projects (Schuurman, 2015). Field trials enable researchers to study the use of the innovation by test users in a natural use context and allow them to discover and understand how technologies are being used and adopted in a real-life setting, which is one of the key principles within living lab research (Ballon et al., 2005; Følstad, 2008; Kjeldskov & Skov, 2014; Schuurman et al., 2013). In contrast to other research methods, participation in field trials requires a prolonged engagement of test users because they are expected to test an innovation during a specific period. Moreover, in most field trials, users are asked to actively provide feedback regarding their usage. However, following participants over a prolonged period also increases the risk of drop-out before the end of a test period (Schuurman & De Marez, 2009).

In previous research on attrition during field trials, some studies have been conducted in the field of eHealth. Within this domain, Eysenbach (2005) introduced the law of attrition, which is "the phenomenon of participants stopping usage and/or being lost to follow-up, as one of the fundamental characteristics and methodological challenges in the evaluation of eHealth applications". Simons, Hampe, and Guldemond (2013) mention time and timing issues as reasons why people stop participating. For eHealth trials on the internet or with self-help applications, high dropout rates "may be a natural and typical feature"; however, it is important to further analyze the attrition data, because it may give an indication of real-life adoption problems (Eysenbach, 2005). Eysenbach (2005) also identified two sorts of attrition, namely *dropout attrition*, which is "the phenomenon of losing participants to follow-up (e.g., participants do not return to fill in follow-up questionnaires)" and *non-usage attrition* in which participants "have lost interest in the application and stopped using it". In a field trial, an example of dropout attrition would be test users continuing to use the innovation but no longer providing feedback, whereas non-usage attrition occurs when test users stop using the innovation but can still give feedback regarding their non-usage. The second type of attrition provides important information for the innovation development process, whereas the first type of attrition generates less information. Therefore, it is especially relevant to minimize the rate of dropout attrition.

Multiple studies have illustrated the occurrence of attrition in the context of eHealth applications, without digging into the causes of the attrition (Grudin, 2002; Korn & Bødker, 2012). Kanstrup, Bjerge, and Kristensen (2010) argue that the stability of the ICT infrastructure and some kind of user support are factors that decrease the rate of attrition, but do not make a distinction between dropout and non-usage attrition.

More in-depth research regarding the attrition within ICT field trials or living lab projects is lacking, despite the specific testing opportunities in multiple real-life contexts of new media innovations and because of their ubiquitous nature (Grudin, 2002; Korn & Bødker, 2012). Therefore, within this paper we want to tackle two main research questions:

1. *To what extent can different types of attrition be distinguished within ICT living lab field trials?*

2. *Which factors play a role in the decision of a test user to continue or stop participating in field trials?*

Methodology

The main goal of this study is to find factors that are related, either positively or negatively, to different types of attrition during field trials. Therefore, we conducted a qualitative analysis within three Living Lab field trials. The field trials were carried out in living lab projects from iMinds Living Labs (tinyurl.com/zqm6qsn), a division of the iMinds ICT research institute of Flanders, Belgium. The attrition rates per field trial (based on project documents) are described in the results section.

In order to find as many factors as possible, we selected three cases that differ in multiple ways, such as sample size, type of innovation, field trial setup, and communication with test users. First, we conducted a quantitative analysis on the attrition rates. The qualitative analysis was done by coding the answers test users gave to open questions from the post-trial survey. Thus, during the analysis and interpretation of the results, we must consider that the survey data will only include information about non-usage attrition, because test users subject to dropout attrition will already have dropped out. The answers to these open questions were analyzed using QSR International's NVivo 10 qualitative data analysis software. When analyzing the factors related to attrition, we coded the factors that the same codes could be used for the three field trials.

Below, the field trials are further described and the answer rates of the post-trial surveys are given. One general finding is that, in all cases, the dropout attrition rate is high. This high dropout attrition rate must be kept in mind when interpreting the results.

Field trial 1

The first field trial was part of a living lab project to develop a location-based service application. The application was tested for seven weeks and participants received weekly emails with updates, tasks related to the innovation, and a feedback form where technical problems could also be reported. At the end of the trial, a survey was sent to 558 test users to receive feedback regarding their experience during the field trial (Figure 1).

Field trial 2

The second field trial was part of a living lab project in which an application to meet up with friends was co-created. The application was tested for five weeks and the participants received weekly emails to give feedback about the innovation, and they were given a weekly assignment. At the end of the field trial, a post-trial survey about the innovation was sent to the 55 participants (Figure 2). The test users could also send the survey to their friends or family that also tested the application. In total 35 test users and eleven contacts of the test users filled in the survey completely.

Field trial 3

The third field trial was more data-driven. Participants had to read 30 news articles for the first try-out and 60 for the second and third try-outs. They could choose to participate in one or more try-outs. Within this field trial, participants did not co-create the innovation, but instead received different assignments that generated data that was needed to test an underlying technology. They did not know the exact intention of the field trial. Participants were rewarded with a cinema ticket for each finished assignment. At the end of the field trial, a survey was sent to 350 participants to get feedback about the trial (Figure 3).

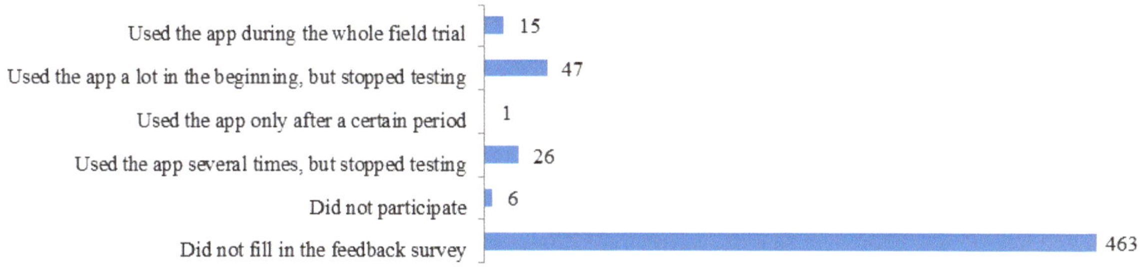

Figure 1. Field trial 1: Summary of responses to the post-trial survey (n= 558)

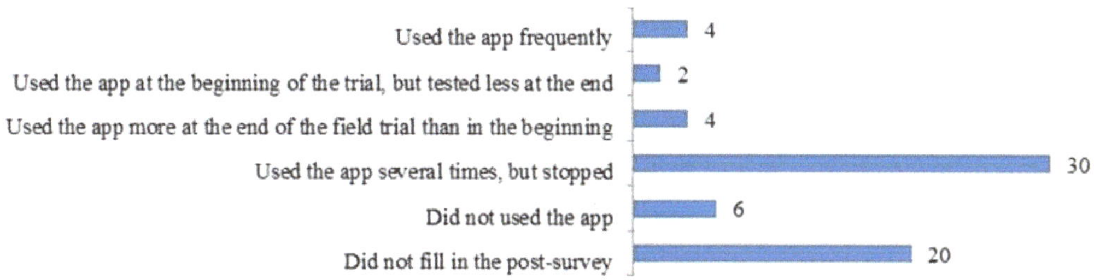

Figure 2. Field trial 2: Summary of responses to the post-trial survey (n= 55 + 11 friends/family of test users)

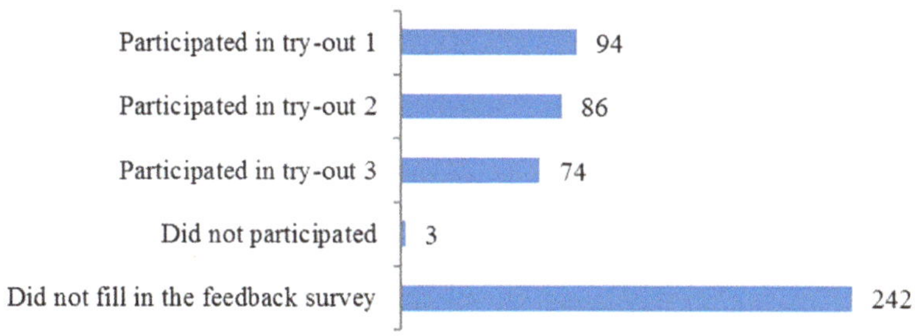

Figure 3. Field trial 3: Summary of responses to the post-trial survey (n= 350)

Results

Attrition rates during field trials
Within this section, we dig deeper into the first research question: To what extent can different types of attrition be distinguished within ICT living lab field trials? Within the first field trial (Figure 4), we see two dips in the attrition rate: i) when respondents have to fill out an intake survey to participate during the trial (dropout attrition) and ii) at the end of the field trial, when many test users stopped using the innovation (non-usage attrition). Thus, many test users did not participate for the entire duration of the field trial.

For the second field trial, we see that the pattern of attrition (Figure 5) is similar to the first field trial. There is high dropout attrition when people have to complete the intake survey, and then there is further (non-usage) attrition during the field trial. However, at the end of the trial, 35 participants filled out the post-trial survey.

Concerning the third field trial, the highest attrition rate was observed when test users had to complete the assignments (Figure 6). For the first assignment, for which the users were asked to read 30 news articles, the non-usage attrition rate was approximately 10% lower than for the two subsequent assignments, each of which required them to read 60 articles. Thus, the lower attrition rate in the first assignment may be explained by it being less cumbersome than the other two assignments. Because it was not expected from the test users to give feedback about an innovation via several research methods, the dropout attrition during this field trial was rather low.

In general, we can conclude that, within living lab field trials, *dropout attrition* occurs during different phases of the trial. A crucial moment for dropout attrition seems to be the intake survey. This increased attrition is most pronounced in the first and second field trial, which seems to be caused by the fact that these surveys had more than 20 questions, whereas in the third field trial, users only had to fill in five questions. Within the first and the third case, there was a delay of several days between the intake survey and the start of testing. However, compared to the second trial, in which the participants received a link to test the application immediately after filling in the survey, there were no substantial differences in attrition.

Non-usage attrition occurs especially after the first time the test-users are confronted with the innovation. Within the next section we will dig deeper into the reasons why participants dropped out during living lab field trials.

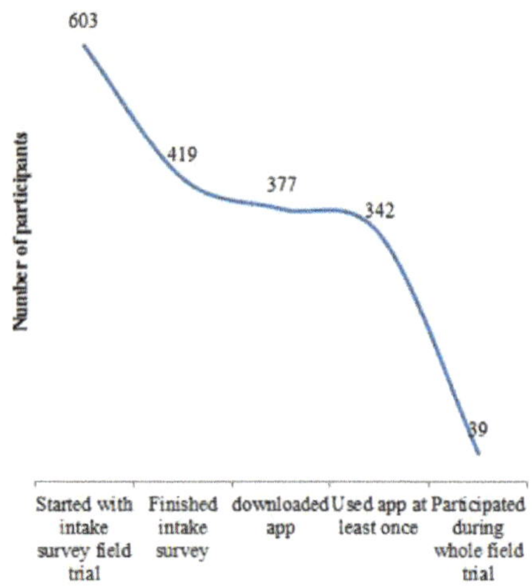

Figure 4. Attrition within field trial 1

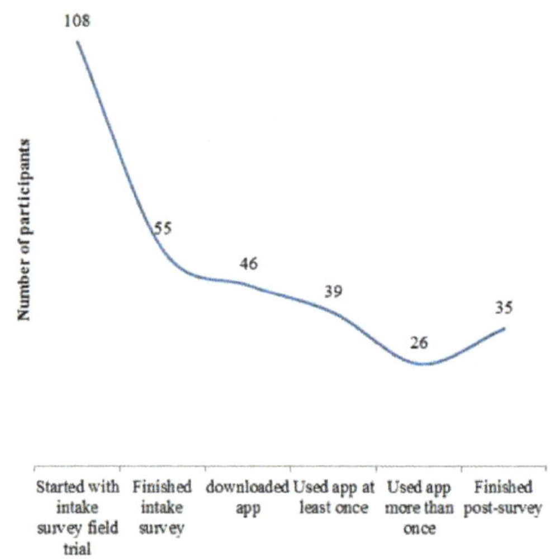

Figure 5. Attrition within field trial 2

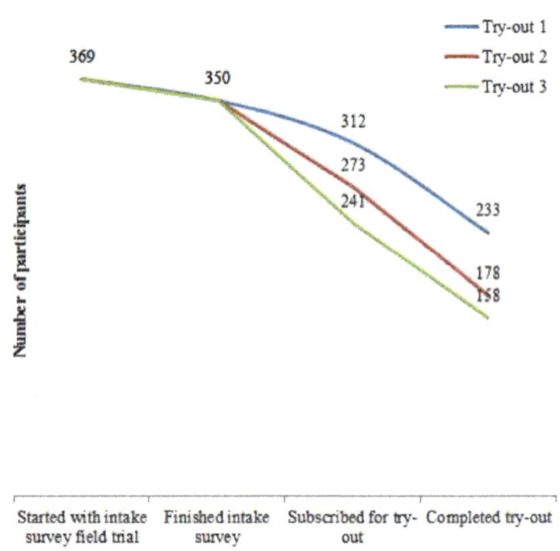

Figure 6. Attrition within field trial 3

Factors related to participation in field trials

Next, we examine factors that can play a role in the attrition during field trials. The data used for this analysis is based on the post-trial surveys at the end of the field trials. The respondents were asked to explain why they stopped using the application or why their use decreased, increased, or stayed constant throughout the entire field trial. These answers were coded according to the different factors related to attrition.

First, we analyzed *the factors that are positively related to participation* in field trials (Table 1). When analyzing the first field trial, the assignments that were given to test the innovation were seen as particularly positive, likely because of the users' curiosity: they wanted to know what the innovation was about. For the second field trial, only a few people kept on testing the innovation during the field trial. Therefore, only three factors were mentioned: i) the "fun factor" of testing the innovation, ii) the added value of the app, and iii) the fact that friends also started to use and test the innovation. For the third field trial, extrinsic motivation (incentives) and intrinsic motivation to participate in scientific research played a role as a factor that motivated test users to finish the assignments.

When comparing the factors that are related positively to participation in field trials with the motivational factors mentioned by Ståhlbröst and Bergvall-Kåreborn (2011, 2013), we see that learning new things (e.g., increasing one's own skills), being entertained (e.g., fun), and stimulating curiosity were also mentioned by the participants in the field trials. During the three field trials, the fun factor played a motivating role.

Next, we analyzed *the factors that are negatively related to participation* in field trials (Table 2) and found that different factors are of importance for each field trial. Within the first field trial, users stopped using the innovation because they did not see the benefit of using it. There were only a limited amount of features available, which made the innovation less interesting to test and made it less likely that test users would test it for a longer period, keeping in mind their time restrictions. For the second field trial, participants mentioned that the innovation did not satisfy their needs. Furthermore, technical issues and a small user base generated dropout among the participants. This finding is in line with Kanstrup, Bjerge, and Kristensen (2010), who argue that users dropout when the technology is unstable. Finally, for the third field trial, time restrictions caused non-participation in the assignments, as was also found by Simons, Hampe, and Guldemond (2013).

When analyzing the factors across the three field trials (Table 3) and when digging deeper into the difference between dropout and non-usage attrition, our studied cases suggest that dropout attrition is mainly linked to the research setup, whereas non-usage attrition is mainly linked to factors related to the innovation itself.

When comparing the *non-usage attrition* over the three field trials, we see that it is high for the first and second field trial because these projects focused more on user co-creation of the innovation, which corresponds with the active user involvement characteristic of living lab research. In the third field trial, the non-usage attrition was lower. The focus in this project was more on the users generating data that allowed testing of the underlying technology, which made the co-creation aspect less important. Next to this, the participants also received cinema tickets after completing their assignments. In the first and second field trial, the participants were not certain they would receive a material incentive, however they did not mentioned this as a factor in their decision to participate in the field trial. Thus, incentives helped when participants had to finish a certain assignment, but when test users had to co-create, intrinsic motivations became more important.

The higher non-usage attrition for the first and second field trial is interesting for the instigator of the project: it points to factors related to the innovation (e.g., usability problems or users not seeing the benefit of the application), which should lead to iteration of the innovation or of the use cases. This finding is in line with Eysenbach (2005), who argues that attrition data can give clues about real-life adoption problems.

Also, network externalities, or the nature of the innovation itself, can cause non-usage attrition. For example, during the second field trial, the testing involved an application for meeting up with friends, which implied that the friends of the test users also had to use the application. These network externalities related to the innovation had a negative influence on the sustained usage of the innovation as the factor "not enough users" scored very high for this field trial.

Also, differences in the *dropout attrition* are noticeable between the trials. These factors are mostly related to the design of the field trial. For the first field trial, we see, for example, high interest among participants to start the field trial, but a very high attrition rate subsequently. This high level of interest in participating can be explained by the communication strategy that was used. A narrative was generated for the field trial,

Table 1. Factors positively related to participation in field trials

		Field trial 1 (n= 23)	Field trial 2 (n= 7)	Field trial 3 (n= 142)
Factors Related to the Innovation	Challenge to conduct the whole field trial			3 (2.1%)
	Curiosity about the innovation	1 (4.4%)		2 (1.4%)
	Increase own skills			4 (2.8%)
	Interesting study			8 (5.6%)
	Like the concept/idea			14 (9.9%)
	Like the app	1 (4.4%)		
	Something new to do			5 (3.5%)
	First one to test it	1 (4.4%)		
	Friends test the app		2 (28.6%)	
	App helps in daily life		3 (42.9%)	
Factors Related to the Field Trial Setup	Anonymous	1 (4.4%)		
	Incentives			46 (32.4%)
	Mailing	1 (4.4%)		
	Tasks	7 (30.4%)		
	Test when and where you want			4 (2.8%)
Other Factors	Conducted by iMinds			3 (2.1%)
	Like to participate in research	2 (8.7%)		18 (12.7%)
	See the evolution of the app			1 (0.7%)
	Help in the development of an innovation	1 (4.4%)		9 (6.3%)
	Curious to know what the innovation was about	5 (21.7%)		
	Try something new			5 (3.5%)
	Fun	2 (8.7%)	2 (28.6%)	20 (14.1%)
	Engagement	1 (4.4%)		

n= total number of reasons to participate in field trials mentioned by the participants
%= number of times a reason was mentioned by participants, divided by n

Table 2. Factors negatively related to participation in field trials

		Field trial 1 (n= 89)	**Field trial 2** (n= 44)	**Field trial 3** (n= 17)
Factors Related to the Innovation	Not interested	21 (23.6%)		
	Did not like the design		3 (6.8%)	
	Did not see the benefit of the app	32 (36.0%)	3 (6.8%)	1 (5.9%)
	Did not trust the app (would not install)	1 (1.1%)		
	Lack of features	8 (9.0%)	2 (4.6%)	
	Not innovative enough	3 (3.4%)		
	Innovation did not satisfy needs	2 (2.3%)	5 (11.4%)	
	Technical issues	4 (4.5%)	6 (13.6%)	3 (17.7%)
Factors Related to the Field Trial Setup	Problems with installing the app	3 (3.4%)		
	No incentive to participate	2 (2.3%)		
	Did not like the tasks	1 (1.1%)		
	Not enough triggers to test	1 (1.1%)		
	Unclear what was expected			2 (11.8%)
Other Factors	Forgot to test	3 (3.4%)		2 (11.8%)
	Lack of motivation	3 (3.4%)		1 (5.9%)
	Not enough users		17 (38.6%)	
	Unforeseen circumstances		1 (2.3%)	1 (5.9%)
	Time restrictions	5 (5.6%)	2 (4.6%)	7 (41.2%)
	No opportunities to test the app		5 (11.4%)	

n= total number of reasons to participate in field trials mentioned by the participants
%= number of times a reason was mentioned by participants, divided by n

Table 3. Comparison across field trials

	Number of Users at Start	**Non-Usage Attrition**	**Dropout Attrition**	**Incentive**	**Duration**
Field trial 1	High	High	Highest	Intrinsic	6 weeks
Field trial 2	Medium	High	Lowest	Intrinsic	4 weeks
Field trial 3	Medium	Low	Medium	Extrinsic + intrinsic	1 week

which asked the test users to help as "undercover agents" and to go on missions to test a new secret application. The mysterious nature of the narrative seemed to have a positive influence on the willingness to participate by triggering the curiosity of the test users. The long intake survey, which was cumbersome, and the lack of perceived added value of the application caused the highest attrition rate.

For the second field trial, the participants were clearly briefed regarding the innovation and were stimulated to provide feedback that would be taken into account by the project instigator. This trial attracted a lower number of test users, but we noticed a lower rate of dropout attrition: some test users kept on giving feedback although they stopped testing the innovation itself. This relatively low rate of dropout attrition seems to be caused by the intrinsic motivations of the participants, who were involved in active co-creation, coupled with reminders that were sent for filling in the feedback surveys.

Concerning the non-usage attrition, the duration of the trial also can play a role in attrition. For example, the first field trial lasted for six weeks, the second trial lasted for four weeks, and the third trial lasted one week per assignment. When comparing the trials, we see a bigger attrition rate for longer field trials.

Guidelines for Project Instigators and Managers

Although the results presented here are exploratory in nature, and further research is needed, we have summarized the main lessons learned in the form of practical guidelines related to: i) the innovation and non-usage attrition and ii) the field trial setup and dropout attrition.

Guidelines related to the innovation and non-usage attrition

1. Introduce the innovation clearly and underline its benefits.

2. Stress the co-creation aspect: test users can be motivated by knowing that their contributions can impact the innovation.

3. Conduct usability testing before the start of the field trial so that any technical issues can be solved beforehand. If there are still technical issues during the field trial, then provide a clear help channel and manage the expectations of test users by, for example, reminding them that the innovation is still in its development phase.

4. Try to anticipate network externalities, because the number of test users can impact the relevance of certain functionalities of an innovation.

5. Communicate clearly at the beginning of the trial what is expected from the test users. Define tasks for the test users to stimulate usage.

6. Remind test users to perform the requested tasks. Some may not otherwise set aside time for testing or they may not remember that a task is to be completed.

Guidelines related to the field trial set-up and dropout attrition

1. Create an accessible helpdesk and make it clear who is responsible for operating it. By including a helpdesk, test users can always give useful feedback when they have the time.

2. Ensure that the testing initiation process is clear and straight forward (e.g., by providing a clear test link at the start).

3. Provide incentives to encourage test users to complete tasks. However, note that incentives do not trigger test users to give valuable feedback.

4. Include some fun (or even funny) tasks or assignments that challenge the users or trigger their curiosity. Appeal to the motivating factor that encourages participation just for fun.

Conclusion

Within living labs, field trials help researchers study the extent to which innovations are being used by test users in a real-life environment. However, several authors have highlighted the difficulty in finding motivated and engaged (long-term) users (Ebbesson & Eriksson, 2013; Kaasinen et al., 2013; Schuurman & De Marez, 2009). This challenge can be problematic, because the setup of a field trial is very time consuming and expensive. Currently, the literature on user participation in field trials during living labs is scarce. In the research domain of eHealth, Eysenbach (2005) explained the law of attrition within field trials and the difference between non-usage attrition and dropout attrition. Although it is difficult to extrapolate these results to field trials in a living lab context, we used this framework to analyze attrition within living lab field trials. With this study, we conducted a qualitative analysis of open questions in post-trial surveys of three living lab field trials and an analysis of attrition data from project documents.

This research has some limitations, including for example, that dropout attrition occurred when the post-trial surveys were sent to the test users. Future research could elaborate on this aspect by exploring how to minimize the dropout attrition so that there is information about why test users dropped out. Future studies should also ask why people stop testing an innovation and how many people stop testing an innovation. Although data logging can be used to measure attrition during field trials, it does not help researchers understand why the users stopped. There are thus many opportunities within this domain for quantitative as well as qualitative research. Although the results of this research are exploratory and difficult to generalize to other field trials, we believe the results are valuable for other researchers, practitioners, and idea owners of new products and services to organize and follow-up field trials. Researchers can pro-actively take into account the factors that play a role in the attrition of test users during the preparations for these trials. The idea owners can also practically gain from these findings because some attrition factors relate directly to the innovation itself.

Within this exploratory research, we can conclude that non-usage attrition as well as dropout attrition occurs. Whereas dropout attrition is mainly linked to the research setup, non-usage attrition is mainly linked to the innovation itself. The factors that affect attrition differ for each field trial because of the differences in the innovation and design of the trial. In this study, the main factors why participants stopped testing is because of time restrictions, because they did not see the benefits of using the application, or the application did not address the user's need as well as intended. We also provided practical guidelines to help instigators and managers reduce attrition in their living lab field trials. Here, the main outcome is that communication with test users plays an important role in minimizing dropout attrition, which in turn yields valuable information regarding non-usage attrition. Project instigators and managers should take care to recognize the factors that affect attrition and consider how they can predict future adoption behaviour.

About the Authors

Annabel Georges is a Junior Researcher in the research group iMinds – MICT – Ghent University in Belgium. She holds a Master's degree in Communication Sciences from Ghent University, with a specialization in New Media and Society. In her master's thesis, *Social Media from A to Z: The Role of Media Coaches in the Diffusion of Social Media Literacy within the Library as an Organization,* she used social network analysis to study the diffusion of social media literacy with library staff. At iMinds, her main interests are social innovation and the factors that motivate test users to participate in field trials.

Dimitri Schuurman holds a PhD (2015) and Master's degree in Communication Sciences (2003) from Ghent University in Belgium. He joined the research group iMinds – MICT – Ghent University in Belgium in 2005 and started working at iMinds Living Labs in 2009. Together with his iMinds colleagues, Dimitri developed a specific living lab offering targeted at startups and SMEs, in which he has managed over 50 innovation projects. As a senior researcher, Dimitri is currently responsible for the methodology and academic valorization of living lab projects. He also coordinates a dynamic team of living lab researchers from iMinds – MICT – Ghent University. His main interests and research topics are situated in the domains of open innovation, user innovation, and innovation management. In early 2015, he finished his PhD entitled *Bridging the Gap between Open and User Innovation? Exploring the Value of Living Labs as a Means to Structure User Contribution and Manage Distributed Innovation.*

Koen Vervoort manages and coordinates panels within living lab settings within one of the first living lab organizations in Europe: iMinds Living Labs in Belgium. He also represents iMinds within large Flemish and European living lab projects, hosts workshops, benchmarks internal processes within the entire living lab community (and beyond), oversees an internal quality survey, and organizes fieldwork for iMinds' flagship, Digimeter (digimeter.be), a report that tracks the ownership and use of media (technology) among the Flemish population.

Acknowledgements

An earlier version of this article was presented at the XXVI International Society for Professional Innovation Management (ISPIM) Conference – Shaping the Frontiers of Innovation Management, Budapest, Hungary, June 14–17, 2015.

References

Baccarne, B., Logghe, S., Veeckman, C., & Schuurman, D. 2013. Why Collaborate in Long-Term Innovation Research? An Exploration of User Motivations in Living Labs. In *4th ENoLL Living Lab Summer School 2013*. European Network of Living Labs.

Ballon, P., Pierson, J., & Delaere, S. 2005. *Test and Experimentation Platforms for Broadband Innovation: Examining European Practice*. Brussels: Vrije Universiteit Brussel. http://dx.doi.org/10.2139/ssrn.1331557

Dutilleul, B., Birrer, F. A. J., & Mensink, W. 2010. Unpacking European Living Labs: Analysing Innovation's Social Dimensions. *Central European Journal of Public Policy*, 4(1): 60–85.

Ebbesson, E., & Eriksson, C. I. 2013. Co-creating Innovative UGC Services with the Media Industry. In *46th Hawaii International Conference on System Sciences (HICSS)*: 3057–3066. New York: IEEE. http://dx.doi.org/10.1109/HICSS.2013.133

European Commission. 2009. *Living Labs for User-Driven Innovation: An Overview of the Living Labs Methodology, Activities and Achievement*. Brussels: European Commission. http://dx.doi.org/10.2759/34481

Eysenbach, G. 2005. The Law of Attrition. *Journal of Medical Internet Research*, 7(1): e11. http://dx.doi.org/10.2196/jmir.7.1.e11

Følstad, A. 2008. Living Labs for Innovation and Development of Communication Technology: A Literature Review. *The Electronic Journal for Virtual Organisations and Networks*, 10: 99–131.

Grudin, J. 2002. Group Dynamics and Ubiquitous Computing. *Communications of the ACM*, 45(12): 74–78. http://dx.doi.org/10.1145/585597.585618

Kaasinen, E., Koskela-Huotari, K., Ikonen, V., & Niemeléi, M. 2013. Three Approaches to Co-Creating Services with Users. In G. Salvendy & W. Karwowski (Eds.), *Advances in the Human Side of Service Engineering*: 286–295. Boca Raton, FL: Taylor & Francis Group.

Kanstrup, A. M., Bjerge, K., & Kristensen, J. E. 2010. A Living Laboratory Exploring Mobile Support for Everyday Life with Diabetes. *Wireless Personal Communications*, 53(3): 395–408. http://dx.doi.org/10.1007/s11277-010-9953-3

Kjeldskov, J., & Skov, M. B. 2014. Was It Worth the Hassle?: Ten Years of Mobile HCI Research Discussions on Lab and Field Evaluations. In *Proceedings of the 16th International Conference on Human-Computer Interaction with Mobile Devices & Services*: 43–52. New York: ACM. http://dx.doi.org/10.1145/2628363.2628398

Korn, M., & Bødker, S. 2012. Looking Ahead: How Field Trials Can Work in Iterative and Exploratory Design of Ubicomp Systems. In *Proceedings of the 2012 ACM Conference on Ubiquitous Computing*: 21–30. New York: ACM. http://dx.doi.org/10.1145/2370216.2370221

Schuurman, D., & De Marez, L. 2009. User-Centered Innovation: Towards a Conceptual Integration of Lead Users and Living Labs. In *Proceedings of COST 298 Conference*: 13–15. Ljubljana, Slovenia: University of Ljubljana.

Schuurman, D., Baccarne, B., Kawsar, F., Seys, C., Veeckman, C., De Marez, L., & Ballon, P. 2013. Living Labs as Quasi-Experiments: Results from the Flemish LeYLab. In *Proceedings of The XXIV ISPIM Conference*, Helsinki, Finland, June 16–19, 2013.

Schuurman, D. 2015. *Bridging the Gap between Open and User Innovation? Exploring the Value of Living Labs as a Means to Structure User Contribution and Manage Distributed Innovation.* Doctoral dissertation, Ghent University, Belgium.

Simons, L. P., Hampe, J. F., & Guldemond, N. A. 2013. Designing Healthy Living Support: Mobile Applications Added to Hybrid (E) Coach Solution. *Health and Technology*, 3(1): 85–95. http://dx.doi.org/10.1007/s12553-013-0052-9

Ståhlbröst, A., & Bergvall-Kåreborn, B. 2011. Exploring Users Motivation in Innovation Communities. *International Journal of Entrepreneurship and Innovation Management*, 14(4): 298–314. http://dx.doi.org/10.1504/IJEIM.2011.043051

Ståhlbröst, A., & Bergvall-Kåreborn, B. 2013. Voluntary Contributors in Open Innovation Processes. In J. S. Z. Eriksson Lundström, M. Wiberg, S. Hrastinski, M. Edenius, & P. J. Ågerfalk (Eds.), *Managing Open Innovation Technologies*: 133–149. Berlin: Springer Berlin Heidelberg. http://dx.doi.org/10.1007/978-3-642-31650-0_9

Keywords: living lab, field trial, drop-out, attrition, user involvement, user engagement, open innovation

Heading for a FALL: A Framework for Agile Living Lab Projects

Tanguy Coenen and Sarie Robijt

> *"Our greatest glory is not in never falling, but in rising every time we fall."*
>
> Confucius (551–479 BC)
> Teacher, editor, politician, and philosopher

Living lab methodologies need to enhance reactivity to changing requirements as these appear in a project. Agile methods allow for quick reactivity, but have been critiqued for not sufficiently taking into account the end-user perspective. In this article, we describe how to blend living lab methodologies with agile methods and, to this end, we present a Framework for Agile Living Lab projects (FALL). To make the framework actionable, we propose a number of actor roles. With concrete examples from living lab practice and a discussion of the theoretical basis, this article is relevant to both academics and practitioners.

Introduction

Agile development and living labs have separately received much attention from innovation-driven practitioners and academics over the last decade (Almirall et al., 2008; Dybå & Dingsøyr, 2008; Følstad, 2008). Despite different backgrounds and foci, both concepts share some commonalities. As major goals common to both approaches, we identify shared ambitions to: i) increase cost efficiency, ii) augment stakeholder collaboration, and iii) cooperate with users. However, both approaches are characterized by some weaknesses. Although living labs champion end-user involvement in both design and development, results from user co-creation often fail to become incorporated in ongoing technological development cycles (Sauer, 2013). Given that innovation frequently has unintended outcomes (Sveiby et al., 2009) – such as unforeseen shifts in requirements – living labs react more rapidly to such shifts in scope. Agile methodologies, on the other hand, have been developed specifically to tackle shifting requirements, yet they lack a structured focus on users and do not target collaboration with them (Cajander, 2013; Singh, 2008).

The question that we aim to answer here is:

How can we integrate agile methodologies, with their advantages for structuring flexible work processes, with living lab methodologies, which are known to be user-driven?

Addressing this question should yield novel insights into how to conduct living lab projects, both from a theoretical and practical perspective, by means of a framework, which we have named the Framework for Agile Living Labs (FALL).

The article is structured as follows. First, we introduce FALL and its component phases. Next, we describe FALL's various actor roles and their associated tasks. Then, we illustrate how SCRUM acts as a backbone for this agile framework. Finally, we highlight the contribution of the study and its limitations, and we offer conclusions.

FALL: Framework for Agile Living Lab projects

In this section, we present the essence of the FALL model, which is the result of an inductive (from practice to model) and deductive (from existing theory to model) process. The inductive part was done by participatory action research through hands-on experience in a variety of Flemish living lab projects run by imec.livinglabs (www.iminds.be/livinglabs/) and the "VRT Proeftuin" (deproeftuin.vrt.be). The theoretical foundations for the deductive part are rooted in design science research, a scientific discipline that aims to contribute to the scientific body of knowledge by building information systems. The main phases of FALL are derived from the action design research (ADR) method (Sein et al., 2011). However, in order to make a research-oriented framework more practical and applicable in daily living lab practice, we implemented FALL using agile techniques.

Figure 1 illustrates FALL, which focuses on living lab projects running from the early stages of a project idea to the real-world evaluation of a working software prototype. This framework was created because living labs need a robust methodology to structure and value user feedback. Because innovation frequently has unintended or unexpected effects (Sein et al., 2011), living labs must adjust rapidly to user feedback. This need for rapid adjustment is further underlined by variability in the timelines of project objectives and the diversity of control points, as commonly experienced in living labs (Leminen & Westerlund, 2011). Each of FALL's component phases is described in the subsections that follow.

Phase 1: Problem Formulation

The first FALL phase, Problem Formulation, aims to produce a concise statement to scope the effort of the team, which should reflect the perspectives of relevant end users and stakeholders. There are different ways of gathering information on what the problem statement should contain. The living lab and user research literature has much to say on how to leverage insights and domain knowledge from representative end users and stakeholders, for example through co-creation techniques and focus groups. Another way is to create online crowdsourced ideation campaigns on general topics that representative end users can participate in. Outputs of such efforts (e.g., story boards, user scenarios, Lego Serious Play models) in the problem formulation phase are by definition abstract, guiding the overall design and development efforts in the subsequent phases, where the outputs of co-creation become more concrete.

In addition to the co-creation of problem formulations, and as is customary in design science research, it is important to map the existing state of the art on the type of system being created. This step is often forgotten or not accounted for in living lab literature, yet creates an important baseline against which to gauge the innovation potential of the project.

From the problem definition, a first solution can be devised in the form of a set of assumptions to be tested by building minimum viable products (MVPs). However, these assumptions are often uncertain statements that should be verified. As in lean UX (Gothelf & Seiden 2013), selecting what assumption to test first can be done by prioritizing them in terms of high risk and low maturity. Risk refers to the consequences of the assumption being false while the project still holds the assumption to be true. Maturity is the amount of knowledge that the project team has regarding the assumption. Testing high-priority (i.e., high-risk and low-maturity) assumptions should be the focus of MVP 0 (see Example 1). The functionality of the MVP can be described using a SCRUM backlog that bundles and de-

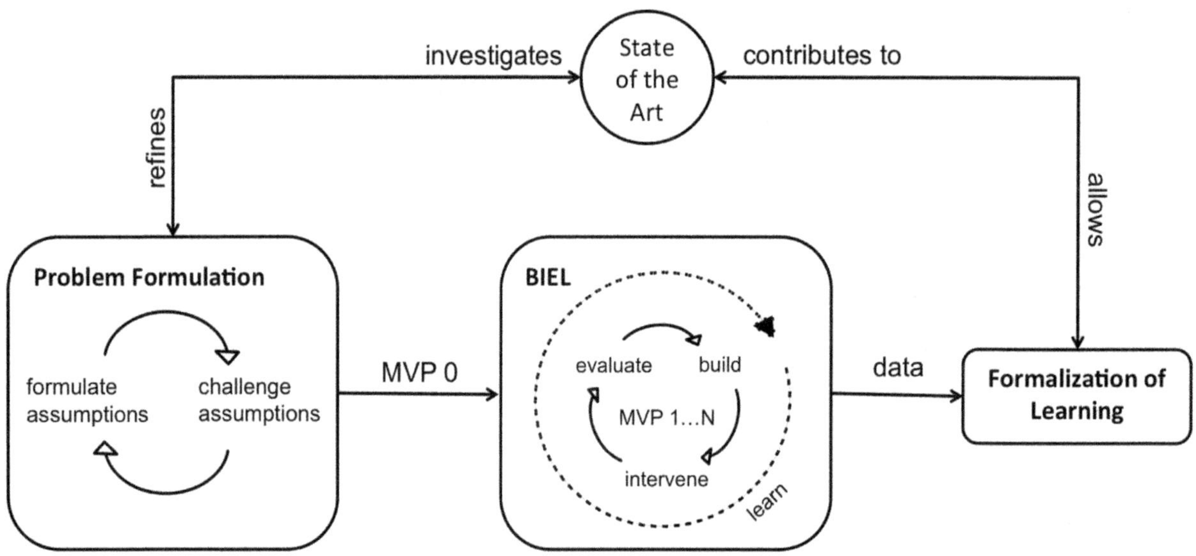

Figure 1. Framework for Agile Living Lab projects (FALL)

Example 1. Testing high-priority assumptions with the first MVPs

In the ZWERM (www.zwermgent.be) project, we started from a very general problem statement: to engage smart citizens with the city through mobile applications. However, this problem statement was too broad to be actionable. By gathering feedback from different stakeholders in the project (including a great number of citizen inhabitants, which were the prospective end users of the project), we arrived at the following problem statement: "How can we build a system that allows neighbourhood citizens to play a game through which they increase social cohesion and use this social cohesion to take actions that are important to the neighbourhood?" A conceptual solution was imagined to answer this question, which could be expressed as a number of assumptions:

1. The question can be answered through Internet-of-Things-enabled public space furniture.

2. The question can be answered by building a game in which a "check-in" (a person swiping an RFID card on a card reader) will be the central game mechanism to engage players with the system and to get them to know each other.

3. The question can be answered by creating a number of missions that can be played with the user's own device and inciting the user to take positive action.

Of these three assumptions, the second and third one were identified as being the highest risk and least mature. Therefore, they became the object of the first MVPs.

scribes the functionality of the MVP as user stories. A user story has the form "as a <role> I can <functionality>". Once MVP 0 has been defined in a backlog, it needs to be built, evaluated and tested, which happens in the next phase.

Phase 2: BIEL (Build - Intervene - Evaluate - Learn)
The creation of the successive MVP prototypes takes place in the "build - intervene - evaluate - learn" (BIEL) phase, in which the MVPs are always created (built), presented for feedback (intervention), and evaluated. The BIEL phase is derived from Sein and colleagues' (2011) action design research method. The reason why these four activities are bundled into one phase is because they take place concurrently and not necessarily in sequence. Indeed, it is often the case in a living lab that an intervention in a real-world environment takes place over a longer period of time. In such a situation, building goes on while the intervention is taking place, as for example, bug fixes and change requests are addressed and integrated in the live functional prototype. Similarly, evaluation can take place during the intervention, for instance using qualitative observations.

Learning has been added as a separate loop in Figure 1 to indicate that it is highly ingrained in all activities taking place in the BIEL phase. Indeed, living lab practitioners, like other people, do not learn as a separate activity, but learn by performing all the necessary activities in the BIEL phase. For example, the actual building of MVPs yields extensive learning on what will work and what will not.

Such MVPs, in the context of FALL, would better be termed minimum viable *prototypes*. Indeed, all MVPs submitted for feedback are intermediate prototypes on the road to the outcome of a living lab project. Examples of prototypes often used in living labs include:

- paper prototypes (Snyder, 2003), which are sketchy representations of the graphical user interface (GUI)

- graphical user interface (GUI) mock-ups or extended paper prototypes with graphic style added (see Example 2)

- clickable prototypes that allow for a certain degree of interaction

- functional prototypes that can be used on the device of the user allowing real-world intervention and testing

In living labs, all participating stakeholders – including end users – can build or participate in the creation of prototypes. As such, we perceive co-design as a possible process of the BIEL phase. A technique we often use for such co-design is the lean UX "design studio" technique, in which end users are asked to individually draw the GUI for a system, after which a GUI design is made to reflect the group consensus.

Example 2. GUI mockups as MVPs

After identifying the most important assumptions and deciding to make them the subject of MVP 0, the partners of the ZWERM project started working on several MVPs. MVP 0 consisted of GUI mockups for the website used to play the game. Feedback was gathered from representative end users, and a new version of the GUI design was created (MVP 1). Next, a functional version of the website and the check-in system were built and deployed at our research facility during a three-week period (MVP 2). Data was gathered through observations, a survey, and an analysis of the system logs. This allowed us to formulate answers to assumptions 2 and 3 (see Example 1). The feedback on assumption 2 was definitely positive, while the feedback on assumption 3 was more nuanced, with some missions working well and other not working at all. Based on what we learned during the intervention with MVP 2, we created a fully functional prototype (MVP 3) that was tested in real-life environments over a four-week period. Again, data was gathered through observations, a survey, and an analysis of the system logs.

Phase 3: Formalization of Learning

The Formalization of Learning phase is where all that has been learned is reflected upon and placed in some format that is fit for consumption by an academic, a business, or a public audience (see Example 3). In the case of the former, it is important to contextualize the formalization of learning in terms of the existing scientific knowledge base (i.e., the state of the art) and the problem formulation. For a business audience, the formalization of learning may be more oriented towards insights that are useful for market introduction of the concepts, underlying the system.

FALL Roles

In order to use FALL, a number of roles are necessary, because certain tasks must be performed during a FALL project. Assigning these tasks to roles in the living lab team can ensure that all tasks are accounted for. Starting from the main actions to be completed in our experience of running living labs (inductive) and from the literature (deductive), we identify eight key roles, which are borrowed from the literature on living labs, user research, agile methods, and user experience design:

Example 3. Formalizing learnings from the MVPs

After the real-life environment intervention with MVP 3 and the evaluation based on the collected data, we drafted a number of papers, which described the system and formulated guidelines for the future design of similar systems. In addition, the core findings of the entire project were formalized into a project description (vision, architecture, business plan) for a spin off that leveraged the main elements of ZWERM.

1. *Process manager:* as in agile methods, the aim of FALL is to increase the amount of self-organization of the team. However, someone is needed to guide the team with the methodology or process of working with FALL. This is the responsibility of the process manager.

2. *User researcher:* takes the lead in getting input from users at different stages of the project. In addition, the user researcher has the responsibility of keeping the story backlog up to date from the perspective of the end users. They are also responsible for making sure the problem formulation is grounded in the state of the art from a non-technical perspective.

3. *Researcher:* active in an academic domain that is relevant to the design problem in the social or natural sciences, researchers contribute the insights that are needed to create innovations by leveraging knowledge from various research fields and applying them to the design problem at hand.

4. *Architect:* their role is to create the systems architecture and to update and prioritize the backlog in terms of the stories that are not facing towards the user, such as "the server backend should be able to automatically backup the user data that is stored in the database". The architect is also responsible for making sure the problem formulation is grounded in the state of the art from a technical perspective.

5. *UX Designer:* responsible for creating MVPs that represent the GUI of the system. These can be wireframes, clickable prototypes, or GUI mock-ups. It is crucial to note that, although the UX designer holds the skillset to build these artifacts, creating them should never be done solely from only the perspective of the UX designer. Core to the philosophy of

FALL is that the UX designer should work with the feedback that was gathered from the project actors (other team members, representative end users, etc.).

6. *Developer:* responsible for translating the story backlog into functional MVPs.

7. *User:* involved in the project to bring domain-oriented knowledge to the team through co-design and usability testing processes, as guided by the user researcher. It is important that the users involved in the FALL project be as representative as possible of the user group that will eventually use the outcome of the living lab project.

8. *Stakeholder:* like users, stakeholders are also involved in contributing domain-oriented knowledge, but are not necessarily representative of the eventual user population. Stakeholders often hold higher-level interest than users and operate from a policy, commercial, or academic point of view.

SCRUM as a Backbone

The Problem Formulation, BIEL, and Formalization of Learning phases require work to be done by a multitude of people. Within FALL, we propose to facilitate this work through SCRUM, the most widely adopted agile methodology. As a result, work in the living lab will be organized according to sprints, which are time-boxed iterations. At the beginning of each sprint, the objectives and the end-time of that particular sprint are defined. These objectives differ according to the phase of FALL in which the project is situated. In the Problem Formulation phase, the objectives will be focused on scrutinizing the body of knowledge and creating a problem formulation with related assumptions on how to address the defined problem. In the BIEL phase, the objectives will be on creating MVPs, testing them and defining new MVPs based on insights from building and testing previous versions. In the Formalization of Learning phase, the aim will be to contribute to the body of knowledge based on what was learned during the living lab's execution.

At the end of the sprint, a demonstration of the work is given and the whole process is evaluated in a sprint retrospective. Also, the backlog is updated according to the tasks that present themselves in the future milestones of the FALL project. The story backlog therefore constitutes a crucial FALL project management tool, as it keeps an overview of tasks in progress or to be completed. Such a backlog can be established at the start of each new BEIL cycle in which a new MVP is to be developed.

Generating the user stories in the backlog can be done by the project group (including users), for example by allowing group members to generate user stories on sticky notes. Subsequently, user stories can be prioritized using the categories of the MOSCOW method: "must have, should have, could have, won't have" (Figure 2). Most of the time, the former two categories will form the basis for BIEL activity in the upcoming cycle, yet user stories in the latter two should be kept for future use.

Conducting FALL as a SCRUM project provides the agility that is needed in living labs, where requirements are unstable due to ongoing end-user feedback. Having time-boxed iterations, at the start of which the premises of the project are questioned, helps in integrating new insights.

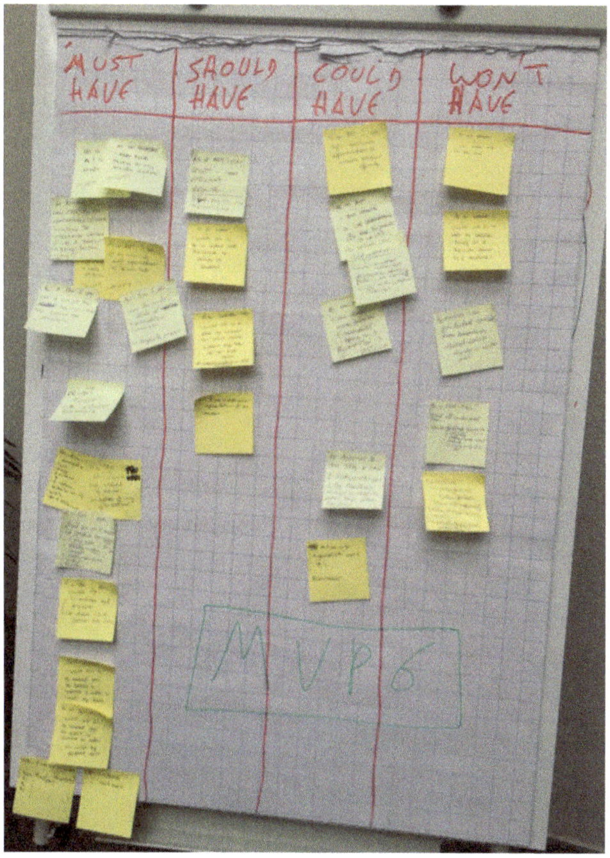

Figure 2. User stories being prioritized using the MOSCOW method
(Photograph courtesy of the D-Pac project: d-pac.be)

Conclusion

With agile methods, it is challenging to take the user perspective into account in a structural manner, whereas living labs often fail to incorporate emergent user feedback into running design and development processes. We proposed to address these issues by creating a framework that allows living labs to be executed in an agile way. FALL – the Framework for Agile Living Labs – draws on lean UX and SCRUM as agile methods. In addition, it takes design science research as a theoretical basis and structures the process along the lines of action design research. Through this article's concrete examples, practical guidelines, and theoretical foundations, we have attempted to address both theoretical and practical implications. FALL can be used as a basis on which to align the research, design, development and evaluation activities that are core to many living lab projects, providing actionable guidelines to researchers and practitioners alike.

This article has made the following contributions. First, we introduced agile methodologies into the theory and practice of living labs. Second, we proposed an actionable, yet theoretically grounded set of constructs (MVP, BIEL, etc.) around which to conduct a living lab project in an agile way. We have placed this into a framework (FALL) and indicated how this framework can be supported methodologically through the different phases of FALL. Third, we proposed principles to be taken into account when performing living lab projects according to FALL: define project roles and use SCRUM as a backbone for project planning. As such, we contributed prescriptive knowledge to living lab theory and made a sep towards overcoming practical hurdles that living lab projects can be confronted with.

A main topic in living lab research is how to gather insights from end users and stakeholders through participative techniques. Although we have hinted on ways to achieve this, we did not cover the participative processes in detail, given our focus on providing a general overview of the processes in FALL. As a key avenue for further research, greater elaboration is needed regarding the participative processes that are appropriate in different phases of FALL and the properties of their outputs.

About the Authors

Tanguy Coenen is head of the software application prototyping team at imec and has a PhD in Business Administration from the Solvay Management School of the Vrije Universiteit Brussel in Belgium. During his career, he has been involved in many living lab projects, both at the national and the European levels. Tanguy fosters a variety of research interests but is focused on a "City of Things" project, which is applying living lab concepts to achieve the vision of a city as a living lab innovation platform.

Sarie Robijt currently works as a Service Design professional in retail business. As a service designer, she takes up several roles from user experience analyst to workshop facilitator and innovation coach. During her time at VRT Proeftuin, she worked on setting up and maintaining a living lab in the media innovation industry.

References

Almirall, E., & Wareham, J. 2008. Living Labs and Open Innovation: Roles and Applicability. *eJOV: The Electronic Journal for Virtual Organization & Networks*, 10: 21–46.

Cajander, Å., Larusdottir, M., & Gulliksen, J. 2013. Existing but Not Explicit: The User Perspective in Scrum Projects in Practice. In *Human-Computer Interaction–INTERACT 2013:* 762–779. Berlin: Springer Berlin Heidelberg.

Dybå, T., & Dingsøyr, T. 2008. Empirical Studies of Agile Software Development: A Systematic Review. *Information and Software Technology*, 50(9): 833–859. http://dx.doi.org/10.1016/j.infsof.2008.01.006

Følstad, A. 2008. Living Labs for Innovation and Development of Information and Communication Technology: A Literature Review. *eJOV: The Electronic Journal for Virtual Organization & Networks*, 10: 99–131.

Gothelf, J., & Seiden, J. 2013. *Lean UX: Applying Lean Principles to Improve User Experience.* Sebastopol, CA: O'Reilly Media Inc.

Leminen, S., & Westerlund, M. 2011. *Open Innovation Paradoxes in Living Labs Networks.* Paper presented at the 13th EURAM Conference, Istanbul, Turkey, June 26–29.

Ries, E. 2011. *The Lean Startup: How Today's Entrepreneurs Use Continuous Innovation to Create Radically Successful Businesses.* New York: Crown Business.

Sauer, S. C. 2013. *User Innovativeness in Living Laboratories: Everyday User Improvisations with ICTs as a Source of Innovation.* The Netherlands: University of Twente.

Sein, M., Henfridsson, O., Purao, S., Rossi, M., & Lindgren, R. 2011. *Action Design Research. Management Information Science Quarterly,* 35(1):37–56.

Singh, M. 2008. U-SCRUM: An Agile Methodology for Promoting Usability. In *Proceedings of AGILE '08:* 555–560. Washington, DC: IEEE.
http://dx.doi.org/10.1109/Agile.2008.33

Snyder, C. 2003. *Paper Prototyping: The Fast and Easy Way to Design and Refine User Interfaces.* Burlington, MA: Morgan Kaufmann.

Sveiby, K. E., Gripenberg, P., Segercrantz, B., Eriksson, A., & Aminoff, A. 2009. Unintended and Undesirable Consequences of Innovation. In *Proceedings of the XXth ISPIM Conference,* Vienna, Austria, June 21–24 June.

Westerlund, M., & Leminen, S. 2011. Managing the Challenges of Becoming an Open Innovation Company: Experiences from Living Labs. *Technology Innovation Management Review,* 1(1): 19–25.
http://timreview.ca/article/489

Keywords: living labs, design science research, agile, methodology, SCRUM, lean UX

21

Creating Brand India: Strategies, Issues, and Challenges

Alok Chakrawal and Pratibha Goyal

> *Brand India has to identify its core product identity... we have to start looking at what we want to stand for.*
>
> Kurush N. Grant
> Former Executive Director of ITC

Just like a product, person, or institution, a country also has an impression in the minds of people. A brand distinguishes the product of one seller from those of the others. With the Indian Prime Minister's recent call to "Make in India", brand building is receiving more attention and momentum in the country. India is the largest democratic country with the highest GDP, and it is becoming a favourable destination for the global investment giants. But, despite various strengths and successes in image building, the country faces challenges in creating a sustainable, conducive environment for making India a production hub of the world. Bureaucracy, corruption, delay in clearance of business proposals, ethical standards and work culture, tax reforms, political interventions, socio-economic barriers, regionalism, etc. are some of the challenges that must be coped with. This article examines these various issues and challenges that must be overcome in creating Brand India.

Introduction

A person, an institution, or even a country leaves an impression in the mind of perceiver. This impression reflects the brand – what distinguishes one person, institution, or country from another. In India and beyond, country-level brand building has been receiving significant attention ever since Indian Prime Minister Narendra Modi's 2014 call to "Make in India" (DIPP, 2016). The initiative is designed to encourage multi-national and national companies to manufacture their products in India, thereby transforming the country into a global design and manufacturing hub. Make in India has became "a rallying cry for India's innumerable stakeholders and partners" (DIPP, 2016) and is a key aspect of the country's future brand. Thus, it is quite pertinent to understand India's image in the minds of the citizen of India and other nationals in the context of the present economic system.

Even in 2011, Kurush Grant, Executive Director of ITC, recognized that "like all successful brands, India as a brand, too, should stand for being a great product in itself. Also, the country and its people have to understand what India symbolises" (Sarkar, 2011). But, if we wish to answer the call to Make in India, we must ask: what is India's brand now? What impression is created in the minds of others when thinking of India today?

In this author's view, a negative impression of India has been created in the minds of foreigners because of perceptions of, for example, poor foreign policies, reports of dishonesty in political parties, unstable relations with neighbouring countries, corruption in the country's administration, and limited trade and cultural relations with the rest of the world (particularly China and fast-growing countries in the region). Even in this age of fast electronic communication and the Internet, where things have become so transparent and easily accessible to people across the globe, people from outside India have a hazy idea about India and its high income disparity, poor health and sanitation conditions, complex caste system, religious and communal conflicts, frenzied approaches of some groups.

India is a diverse country where people live at the greatest of extremes. Around 35 per cent of the urban population lives below the poverty line (SECC, 2015), and yet India is also an atomic power with high-end ballistic missiles and surgical strike capacity. India also has talented people in its information technology sector and many other Indians are working abroad in corporations such as Google, IBM, and Intel. At a projected rate of 7.5% and rising, India is the fastest growing economy in the world (Gil Sandar, 2015). It is indeed time to remake India's brand.

This article elaborates on the image of India, as seen from the inside and outside, with the aim of understanding Brand India today and what brand-related challenges must be faced if the Make in India initiative is to be successful.

Understanding Brand India

Top better understand Brand India and to evaluate its chances for success, we considered the following nine factors, which we consider important based on our reading of the relevant literature:

1. *GDP share as share of the world economy:* At 6.83%, India ranked third among G20 countries in 2014 (Quandl, 2015).

2. *Growth rate of GDP:* As mentioned above, India is now the world's fastest growing economy (Gil Sandar, 2015). And its GDP growth since 2012 is further highlighted by declines in the other four BRICS countries (Brazil, Russia, China, and South Africa) over the same time period (World Bank, 2016). This growth in GDP is fueled by increasing contributions from services and industry relative to agriculture, which has fallen from over 50% of GDP composition in the early 1950s to less than 15% in the early 2010s (Planning Commission, 2014). This shift away from agriculture to services and industry is a sign of a developing economy.

3. *Investment destination:* In the ranking of most attractive destinations for investment, India is ranked first among global investors according to Ernst and Young's India Attractiveness Survey (Ernst and Young, 2015) and is ranked third (behind the United States and the United Kingdom, and just ahead of China and Germany) according to its Capital Confidence Barometer (Ernst and Young, 2016).

4. *Political system and stability:* Among its near neighbours, is considered relatively stable owing to the following factors: i) the absence of military coups in an independent democratic India; ii) its multi-party political system with a national and regional layer; iii) its strong and active judiciary system that has a degree of control over the country's functionary machinery; iv) its effective legal system with a degree of transparency; checks and balances on political system by non-governmental organizations (NGOs); and v) vigilant public participation in political activism.

5. *Corruption perception:* Despite all judiciary strength and peoples' active participation in political activism, corruption is common in India, which is ranked 76th among 168 countries in terms of perceived corruption (Transparency International, 2015). It seems that every day, Indians hear about new methods of corruption innovated by Indians. However, if there are silver linings to be found, they are that these revelations of corruption are possible only because there is now a system to control and prevent; that India ranks favourably among its nearest neighbours; and that its rankings are improving.

6. *Global presence of Indian corporates:* Indian companies are striving hard to make their presence globally and have had some significant achievements. However, there are no Indian companies among the top ten Global 500 companies, whereas China has three (Fortune, 2015). The highest-ranking Indian company is the Indian Oil Corporation (IOC) in the 109th position.

7. *Indians leading global corporations:* Indian workers have good reputations as hardworking, creative, disciplined, and loyal professionals. Accordingly, many large corporations such as Intel, Microsoft, IBM, BMW, Google, Motorola, MasterCard, SanDisk, Nokia, etc. hire talented Indians. In many cases, the global heads of multi-national corporations are of Indian origin (for examples, see Mishra, 2014).

8. *Peace index:* As per the Global Peace Index (Institute for Economics & Peace, 2015), India stands in the 143rd position, behind fellow BRICS countries Brazil (103rd), South Africa (136th), and China (124th), but ahead of Russia (153nd). There is much scope for improvement in this area.

9. *Black money:* In India, "black money" refers to funds earned on the black market, meaning that no taxes have been paid upon such earnings. India may be better in this area than other fast-developing countries, although this is another area where there is scope for improvement.

Besides these nine key factors for projecting the brand image of India, there may many more factors to be considered, such as quality of governance and government effectiveness index, literacy rate and employability quality, adaptability quality, and acceptability.

Boosters to Brand India

A number of initiatives have been launched recently by the Government of India and non-governmental organizations that are likely to give a boost to efforts to develop a strong Brand India:

1. *Clean India Mission* (swachhbharat.mygov.in): Launched in October 2014, the ambitious "Swachh Bharat Abhiyan" (Clean India Mission) was undertaken to clean and maintain the cleanliness of the country's streets, roads, and rivers and to develop key hygiene infrastructure throughout the country, both to improve the attractiveness of the landscape and improve the health of the citizens.

2. *Digital India* (digitalindia.gov.in): This vision of this flagship programme of the Government of India is "to transform India into a digitally empowered society and knowledge economy". The programme focuses on three key areas: i) digital infrastructure for every citizen, ii) governance and services on demand, and iii) digital empowerment of citizens.

3. *Skill India* (skillindia.gov.in): Launched in July 2015, this initiative of the Government of India aims to train over 400 million people in India in different skills by 2022. The initiatives include the National Skill Development Mission, the National Policy for Skill Development and Entrepreneurship 2015, the Pradhan Mantri Kaushal Vikas Yojana (PMKVY) scheme, and the Skill Loan scheme.

4. *Make in India* (makeinindia.com): Launched by the Government of India in September 2014, Make in India is designed to encourage multi-national, as well as national companies to manufacture their products in India.

5. *Startup India* (startupindia.gov.in): This initiative aims at fostering entrepreneurship and promoting innovation by creating an ecosystem that is conducive for growth of startups. The objective is that India must become a nation of job creators instead of being a nation of job seekers. Startup India was launched in January 2016.

In the author's view, an increased focus on the following other activities can also play a role in boosting the current and future Brand India:

1. *Foreign policy:* India's foreign policy is currently focused on improving relations with neighbouring countries in South Asia and on engaging the extended neighbourhood in Southeast Asia and the major global powers.

2. *Foreign strategic alliances:* Recently, India has engaged in many joint military operations with, for example, China, France, Russia, Sri Lanka, and the United Kingdom .

3. *Inviting foreign heads of state:* Among other visits, every year on January 26, India invites the head of state of a leading foreign country to be the chief guest at the Republic Day celebration. Recent guests have included the heads of the states from France, the United States, and Japan.

4. *Global MoUs and business relations:* India is busy in signing important memorandums of understandings (MoUs) in the areas of trade, industry, and the service sector with important countries to encourage economic growth and development in India.

5. *Resolving land issues with neighbours:* A long-pending land issue with Bangladesh was resolved with the efforts of Government of India last year. Such efforts are in full swing for resolving land issues with China, Pakistan, and Myanmar. It is a great image booster element for India in the eyes of foreign countries that India is able to resolve its important land issues with its neighbours in peaceful manner.

6. *Strategic balance in strategic strength:* India is a nuclear power, having great capability to strike most of the countries in the region. This power gives strategic strength to India to maintain the power balance in the region.

7. *United Nations influence:* India seeks a permanent seat in the security council of the United Nations, through which it could increase its global influence.

8. *Special government attention on brand building:* Amongst the leading organizations working on building Brand India is India Brand Equity Foundation (IBEF; ibef.org), a public–private partnership between the Ministry of Commerce and Industry, the Govern-

ment of India, and the Confederation of Indian Industry. It aims to effectively present the Indian business perspective and leverage business partnerships in a globalizing market-place. IBEF regularly tracks government announcements in policy, foreign investment, macroeconomic indicators, and business trends.

Challenges in Creating Brand India

Taking into account the backdrop of initiatives and activities described above, there is much work being done and many challenges to be overcome in building Brand India. And, like a global product, building a country's brand cannot be done overnight or without help. There has to be a comprehensive effort to be initiated from the government as well as from the entrepreneurs and corporate entities of a nation. Some of the key challenges that must be overcome by these groups are described below:

1. *Bureaucratic hassles:* India runs under basically the bureaucratic system established by the British almost one and a half centuries ago. Changes have been made, but many reforms are still needed to reduce the frequency and severity of bureaucratic "hassles". It has happened many times that large investors have pulled back from India due to intolerable delays in the project approvals from the authorities. To help improve Brand India, such hassles need to be reduced substantially. Key steps that need to be taken include adding accountability, incentives, transparency, and freedom in decision making, while removing subjectivity, political interventions, and vindictive approaches.

2. *Corruption:* Corruption is a very common word used in the Indian economy. As we have seen, India has to go a long way to bring down its corruption ranking at a global level. Scams and dishonest behaviour put India's image at stake.

3. *Work culture:* The work cultures of multinational corporations are often quite different than those typically found in India. Multinationals emphasize performance in an informal setting, with options for employees to work from home, for example, as long as output is maintained. Such practices are now becoming more common in the Indian IT sector, but the overall work culture in India tends to overemphasize physical attendance with not enough attention being paid to output. India needs to modernize its work culture to improve its global image.

4. *Quality control:* India needs to improve the quality of both exported and domestic goods and services. Quality control contributes to brand building through resource efficiency, increase standards of living, revenue generation, benchmarks, improved work environments, employee satisfaction, and improved moral values and standards. With exports in particular, India becomes better known through its products across the world. Therefore, Indian companies must ensure that international quality standards are being followed and that their products and services support a positive image of India.

5. *Flagship concept:* India has to take a lead in making and promoting its main products by industry and region. The flagship product concept may give India a big boost in terms of brand building. For example, Japan has a strong brand in technology. Germany is known for its expertise in the automobile sector. Israel is known for new agriculture technology. Similarly, India must decide upon and develop its areas of strength, and its overall brand, as Kurush Grant suggests in the opening quotation of this article.

6. *Technology adoption and implementation:* Indian managers need to be quick to adopt new technology in the primary sector, industry sector, and service sector. To build Brand India, India companies need to be seen as operating on the forefront of innovation and technology

7. *Social and economic welfare:* Indian managers have to make sure that the brand building drive does come at the expense of their basic responsibility for social and economic welfare. Corporate social responsibility as a mandatory clause in some Indian companies has helped, but they need to think beyond this threshold of 2% of net profit. Indian companies have reserves and surpluses that can be used to contribute to the welfare of society.

8. *Research and development (R&D):* India is considered to be a poor investor in R&D activities. The onus lies upon Indian companies and managers to invest in R&D, not only for their own direct benefit but to indirectly contribute to Brand India.

Conclusion

India has a great legacy and is considered to be one of the oldest economies in the world. In modern times, it can achieve distinction if Brand India is created well. Despite a number of strengths, in India, everything is not as per the aspirations of global business houses. Creating sustainable business environment and making India a production hub of the world is a challenging task. Bureaucracy, corruption, delays in clearance of business proposals, ethical standards and work culture, tax reforms, political interventions, socio-economic barriers, and regionalism are some of the challenges. Nonetheless, India is poised to take advantage of its strengths and overcome these challenges so that it can answer the call to Make in India and promote Brand India.

About the Authors

Alok Chakrawal is a Professor of Commerce and Business Administration, and the Coordinator of the Internal Quality Assurance Cell (IQAC) at Saurashtra University in Rajkot, India. He holds MCom and PhD degrees, for which he was awarded a Gold Medal upon graduation followed by a Junior Research Fellowship from the University Grants Commission. He is an accomplished international speaker and award-winning author of academic papers and books, particularly in the domain of stress management. He has been the General Manager and Director of SKSE Securities Ltd., Rajkot, and Controller of Examination of Saurashtra University. He also sits on various boards and committees at national and international levels.

Pratibha Goyal is Professor and Director of the School of Business Studies at Punjab Agricultural University in Ludhiana, India. She received her MBA from Guru Nanak Dev University in, Amritsar, India, and her PhD from Punjabi University in Patiala, India. She began her career as a Deputy Manager of Human Resources in a leading industry. Her immense love for mentoring young minds saw a shift in her career when she joined the faculty of Management in Punajbi University, Patiala, and later in Punjab Agricultural University, Ludhiana. She has been the Principal Investigator of three major research projects of the University Grants Commission and has participated in several national seminars, conferences, and symposia. She is on the Board of Directors of Hero Cycles Limited. She has authored five internationally acclaimed books, besides being Executive Editor of the *Effective Management* journal. She has published several notable research papers in national and international journals and has guided many scholars in their research projects.

References

DIPP. 2016. Make in India: About Us. *Department of Industrial Policy & Promotion (DIPP), Ministry of Commerce, Government of India.* Accessed May 1, 2016: http://www.makeinindia.com/about

Ernst and Young. 2015. *India Attractiveness Survey 2015: Ready, Set, Grow.* New Delhi: Ernst and Young India.

Ernst and Young. 2016. *Global Capital Confidence Barometer,* 14th ed. London: Ernst and Young.

Fortune. 2015. Global 500. *Fortune.com.* Accessed May 1, 2015: http://fortune.com/global500/

Gil Sandar, F., Shome, S,. Seth, S., & Misra, J. 2015. *India Development Update: Fiscal Policy for Equitable Growth.* Washington, DC: World Bank Group.

Institute for Economics & Peace. 2015. *Global Peace Index 2015.* Sydney: Institute for Economics & Peace.

Mishra, R. 2014. Top 10 Global Operations Head of India Origin. *India.com.* February 5, 2014. Accessed May 1, 2016: http://www.india.com/top-n/top-10-global-operations-head-of-indian-origin-10763/

Planning Commission. 2014. Databook for Planning Commission. *Government of India Planning Commission,* December 22, 2014: http://planningcommission.nic.in/data/datatable/data_2312/DatabookDec2014%202.pdf

Sarkar, A. 2011. Is India's Brand Image Still Limited in Global Perception? *afaqs!:* November 2, 2011. Accessed May 1, 2016: http://www.afaqs.com/news/story/32044_AdAsia-2011-Is-Indias-brand-image-still-limited-in-global-perception

Transparency International. 2015. *Table of Results: Corruption Perceptions Index 2015.* Berlin: Transparency International. http://www.transparency.org/cpi2015

Quandl. 2015. GDP as Share of World GDP at PPP By Country. *Quandl.com.* Accessed May 1, 2016: https://www.quandl.com/collections/economics/gdp-as-share-of-world-gdp-at-ppp-by-country

World Bank. 2016. Data: GDP Growth (Annual %). *World Bank.* Accessed May 1, 2016: http://data.worldbank.org/indicator/NY.GDP.MKTP.KD.ZG/countries/1W-IN-CN-RU-ZA-BR?display=graph

Keywords: Brand India, Make in India, innovation, entrepreneurship, government initiatives, challenges, economic growth

22

Comparing the Entrepreneurial Ecosystems for Technology Startups in Bangalore and Hyderabad, India

M H Bala Subrahmanya

> *"The emergence of [the] start-up wave in India is a relatively new phenomenon. India is today undergoing a fundamental shift with entrepreneurship and innovation being primary catalysts in job creation and solving everyday problems."*
>
> Praneeth Rao, Vaibhav Goyal, Agrima Singh, and Simrat Singh
> International Policy Analysis Network Research Team

Technology startups are gaining increasing attention from policy makers the world over because they are seen as a means of encouraging innovations, spurring the development of new products and services, and generating employment. Technology startups tend to thrive when inserted in a conducive entrepreneurial ecosystem. Therefore, ecosystem promotion is being given increasing policy support. However, the emergence and structure of entrepreneurial ecosystems for technology startups have hardly been traced and examined in detail. In India, Bangalore occupies a unique position in the startup world, and Hyderabad is fast emerging as one of the promising startup hubs in the country. Given this background, we set out to explore and examine the structure, evolution, and growth of ecosystems for technology startups in the context of Bangalore and Hyderabad. Both the ecosystems emerged due to the initial foundation laid in the form of government–industry–academia triple helix and their interactions leading to the emergence of a modern industrial cluster followed by an information technology and biotechnology cluster, which then led to R&D cluster serving both the cities. These three clusters together, gradually and steadily, facilitated an entrepreneurial ecosystem for technology startups to emerge. The ecosystem operates within the triple helix model and has a nucleus with two outer layers: i) an inner layer of primary (indispensable) factors and ii) an outer layer of supplementary (secondary) factors. Through the analysis of the experiences of Bangalore and Hyderabad and their ecosystem evolution, its structure, and components, we derive key lessons for others within and beyond India.

Introduction

Entrepreneurial ecosystem development is a process occurring over a period of time. Successful ecosystems are highly developed and matured. The two often-cited benchmarks in this regard are Silicon Valley and Israel, which are uniquely identified for their success in entrepreneurship development and for generating in one year, more successful startups than any other nation would have done in years or decades (Arruda et al., 2013). If the structure of an ecosystem has to be understood for policy implications, it would be appropriate to probe how each ecosystem originated and developed over time, as this has not been adequately explored in the context of entrepreneurial ecosystems for technology startups, as of now (Motoyama & Watkins, 2014).

Today, India has been recognized as one of the key potential sources of technology-based startups in the global economy (Gai & Joffe, 2013), and it currently ranks third globally, in terms of number of existing startups and number of startup exits (NASSCOM, 2015; *Times of India,* 2016). Within India, its "Silicon Valley", Bangalore, was considered to be one of the nine International Startup Hubs outside the United States (Pullen, 2013). Further, recently it ranked 15th among the 20 cities with the best ecosystems for startups worldwide (Startup Genome, 2015). Hyderabad is one of the major upcoming startup locations in the country (NASSCOM, 2015), and it is a close competitor with Bangalore in the spheres of information technology, biotechnology, and R&D-intensive industries (Das, 2015). Given this context, we set out to examine how these two ecosystems have emerged over time and to understand the prevailing structures of the ecosystems for technology startups in these two cities.

The emergence of technology startup hubs in both Bangalore and Hyderabad is a very recent phenomenon. However, the foundation for its emergence was laid decades ago, after India's independence in 1947. These two metropolitan cities host industrial clusters comprising modern manufacturing industries of varying sizes and are also home to several publicly-funded R&D institutions and high-quality public and private educational institutions. Both cities have pro-industry regional governments, which have formulated and implemented regional (state-level) industrial policies periodically for the development of industries and infrastructure. Thus, both have the much needed industry–institute–government combination for the emergence of a conducive ecosystem for startups.

In this article, we trace the evolution of the entrepreneurial ecosystems in the two cities, with a historical perspective since India's independence, and we examine the current structures of these ecosystems. We aim to derive some general lessons from the experiences of these two technology startup hubs of India. But at the outset, we would like to broadly define the two important concepts of this article, namely, startups and ecosystems. A startup is, in general, defined as a new venture with no previous history of operations (Bala Subrahmanya, 2015). Such new ventures suffer from the liability of newness because they are unfamiliar and without any precedence (Certo, 2003). In terms of age, they are age-zero firms or "infants" (i.e., they are less than one-year old) (Kane, 2010). Startups exclude existing enterprises that have been acquired by new ownership, or those inherited by younger generations from the older ones, or "industry spin-offs" where a large firm has a control, directly or remotely, or franchisees of any form (Bala Subrahmanya, 2015). Similarly, an ecosystem in the context of startups, in general, may be defined as a system comprising prospective as well as currently operating startup entrepreneurs, their mentors, financiers, trainers, large firms that provide market support, organizations (universities, institutions, etc.), and government policy makers that support and promote startups, and their interrelationships and interactions.

Objectives and Methodology

This article has two specific research questions:

1. *How did the entrepreneurial ecosystems of Bangalore and Hyderabad emerge?*

2. *What are the structures and components of these entrepreneurial ecosystems?*

To answer these questions, we first reviewed the relevant empirical literature and then conducted personal interviews with ecosystem stakeholders to understand the structure of the ecosystems and the interactions between the various actors.

Our empirical literature survey, which focused on the diverse components of entrepreneurial ecosystems in different parts of the global economy, clearly identified some common elements without which an entrepreneurial ecosystem may not come into existence or may not be effective and successful. The key elements or components of an entrepreneurial ecosystem identified and emphasized by the various empirical studies are presented in Table 1.

Based on the literature survey, and our identification of key stakeholders under different domains of the entrepreneurial ecosystems in both Bangalore and Hyderabad, we prepared an open-ended questionnaire and identified a set of about 50 stakeholders for each city. We approached those identified stakeholders who responded positively to our request and interviewed each one of them personally (51 in Bangalore and 38 in Hyderabad) for about one and half hours during the period from August 2015 to January 2016 (Table 2). An open-ended questionnaire formed the basis of the interviews, which contained 10 questions covering topics such as:

- definition of a high-tech startup
- characteristics of a high-tech startup
- components of an ideal ecosystem for high-tech startups (in India)
- ranking of the identified components

Table 1. The entrepreneurial ecosystem: components and their roles

No.	Components	Role and Importance	Identified Studies
1	Entrepreneurship	At the core of the ecosystem for the birth, growth, and exit of startups.	• Isenberg (2011) • Feld (2012) • Arruda et al. (2013) • Aleisa (2013) • Foster & Shimizu (2013) • Piscione (2013) • Anjum (2014) • Dale & Zell (2014) • WEF (2014) • Bala Subrahmanya (2015) • Stangler & Masterson (201 • Krajcik & Formanek (2015 • Sean (2015) • Haines (2016) • Hemmert et al. (2016) • Aaltonen (2016)
2	Markets	Early product adopters, repeat customers, networks, and multinational corporations.	
3	Finance	Angel investors, private equity firms, venture capitalists, public capital markets for initial public offerings, and debt instruments.	
4	Human Resources	Labour force, technical workforce, and managerial talent.	
5	Education & Research	Institutions that produce non-graduate and graduate workforce, technical and managerial talent, innovations, entrepreneurial training, etc.	
6	Government (regulatory framework & policies)	Favourable laws and regulations for tax incentives, venture capital, bankruptcy, property rights, labour, and exits public research institutions.	
7	Large Companies (including multinational corporations)	As customers, sources of entrepreneurship, nurturers of startups through accelerators, sources of technical and managerial workforce, investors, technology providers, acquirers of startups.	
8	Mentors/Advisors	Technical and managerial advice for startup creation, stability, growth, and exits.	
9	Support Institutions	Accelerators, technology business incubators, soft infrastructure (e.g., professionals such as lawyers and accountants), and hard infrastructure (e.g., telecommunications, transportation, and logistics)	
10	Cultural Support	Tolerance of risk, failures, support for innovation, creativity, drive and hunger for achievements; higher social status for entrepreneurs; respect for wealth creation.	
11	Media	Publicizing: startup creations, its pre-requisites, support available, failure consequences, and achievements of success	
12	Immigration of Talent	Sources of entrepreneurship, human resources, advisors/mentors, finance, support networks, etc.	

Table 2. Entrepreneurial ecosystem stakeholders in Bangalore and Hyderabad interviewed for this study

Stakeholder Type	Bangalore	Hyderabad
Startup Entrepreneurs	17	17
Government Officials	1	1
Large Enterprises (multinational corporations and domestic large firms)	7	2
Professors	3	6
Financiers (angels, venture capitalists, etc.)	7	4
Accelerators, Incubators, and Co-Working Spaces	7	6
Industry Associations	3	1
Mentors	3	1
Media Representatives	3	0
Total	**51**	**38**

- the importance of quality of components, quantity of components, and interaction among components for determining the effectiveness of an ecosystem
- whether Bangalore/Hyderabad has the best ecosystem in India
- factors that led Bangalore/Hyderabad to emerge as one of the best startup hubs in India
- significant components of the ecosystem for high-tech startups in Bangalore/Hyderabad
- components that are lacking or have a weaker presence in Bangalore/Hyderabad
- the level of interaction between the components in the ecosystem of Bangalore/Hyderabad

The author took down notes as he interacted with the stakeholders of the ecosystem. The information gathered through the personal interviews and the review of relevant literature formed the basis of the analyses to answer the two research questions.

Phases of Evolution in the Entrepreneurial Ecosystems of Bangalore and Hyderabad

Both Bangalore and Hyderabad are the capital cities of their respective states, namely, Karnataka and Andhra Pradesh/Telangana in southern India. These are two of the metropolitan cities of India and are industrial/commercial/institutional hubs of the country. According to the Jones Lang LaSalle City Momentum Index (Clarke-Billings, 2017), Bangalore is the most dynamic city in the world whereas Hyderabad is the fifth most dynamic city in the world. Table 3 provides some key statistics on both of these cities.

After India's independence in 1947 (for Bangalore) and after the integration into Indian Union in 1948 (for Hyderabad), the foundation for the evolution of an entrepreneurial ecosystem was laid in both the cities. This was primarily driven by public policies initiated by the national government and were characterized by the location and growth of central public sector undertakings (CPSUs) followed by industrial estates for modern small-scale industries and establishment of national government-funded public R&D institutions, as well as public and private educational institutions (Department of Gazetteer, 1996; Sudhira et al., 2007). This foundation resulted in the emergence of the much-needed regional innovation system, led by the "triple helix" comprising university–industry–government and their interactions, as propounded by Etzkowitz (2003). In the triple helix model, the State (national government) played a dominant role, driving industry and academia for national/regional development, resembling what Etzkowitz and Leydesdorff have described as a "statist" model (2000). The development of such a regional innovation system occurred over a period comprising almost four decades, ranging from the late 1940s to the mid-1980s. This marked the first phase in the evolution of an entrepreneurial ecosystem in both the cities (Figures 1 & 2).

The transition from the first phase to the second phase occurred rather smoothly and spontaneously beginning in the mid-1980s in Bangalore, with the entry of Texas Instruments and the origin of Bangalore-based information technology firms. The process was stimulated by the introduction of national economic reforms in 1991 and their acceleration since then, which virtually removed the entry barriers for multinational corporations in the information technology industry. This, coupled with the initiatives of regional government to strengthen the physical and virtual infrastructure as well as educational institutions, Bangalore received a

Table 3. Bangalore and Hyderabad: comparative statistics

Characteristics	Bangalore	Hyderabad
Year of Founding	1537	1591
Area	741 km^2	650 km^2
Altitude Above Sea Level	920 m	550 m
Climate	18° C – 38° C	15° C – 40° C
Population	8.4 million	7.7 million
Literacy Rate	89%	83%
Major Industries	Information technology and biotechnology industries, manufacturing, and textiles	Information technology and biotechnology industries, and manufacturing
Universities and Institutes of National Importance	13	6
Central Government Institutes and Central Public Sector Undertakings	57	47
State Government Institutes and State Public Sector Undertakings	20	13
Scientific & Industrial Research Organizations (SIROs)	35	32
DSIR-Registered Private Sector In-House R&D Units	178	185
Private Sector R&D Units Not Registered with DSIR	104	79

Notes:
1. Population numbers and literacy rate percentages: Office of the Registrar General & Census Commissioner, India (2011) .
2. Bangalore year of founding, area, altitude, and climate: Sudhira et al. (2007).
3. Hyderabad year of founding, area, altitude, and climate: Das (2015).
4. Data on universities, government institutes, public sector undertakings, SIROs, DSIR registered R&D units, and private sector R&D units: DST (2015).
5. DSIR – the Department of Scientific & Industrial Research – is a part of the Ministry of Science & Technology, Government of India. The primary endeavour of DSIR is to promote R&D by the industries, and it grants recognition to in-house R&D units in industrial enterprises, provided they satisfy terms and conditions as stated by the DSIR.

boost to emerge as an information technology industry cluster, soon compounded by a biotechnology industry cluster, by the late 1990s. With the introduction and acceleration of national economic reforms, the role of the State (national government) declined in intensity, though regional government continued to be important, and the role of industry and academia became more important. Thus, in the triple helix model, all the three partners, namely, State, Academia, and Industry seemed to emerge as equally important, thereby resembling the "laissez-faire" model described by Etzkowitz and Leydesdorff (2000). The role of the industry shifted increasingly from central public sector undertakings to multinational corporations and domestic information technology industry firms (Figures 1 & 2).

However, in Hyderabad, the transition from the first phase to the second phase was neither smooth nor spontaneous. Though Hyderabad, similar to Bangalore, had emerged as a cluster of modern manufacturing industries by the 1980s, it was the explicit initiative taken by the State (regional government) since the late 1990s for the transformation of the city towards the creation of a knowledge society and the positive response from the multinational corporations in the information technology industry, duly supported by domestic informa-

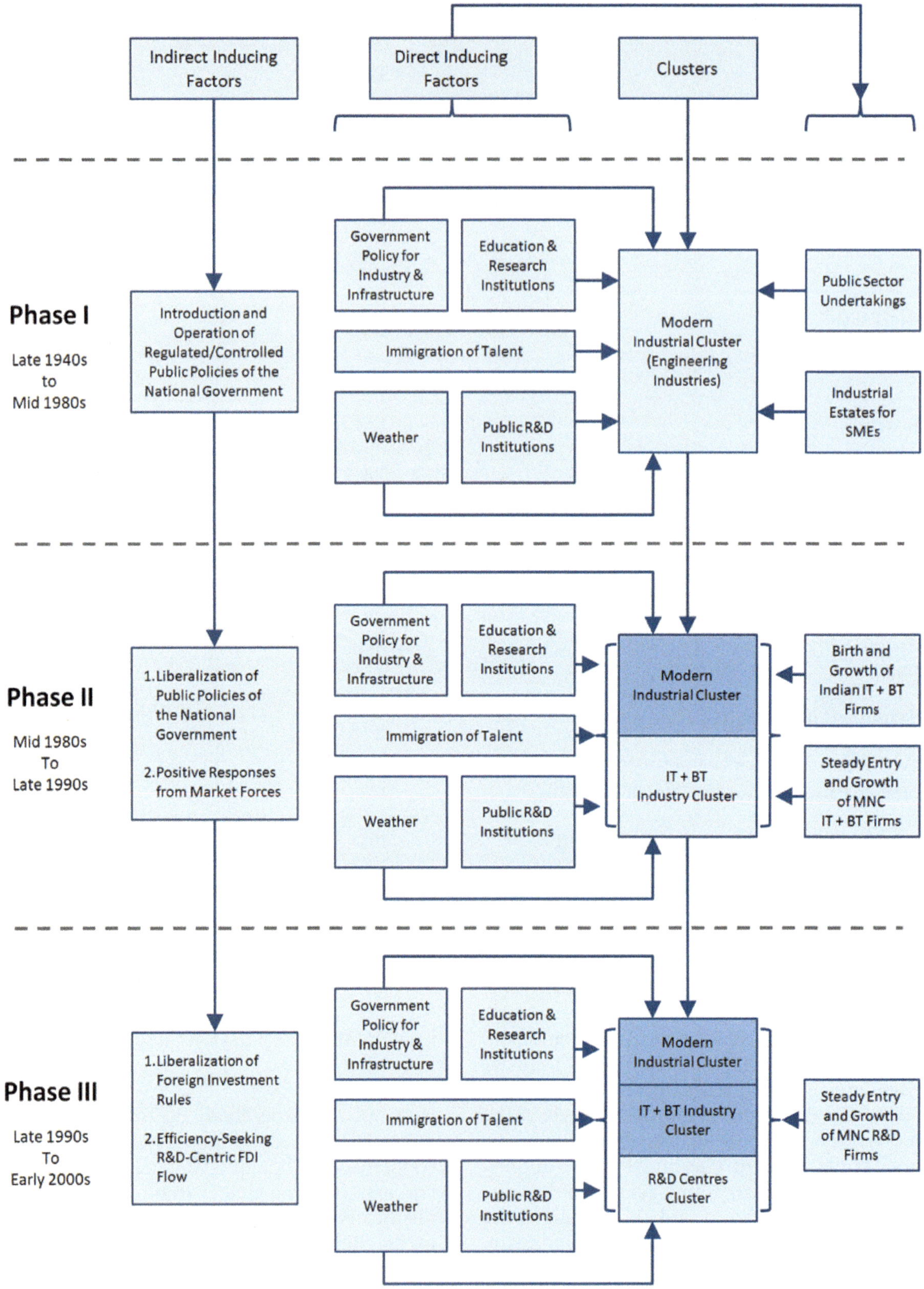

Figure 1. Bangalore: Phases in the evolution of its entrepreneurial ecosystem for technology startups

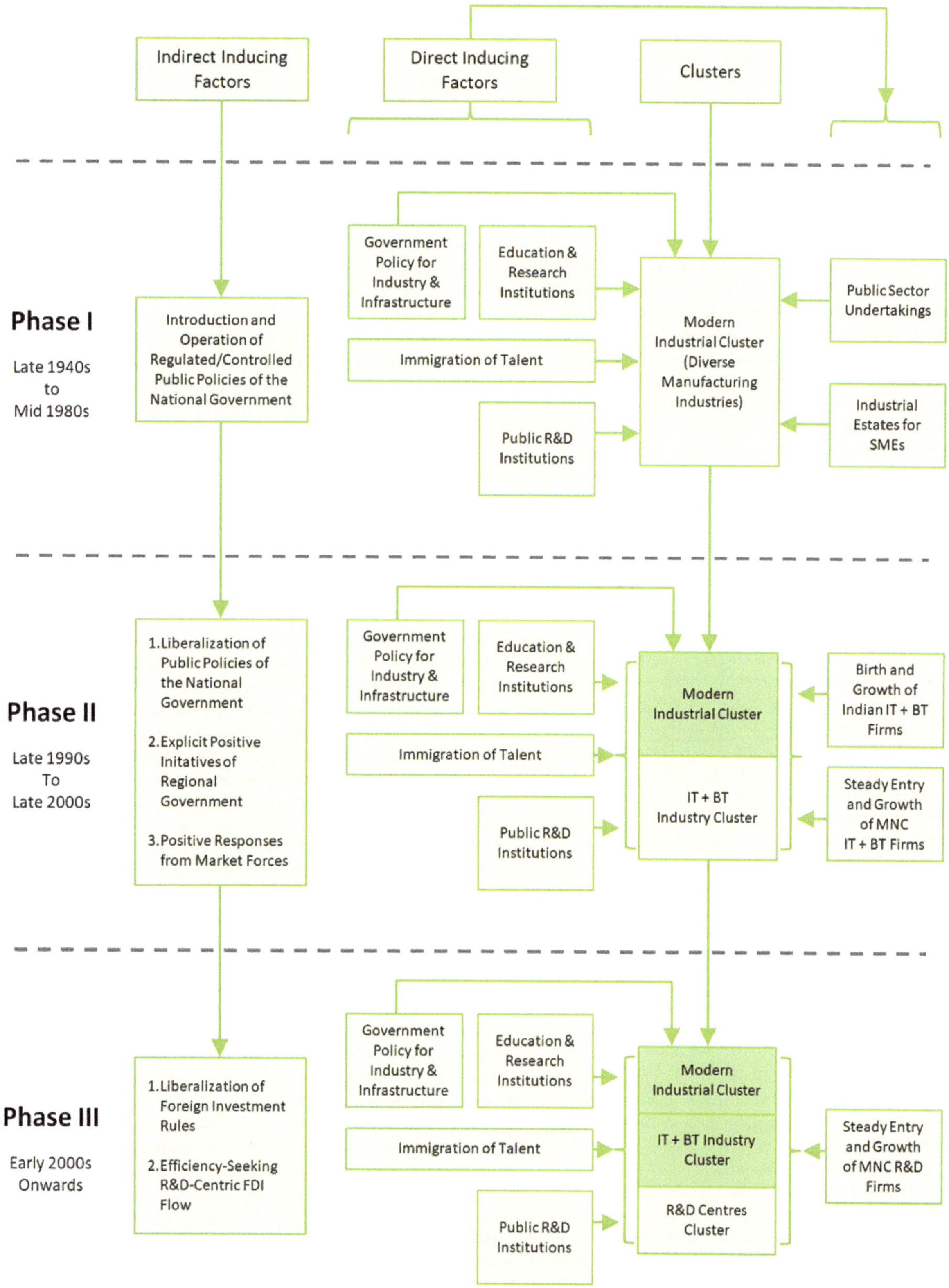

Figure 2. Hyderabad Phases in the evolution of its entrepreneurial ecosystem for technology startups

tion technology industry firms, which led to the emergence and growth of an information technology industry cluster in the subsequent decade (Das, 2015). Further, unlike Bangalore, Hyderabad had a heavy concentration of pharmaceutical industries and a higher concentration of biotechnology industries; both had come up (in 1980s as well as 1990s) prior to the emergence and growth of the information technology industries. The State (regional government in the place of national government) continued to be the dominant partner with industry and academia, implying that the "statist model" continued to be relevant for Hyderabad, unlike Bangalore, even in the second phase. Only the role of industry shifted from central public sector undertakings to multinational corporations and domestic private firms, and the role of academia was strengthened by the addition of private sector academic institutions (Figures 1 & 2).

The transition from the second phase to the third phase, again, has taken place smoothly and spontaneously since the late 1990s in Bangalore, with more and more multinational corporations locating their R&D affiliates in the city. This, coupled with the already established publicly funded R&D institutions, facilitated the emergence of an R&D cluster. In the process, interactions through partnerships and networking initiatives between R&D affiliates of multinational corporations and information technology industry development centres, academia and government grew. Perhaps all this led to the emergence of Bangalore as one of the 46 Global Hubs of Technological Innovation (UNDP, 2001) and the more recent identification of the city by the *MIT Technology Review* as one of the eight largest technology innovation clusters in the world (*Times of India*, 2013). Considering this, Bangalore might be in the process of witnessing what Etzkowitz and Leydesdorff (2000) have described as a "balanced" triple helix model (Figure 1). Though Hyderabad, similar to Bangalore, started to attract multinational corporations to locate their R&D affiliates, it is still in the process of catching up with Bangalore as a cluster for R&D centres (Figures 1 & 2). Given this, Hyderabad might take some more time to witness a "balanced" triple helix model.

The question is, what is the relevance of the triple helix model and its different configurations for an entrepreneurial ecosystem for technology startups? It is the triple helix of university–industry–government interactions that generates knowledge and innovations (Ranga & Etzkowitz, 2013), which could lead to the emergence and growth of technology-based startups. In addition, along with technology startups, triple helix interactions provide or influence all the required components of an entrepreneurial ecosystem for technology startups such as: i) sources of entrepreneurship; ii) sources of finance; iii) market; iv) human resources; v) support structure comprising accelerators, technology business incubators, and co-working spaces, among others; vi) mentorship; vii) policy; viii) conducive culture; and ix) supportive media. From the triple helix interactions, both Bangalore and Hyderabad have witnessed the emergence of these components in varying degrees.

In addition, weather has played a distinctly positive role in the case of Bangalore. Bangalore's strong point is its favourable weather, which helps foster a culture of optimism and openness, as stated by the MIT Technology Review (Times of India, 2013). Its location over the Deccan Plateau at an altitude of more than 3000 ft above the sea level keeps it cooler than most other Indian cities, and gives it possibly the best climate among all the metropolitan cities of India. It is often stated that Bangalore enjoys a salubrious climate all through the year (Sudhira et al., 2007). This favourable weather has contributed significantly in attracting investments and technology workforce to the city on a continuous basis. Given this, it is appropriate to understand the structure of entrepreneurial ecosystems in the two technology startup hubs.

Structure and Components of the Entrepreneurial Ecosystems of Bangalore and Hyderabad

The more than six decades' growth of Bangalore, which saw the setting up of central public sector undertakings comprising modern manufacturing industries in machinery and electronics, coupled with the growth of much needed SMEs, educational institutions, public R&D institutions, followed by information technology and biotechnology industries, and then by R&D affiliates of multinational corporations, ensured the emergence and growth of triple helix model partners, namely, government, industry, and academia. This gradually but steadily led to the emergence of different components of the entrepreneurial ecosystems for technology startups. Bangalore's current entrepreneurial ecosystem for technology startups is summarized in Figure 3. The ecosystem can be broadly viewed as a system within the triple helix model comprising: i) a nucleus of technology startups and prospective technology startups, surrounded by the existence of ii) indispensable (primary) factors consisting of sources of finance including seed funds, angel investors, venture capital funds, and private equity firms; market support; human

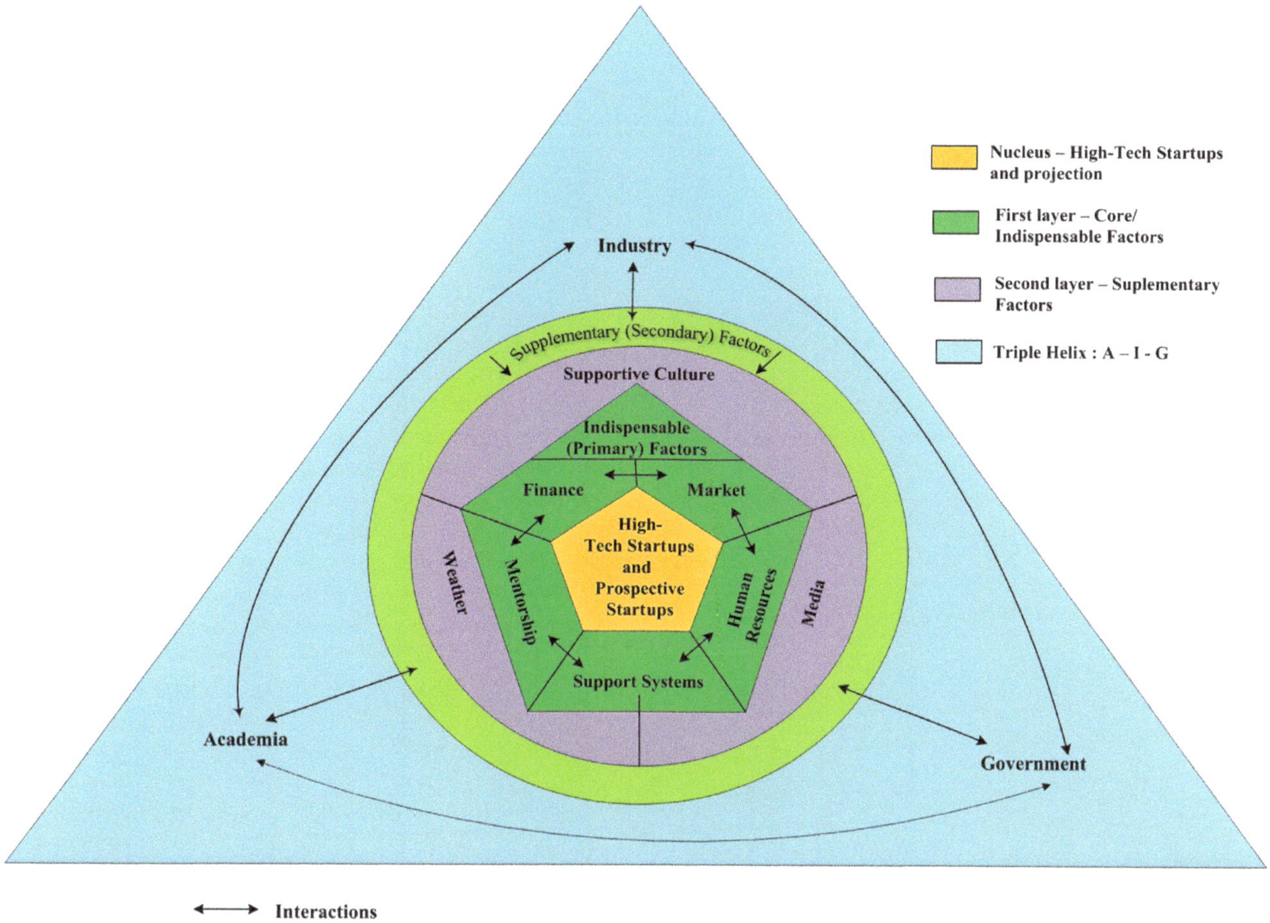

Figure 3. Bangalore: Structure and components of the entrepreneurial ecosystem for technology startups in relation to the triple helix model

resources, support systems in the form of accelerators, incubators, and co-working spaces; and technology and business mentors, and iii) supplementary (secondary) factors consisting of supportive culture (induced by constant talent immigration, supportive media and good weather).

The large base of education and research institutions in Bangalore play multiple roles in supporting and promoting the ecosystem. Karnataka state, of which Bangalore is the capital, has the third-largest concentration of professional (technical and management) education institutions and has one of the highest concentrations of universities and university-level institutes in the country (Government of India, 2016), the majority being located in Bangalore. First and foremost, these institutions generate entrepreneurial as well as human resource talent for technology startups. Second, faculty members in these institutions provide mentorship and sometimes even facilitate market support as the early adopters of startup products/services. Third, some of these institutions (such as the Indian Institute of Science, the Indian Institute of Management Bangalore, the International Institute of Information Technology Bangalore, the Institute for Bioinformatics and Applied Biology, and the National Centre for Biological Sciences) have their own technology business incubators and arrange for other support services, such as intellectual property consultancy, to nurture entrepreneurship through technology innovation commercialization, particularly among their own faculty/would-be graduates.

Large firms, similar to education and research institutions, form an indispensable part of the ecosystem for various reasons. These include central public sector undertakings, domestic private firms, and multinational corporations. First, similar to education and research institutions, they generate entrepreneurial talent as

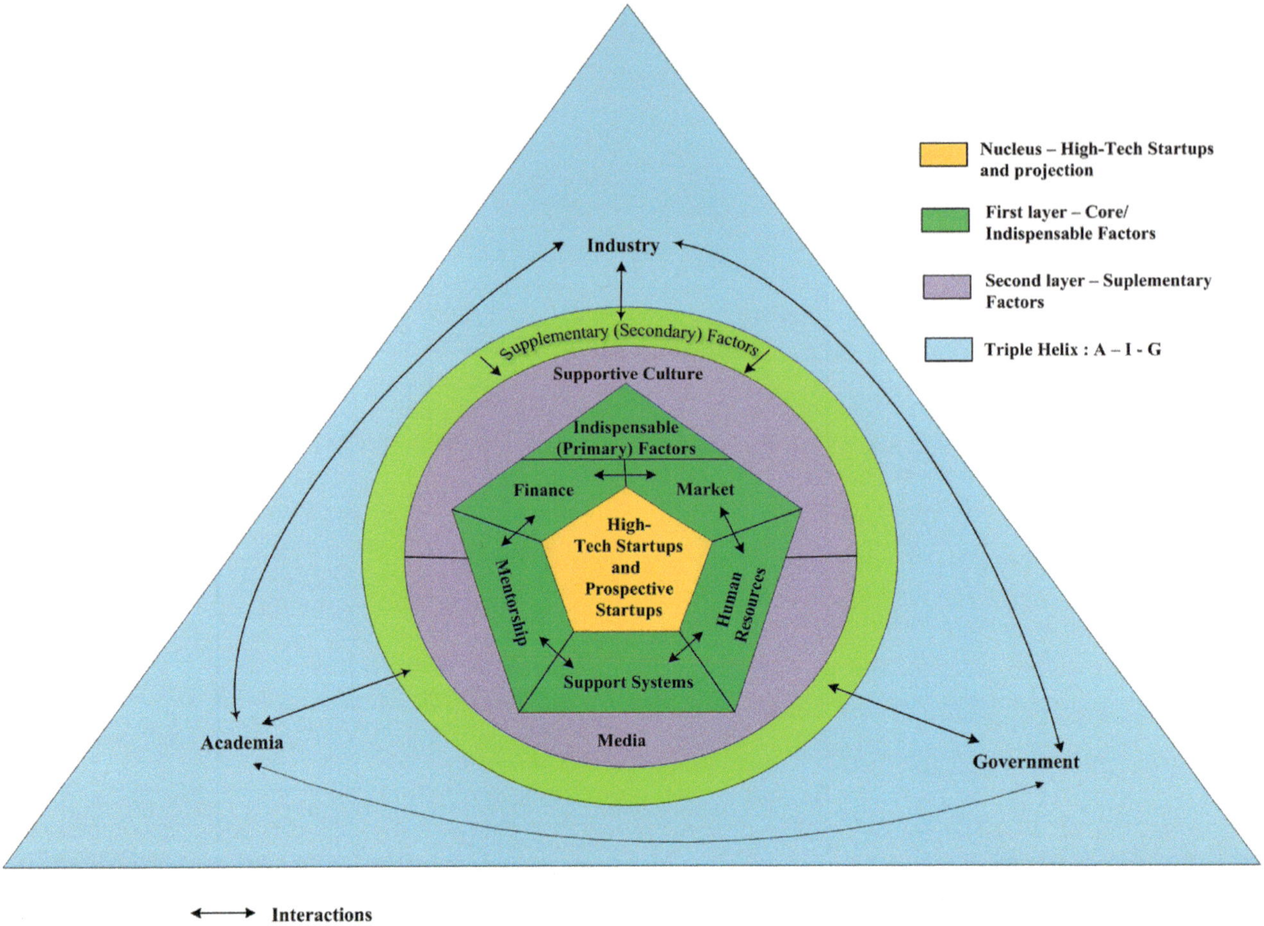

Figure 4. Hyderabad: Structure and components of the entrepreneurial ecosystem for technology startups in relation to the triple helix model

well as human resources for technology startups. The employees of large firms quit their jobs and enter the entrepreneurial world to implement their innovative ideas. Some of the multinational corporations explicitly encourage entrepreneurship among their employees by providing them sabbatical leave to experiment with their entrepreneurial ideas, with the option of returning to the company if the startup fails (ETB, 2016). Second, large firms provide early market support in the form of early adopters of startup products/services. Third, at times, they consciously nurture and promote startups through their own accelerators, with the objective of expanding market base for their own final products and services. Bangalore is home to many accelerators such as Kyron, Microsoft, SAP Labs, and Target. At a later stage, depending upon the "complementarity of business" and "compatibility of culture", they even acquire some of these startups.

India being a quasi-federal state, both the national government and the state government can play a direct or indirect role in the ecosystem for technology startups. The role of the national/regional government in startup promotion can be either passive or active. It can be passive if the government does not play any facilitating role. However, to begin with, immediately after independence, the national government located several key central public sector undertakings, public R&D institutions, and universities in Bangalore. In the meantime, the regional government created the much-needed industrial infrastructure (in the form of power, industrial area development, communication, water, etc.) and stimulated modern industrialization. This was followed by the growth of information technology and biotechnology industries, and multinational corporations setting up R&D centres, apart from facilitating the establishment of education and research institutes. All this led to an incessant inflow of talent from diverse corners of India.

Now, regional government appears to be an active facilitator of technology startups through the creation of support infrastructure such as venture capital funds and technology business incubators, though not on a considerable scale. But, the role of the government as an early adopter of startup products/services is hardly visible. More importantly, the Government of Karnataka (2015) has recently come out with an exclusive policy for startups to give them a further boost. It is the triple helix consisting of the academia–industry–government combination and its interactions which led to the gradual and steady emergence of a nucleus with a set of indispensable (primary) ecosystem components and supplementary (secondary) components.

The more than six decades' growth of Hyderabad, similar to Bangalore, saw the setting up of central public sector undertakings comprising a wide range of modern manufacturing industries such as machinery, electronics, and pharmaceuticals, coupled with the growth of much needed SMEs, educational institutions, and public R&D institutions. This was followed by the emergence and growth of information technology and biotechnology industries, and of late, by R&D affiliates of multinational corporations. This has resulted in the emergence and growth of triple helix model partners, namely, government, industry, and academia. All this has gradually and steadily led to the nurturing of different components of entrepreneurial ecosystem for technology startups. Hyderabad's current entrepreneurial ecosystem for technology startups is summarized in Figure 4. The ecosystem can be broadly viewed, similar to Bangalore, as a system within the triple helix model comprising: i) a nucleus comprising technology startups and prospective technology startups, surrounded by the existence of ii) indispensable (primary) factors (finance, market, human resources, support systems, and mentors), and iii) supplementary (secondary) factors involving supportive culture and media. The role of weather, unlike in Bangalore, is not explicitly visible in the context of Hyderabad.

Hyderabad is the home for several national-level education and research institutions, which play multiple roles in supporting and promoting the ecosystem. First, these institutions are the sources of entrepreneurship and human resources for technology startups. Second, these institutions provide mentorship and sometimes even facilitate market support for product testing as well as acting as early adopters of startup products/services. Third, some of these institutions (such as the International Crops Research Institute for the Semi-Arid Tropics, the International Institute of Information Technology Hyderabad, the University of Hyderabad, the IKP Knowledge Park, and the Birla Institute of Technology and Science Pilani) have their own technology business incubators and arrange for other support services including intellectual property consultancy, to nurture innovation and entrepreneurship through technology commercialization, particularly among their own faculty/would-be graduates. More recently, the Indian Institute of Technology Hyderabad (which was started in 2008) initiated its own technology business incubator.

Large firms, particularly multinational corporations, form another indispensable part of the ecosystem for various reasons. These include central public sector undertakings, domestic private firms, and multinational corporations. First, similar to Bangalore, they generate entrepreneurial talent as well as human resources for technology startups. The employees of large firms quit their jobs and enter the entrepreneurial world to implement their innovative ideas or they fund promising entrepreneurial ideas as angel investors. Second, large firms provide early market support in the form of early adopters of startup products/services that have business compatibility. Third, at a later stage, depending upon the "complementarity of business" and "compatibility of culture", they undertake mergers and acquisitions of some of these startups. Thus, large firms provide the much-needed direct and indirect support for technology startups such as entrepreneurship, market, finance and human resources.

Like in Bangalore, to begin with, immediately after the independence, the national government located several key central public sector undertakings, public R&D institutions, and universities in Hyderabad. During this time, the regional government was not visibly active in promoting industrialization. However, since the late 1990s, the state government has taken a pro-active stand for the promotion of information technology industries through the explicit development of specific infrastructure, while pharmaceutical and biotechnology industries have already registered their presence along with the growth of other manufacturing industries (Das, 2015). The process of setting up of R&D centres by multinational corporations began much later. Slowly and steadily, this led to an incessant inflow of talent, though not to the extent of Bangalore, from diverse corners of India.

As of now, the regional government is an active facilitator of technology startups (much more than that of Bangalore) through the creation of support infrastructure. The setting-up of T-Hub (t-hub.co) in 2015, the largest technology business incubator in the country, represents just one example of the positive initiatives taken by the state government. T-Hub is designed for technology startups with the mission to catalyze the creation of one of the most vibrant entrepreneur communities in the world to encourage and fuel more startup success stories. It is a unique public–private partnership between the Government of Telangana and three premier institutes, namely, the International Institute of Information Technology Hyderabad (IIITH), the Indian School of Business (ISB), Hyderabad, and the National Academy of Legal Studies and Research (NALSAR) University, Hyderabad. It stands at the intersection of the startup, academic, corporate, research, and government sectors (T-Hub, 2016). The regional government has recently brought out an Innovation Policy and an Information & Communication Technology Policy, respectively which, among others, focused on the promotion of technology startups (Government of Telangana, 2016a; Government of Telangana, 2016b). Thus, similar to Bangalore, it is the triple helix consisting of academia–industry–government combination and its interactions that resulted in the gradual and steady growth of a nucleus with a set of indispensable (primary) ecosystem components and supplementary (secondary) components.

A summary of the similarities and the differences between Bangalore and Hyderabad, the two promising but evolving technology startups hubs of India, is presented in Table 4.

Table 4. Similarities and differences between the Bangalore and Hyderabad ecosystems

Similarities	Differences
The ecosystem foundation was laid after India's independence in 1947 for Bangalore and after the accession of Hyderabad to India in 1948 for Hyderabad. The public policies of the national government, supplemented by that of regional governments, subsequently played a key role in each ecosystem.	The period of emergence of the second phase as well as that of the third phase in Bangalore is distinctly different from that of Hyderabad.
Three distinct phases are ascertained in the evolution of entrepreneurial ecosystems for technology startups in both the cities.	The positive response of market forces to economic policy reforms played a major role in the early emergence of an information technology and biotechnology cluster followed by an R&D cluster in Bangalore, whereas pro-active regional government initiatives followed by market force responses played a major role in the subsequent emergence of the information technology and biotechnology cluster followed by an R&D cluster in Hyderabad.
The modern industrial cluster followed by the information technology and biotechnology cluster and then by the R&D centre cluster led to the gradual evolution of an ecosystem structure and its components.	Weather played a distinct positive role in the emergence of an entrepreneurial ecosystem in Bangalore, whereas weather has not played any explicit role in Hyderabad.
The structure and components of entrepreneurial ecosystems in both the cities are observed to be largely similar, but for the role of weather which stands apart in the context of Bangalore.	Bangalore accounts for the highest number of unicorn companies in the country whereas Hyderabad has none, at present.
Both the ecosystems are in the process of emerging and therefore in the "phase of evolving".	Bangalore accounts for a higher concentration of early stage startups and "cockroaches" relative to Hyderabad due to a higher level of vibrancy in the ecosystem. Accordingly, it is considered to be ahead of Hyderabad in the "quantity and quality" of ecosystem components and their interactions and therefore has a higher level of maturity relative to the latter.

Conclusion

Entrepreneurial ecosystems for technology startups are unique and they do not emerge overnight. The emergence of such ecosystems takes place over a period of time. Among the economies across the world, India occupies a unique place as an emerging economy, accounting for the third highest concentration of technology startups and startup exits globally. In India, Bangalore has been attracting global attention increasingly, and more recently, as a hub of technology startups. Hyderabad, as one of the fast-emerging technology startup hubs in India, is closely following Bangalore. Given this, two questions are answered in this article: How did the entrepreneurial ecosystems of these two cities emerge over a period of time? and What is the structure of these ecosystems and what are their components?

The seeds for the emergence of the two ecosystems were sown immediately after India's independence in 1947 for Bangalore and after accession into Indian Union in 1948 for Hyderabad. Public policy driven central public sector undertakings, public R&D institutions, educational institutions, supplemented by domestic private enterprises led both the cities to emerge as modern industrial clusters. This was followed by the entry of multinational corporations in the information technology and biotechnology industries in response to economic liberalization (in Bangalore) and in response to regional government initiatives (in Hyderabad), which resulted in the emergence of information technology/biotechnology clusters. Subsequently, a steady liberalization of foreign investment rules and regulations encouraged a steady entry of R&D affiliates of multinational corporations in Bangalore, followed by in Hyderabad, which nurtured clusters of R&D centres. It is the successive addition of these three different but inter-related clusters that gradually and steadily led to the rise of various components of entrepreneurial ecosystems in the two cities with a strong base of government–industry–academia triple helix model.

Today, the entrepreneurial ecosystems of both Bangalore and Hyderabad have all the critical components that have been commonly identified in empirical literature in the international context. In addition, the diverse stakeholders of the two ecosystems enabled us to define the structure of an ecosystem in the Indian context. The triple helix model comprising government, industry, and academia formed the foundation of the ecosystem, within which the system involved a nucleus comprising technology startups and prospective technology startups with two outer layers. The first outer layer consisted of five primary (indispensable) factors (namely, finance, market, human resources, mentors, and support systems), whereas the outer-most layer comprised three secondary (supplementary) factors of culture, media and weather (in the case of Bangalore), culture and media (in the case of Hyderabad).

However, the ways the triple helix model operates in the two cities are different. Whereas in the Bangalore ecosystem, the triple helix interactions appear to have graduated from a "statist" model to a "laissez-faire" model and further to a "balanced" triple helix model, in the Hyderabad ecosystem, a "statist" model seems to be relevant even today. This is primarily because industry and academia have emerged, over a period of time, to be as important as the government, if not more, in Bangalore. But government continues to be the dominant partner in Hyderabad even today, due to its pro-active ecosystem promotional policies. This has been corroborated by our interviewed stakeholders in both the cities. But what does it imply?

Broadly, an ecosystem for startups is assumed to graduate through four different stages of evolution: i) nascent, ii) evolving, iii) mature, and iv) self-sustainable (Cukier et al., 2016). Considering that the vibrancy of both Bangalore and Hyderabad ecosystems is largely confined to a rapidly growing "emergence of technology startups", followed by a number of "stably operating" startups but with few "successful and grown" startups, it would be appropriate to conclude that these two ecosystems are still "evolving", and therefore, have only "moderately matured". Recently, in terms of the growth index for startup hubs, Bangalore has been ranked at the top internationally, followed by London and Tel Aviv (Velayanikal, 2015). The balanced triple helix model would prove beneficial to take Bangalore further towards maturity in the coming years. But the same cannot be assumed to happen in the case of Hyderabad. Only the continued active involvement of the government might ensure its further evolution towards maturity. In other words, any withdrawal or reduction in the role of the government might affect the further evolution of the Hyderabad ecosystem towards maturity.

Given the above, what are the key lessons/insights that we can derive from the experiences of Bangalore and Hyderabad for elsewhere in India and the world? The findings of our analysis yield three important lessons. First, the evolution of an entrepreneurial ecosystem in a technologically vibrant innovation cluster (Bangalore, which is often compared with a developed economy cluster) and an emerging innovation cluster (Hyderabad) within an emerging economy (India) is historically explored and analyzed. The critical factors that led to the laying of an ecosystem foundation gradually and steadily over a period of almost six decades in three different phases, along with other supportive factors are brought out. This finding clearly demonstrates that ecosystem development is an evolving process. But, this process will neither evolve naturally nor can it be built by design. As argued by Isenberg (2011), ecosystems are usually the result of intelligent evolution (a process that blends the invisible hand of markets and the deliberate helping hand of public policies).

Second, an entrepreneurial ecosystem is essentially regional in character. The present study has revealed the unique features and elements involved in the development of the Bangalore and Hyderabad ecosystems. An entrepreneurial ecosystem is unique as it is the result of multiple elements interacting in highly complex and idiosyncratic ways, as brought out by these two cases. Though the Bangalore and Hyderabad ecosystems have some similarities, they are not identical. This finding highlights that it is neither feasible nor desirable to replicate either Bangalore or Hyderabad, even within India. What is essential is to ascertain and understand the structure and critical components of an entrepreneurial ecosystem, without which an ecosystem might not emerge, even if it emerges it may not survive, and even if it survives it may not prove to be effective or successful.

Third, the structure and critical components of Bangalore and Hyderabad entrepreneurial ecosystems are ascertained. The triple helix comprising government, industry, and academia form the basis for an entrepreneurial ecosystem as it supports or generates the critical components of an entrepreneurial ecosystem. An entrepreneurial ecosystem for technology startups is defined as a structure with the triple helix as the base, and it comprises a nucleus (startup entrepreneurs and prospective entrepreneurs) with two outer layers consisting of some core and indispensable components as the first outer layer, which include: i) financiers (in the form of seed funds, angel investors, venture capitalists, and private equity firms), ii) markets, iii) human resources, iv) nurturers (in the form of accelerators, business incubators, and co-working spaces), and v) business and technology mentors. The second outer layer consists of supplementary factors such as supportive culture, effective media, and, if possible, good weather. Thus, the nucleus and the core components, supplemented adequately by the secondary factors, and with the triple helix as the base, define the structure of an ecosystem for technology startups. Their adequate and growing presence and interactions will generate vibrancy in the ecosystem for its further growth. This would lead to an increasing scale of technology startup emergence, nurturing for survival and stability, and growth of the successful ones, thereby contributing to national economic growth, innovation achievements, and employment generation.

Acknowledgements

This study forms a part of the research project titled "A Study on the Ecosystem for High-Technology Startups in Bangalore and Hyderabad", which is sponsored by the Department of Science and Technology (DST), Government of India, New Delhi. DST Project F.No. DST/NSTMIS/05/178/2014-15.

About the Author

M H Bala Subrahmanya is a Professor of Economics in the Department of Management Studies at the Indian Institute of Science (IISc), Bangalore, India. He joined Indian Institute of Science, Bangalore as an Assistant Professor in 1996, was promoted to Associate Professor in 2002, and further to a Professor in 2008. He has more than 27 years of professional experience, of which more than 21 years are at the IISc. During his professional career, he received a Commonwealth Fellowship (1999–2000), a Japan Foundation Fellowship (2004/2005), and a Fulbright-Nehru Senior Research Fellowship (2009/2010). He holds a PhD in Economics from the Institute for Social & Economic Change (ISEC) in Bangalore, and his field of specialization is Industrial Economics.

References

Aaltonen, A. .2016. *Factors Shaping Entrepreneurial Ecosystems and the Rise of Entrepreneurship: A View from Top Management Journals.* Helsinki: Demos Helsinki.

Aleisa, E. 2013. *Startup Ecosystems: Study of the Ecosystems Around the World; Focusing on Silicon Valley, Toronto and Moscow.* Project Report. http://www.janrecker.com/wp-content/uploads/2013/02/20130213_FinalReport_Startup-Ecosystems.pdf

Anju, Z. 2014. *Startup Capitals: Discovering the Global Hotspots of Innovation.* Gurgaon, India: Random House.

Arruda, C., Nogueria, V. S., & Costa, V. 2013. The Brazilian Entrepreneurial Ecosystem of Startups: An Analysis of Entrepreneurship Determinants in Brazil as Seen from the OECD Pillars. *Journal of Entrepreneurship and Innovation Management,* 2(3): 17–57.

Bala Subrahmanya, M. H. 2015. New Generation Startups in India: What Lessons Can We Learn from the Past? *Economic and Political Weekly,* 50(12): 56–63.

Certo, S. T. 2003. Influencing Initial Public Offering Investors with Prestige: Signalling with Board Structures. *Academy of Management Review,* 28(3): 432–446. http://www.jstor.org/stable/30040731

Clarke-Billings, L. 2017. Bangalore Named Fastest-Changing and Most Dynamic City in the World. *Newsweek,* January 18, 2017. Accessed March 10, 2017: http://europe.newsweek.com/bangalore-fastest-changig-dynamic-innovative-city-world-544166

Cukier, D., Kon, F., & Lyons, T. S. 2016. Software Startup Ecosystems Evolution: The New York City Case Study. In *Proceedings of 2nd International Workshop on Software Startups, IEEE International Technology Management Conference,* Trondheim, Norway.

Dale, R., & Zell, Z. 2014. *Startup Ecosystems: A Look at Entrepreneurship in London and Boulder.* Working Paper. London: FNCE.

Das, D. 2015. Hyderabad: Visioning, Restructuring and Making of a High-Tech City. *Cities,* 43: 48–58. https://doi.org/10.1016/j.cities.2014.11.008

Department of Industries & Commerce. 1996. *Industrial Policy 1996–2001.* Bangalore, India: Government of Karnataka.

DST. 2015. *Directory of R&D Institutions 2015.* New Delhi, India: National Science and Technology Management Information System, Department of Science and Technology, Government of India.

Etzkowitz, H. 2003. Innovation in Innovation: The Triple Helix of University-Industry-Government Relations. *Social Science Information,* 42(3): 293–337. https://doi.org/10.1177/05390184030423002

Etzkowitz, H., & Leydesdorff, L. 2000. The Dynamics of Innovation: From National Systems and "Mode 2" to a Triple Helix of University–Industry–Government Relations. *Research Policy,* 29(2): 109–123. https://doi.org/10.1016/S0048-7333(99)00055-4

Foster, G., & Shimizu, C. 2013. *Entrepreneurial Ecosystems Around the Globe and Company Growth Dynamics.* Cologny, Switzerland: World Economic Forum.

Gai, B., & Joffe, B. 2013. *India Startup Report.* World Startup Report. http://www.worldstartpreport.com

Government of Karnataka. 2015. *Karnataka Startup Policy 2015–2020.* Department of Information Technology, Biotechnology and Science & Technology. Bangalore, India: Government of Karnataka.

Government of Telangana. 2016a. *Innovation Policy 2016.* Telangana, India: Department of Information Technology, Electronics and Communications, Hyderabad.

Government of Telangana. 2016b. *ICT Policy Framework 2016.* Department of Information Technology, Electronics and Communications, Hyderabad.

Haines, T. 2016. Developing a Startup and Innovation Ecosystem in Regional Australia. *Technology Management Innovation Review,* 6(6): 24–32. https://timreview.ca/article/994

Hemmert, M., Cheng, Y., Kohlbacher, F., Kotosaka, M., Loh, C. T., & Waldnberger, F. 2016. *High-tech Startup Ecosystems in East Asian Agglomerations: Are They Different from the West?* Tokyo: German Institute of Japanese Studies.

Isenberg, D. 2011. *The Entrepreneurship Ecosystem Strategy as a New Paradigm for Economic Policy: Principles for Cultivating Entrepreneurship.* Babson Park, MA: Babson Entrepreneurship Ecosystem Project, Babson College.

Kane, T. 2010. *The Importance of Startups in Job Creation and Job Destruction.* Working Paper. Kansas City, MI: Kauffman Foundation Research Series: Firm Formation and Economic Growth.

Krajcik, V., & Formanek, I. 2015. Regional Startup Ecosystem. *European Business & Management,* 1(2): 14–18. http://dx.doi.org/10.11648/j.ebm.20150102.12

Motoyama, Y., & Watkins, K. 2014. *Examining the Connections within the Startup Ecosystem: A Case Study of St. Louis.* Kansas City, MI: Kauffman Foundation.

NASSCOM. 2015. *Startup India: Momentous Rise of the Indian Startup Ecosystem.* Bangalore, India: Zinnov Consulting.

OECD. 2013. *Startup Latin America: Promoting Innovation in the Region.* Paris: Development Centre Studies.

Office of the Registrar General & Census Commissioner, India. 2011. *Census of India.* New Delhi, India: Government of India.

Piscione, D. P. 2013. *Secrets of Silicon Valley: What Everyone Else Can Learn from the Innovation Capital of the World.* London: Macmillan.

Pullen, J. P. 2013. Emerging Tech: 9 International Startup Hubs to Watch. *Entrepreneur,* May 7, Business Daily, USA.

Ranga, M., & Etzkowitz, H. 2013. Triple Helix Systems: An Analytical Framework for Innovation Policy and Practice in the Knowledge Society. *Industry and Higher Education,* 27(4): 237–262. https://doi.org/10.5367/ihe.2013.0165

Sean, TAN C S. 2015. *Startup Ecosystem in Singapore.* Working Paper. The Netherlands: Rijksdienst voor Ondernemend Nederland.

Stangler, D., & Masterson, J. B. 2015. *Measuring an Entrepreneurial Ecosystem*. Kansas City, MI: Kauffman Foundation.

Startup Genome. 2015. *Startup Ecosystem Report 2015*. San Francisco, CA: Startup Genome.

Sudhira, H. S., Ramachandra, T. V., & Bala Subrahmanya, M. H. 2007. City Profile: Bangalore, Cities. *The International Journal of Urban Policy and Planning*, 24(5): 379–390.

T-Hub. 2016. *T-Hub*. Accessed July 1, 2017:
http://t-hub.co

Times of India, 2013. Bengaluru Among Top 8 Technology Innovation Clusters. *Times of India*, August 2, 2017. Accessed July 1, 2017:
http://economictimes.indiatimes.com/tech/ites/bangalore-among-top-8-technology-innovation-clusters/articleshow/21546192.cms

Times of India. 2016. India Ranks Third in Tech Startup Exits. *Times of India*, September 22, 2017. Accessed July 1, 2017:
http://timesofindia.indiatimes.com/business/india-business/India-ranks-3rd-in-tech-startup-exits/articleshow/54453645.cms

UNDP. 2001. *Human Development Report 2001*. New York: Oxford University Press, United Nations Development Programme (UNDP).

World Economic Forum. 2014. *Entrepreneurial Ecosystems around the Globe and Early-Stage Company Growth Dynamics*. Geneva, Switzerland: World Economic Forum.

Keywords: startups, technology, ecosystems, entrepreneurship, Bangalore, Hyderabad, India

Permissions

All chapters in this book were first published in TIM, by Carleton University; hereby published with permission under the Creative Commons Attribution License or equivalent. Every chapter published in this book has been scrutinized by our experts. Their significance has been extensively debated. The topics covered herein carry significant findings which will fuel the growth of the discipline. They may even be implemented as practical applications or may be referred to as a beginning point for another development.

The contributors of this book come from diverse backgrounds, making this book a truly international effort. This book will bring forth new frontiers with its revolutionizing research information and detailed analysis of the nascent developments around the world.

We would like to thank all the contributing authors for lending their expertise to make the book truly unique. They have played a crucial role in the development of this book. Without their invaluable contributions this book wouldn't have been possible. They have made vital efforts to compile up to date information on the varied aspects of this subject to make this book a valuable addition to the collection of many professionals and students.

This book was conceptualized with the vision of imparting up-to-date information and advanced data in this field. To ensure the same, a matchless editorial board was set up. Every individual on the board went through rigorous rounds of assessment to prove their worth. After which they invested a large part of their time researching and compiling the most relevant data for our readers.

The editorial board has been involved in producing this book since its inception. They have spent rigorous hours researching and exploring the diverse topics which have resulted in the successful publishing of this book. They have passed on their knowledge of decades through this book. To expedite this challenging task, the publisher supported the team at every step. A small team of assistant editors was also appointed to further simplify the editing procedure and attain best results for the readers.

Apart from the editorial board, the designing team has also invested a significant amount of their time in understanding the subject and creating the most relevant covers. They scrutinized every image to scout for the most suitable representation of the subject and create an appropriate cover for the book.

The publishing team has been an ardent support to the editorial, designing and production team. Their endless efforts to recruit the best for this project, has resulted in the accomplishment of this book. They are a veteran in the field of academics and their pool of knowledge is as vast as their experience in printing. Their expertise and guidance has proved useful at every step. Their uncompromising quality standards have made this book an exceptional effort. Their encouragement from time to time has been an inspiration for everyone.

The publisher and the editorial board hope that this book will prove to be a valuable piece of knowledge for researchers, students, practitioners and scholars across the globe.

List of Contributors

Bastiaan Baccarne

Sara Logghe

Dimitri Schuurman

Lieven De Marez

Soile Juujärvi

Virpi Lund

Chantal Trudel

Sue Cobb

Kathryn Momtahan

Janet Brintnell

Ann Mitchell

Edward D. Lemaire

Seppo Leminen

Mervi Rajahonka

Mika Westerlund

Anna-Greta Nyström

Miia Mustonen

Seppo Yrjölä

Rens Brankaert

Elke den Ouden

David Guimont

Dominic Lapointe

Sonja Pedell

Alen Keirnan

Gareth Priday

Tim Miller

Antonette Mendoza

Antonio Lopez-Lorca

Leon Sterling

Louise Savelkoul

Murk Peutz

Nicole Coviello

Stoyan Tanev

Kris Steen

Ellen van Bueren

Rowena Barrett

Kaisa Still

Louna Hakkarainen

Sampsa Hyysalo

Ruben D'Hauwers

Aron-Levi Herregodts

Lynn Coorevits

Olivier Rits

Pieter Ballon

Annabel Georges

Koen Vervoort

Tanguy Coenen

Sarie Robijt

Alok Chakrawal

Pratibha Goyal

M H Bala Subrahmanya

Index